Brief Contents

KT-117-305

Contents

ELECTRONIC COMMERCE

Second Edition

Gary P. Schneider, Ph.D., CPA
University of San Diego

and

James T. Perry, Ph.D.
University of San Diego

**COURSE
TECHNOLOGY**
— ✶ — ™
THOMSON LEARNING

Australia • Canada • Mexico • Singapore • Spain • United Kingdom • United States

COURSE
TECHNOLOGY
™
THOMSON LEARNING

Electronic Commerce
is published by Course Technology.

Publisher
Kristen Duerr

Managing Editor
Jennifer Locke

Product Manager
Margarita Donovan

Development Editor
Amanda Brodkin

Production Editor
Elena Montillo

Marketing Manager
Toby Shelton

Editorial Assistant
Janet Aras

Text Designer
Books by Design

Cover Designer
Abby Scholz

Manufacturing Coordinator
Alec Schall

Chapter 3
Web-Based Tools for Electronic Commerce

Chapter 4
Electronic Commerce Software

Chapter 5
Security Threats to Electronic Commerce

Chapter 6
Implementing Security for Electronic Commerce

Chapter 7
Electronic Payment Systems

Chapter 8
Strategies for Marketing, Sales, and Promotion

Chapter 9
Strategies for Purchasing and Support Activities: From Electronic Data Interchange to Electronic Commerce

Chapter 10
Strategies for Web Auctions, Virtual Communities, and Web Portals

Chapter 11
The Environment of Electronic Commerce: International, Legal, Ethics, and Tax Issues

Chapter 12
Planning for Electronic Business: Resource and Implementation Issues

Preface

Electronic Commerce, Second Edition provides complete coverage of the constantly changing field of electronic commerce. We assume that readers have no previous electronic commerce knowledge or experience. This book is designed to help you learn about the key business and technology elements of electronic commerce from the ground up.

Having worked in research, consulting, and corporate training in electronic commerce, we began developing both an undergraduate business school course and an MBA-level course in electronic commerce. Although we had used a variety of materials in our corporate training work, we were concerned that those materials would not work well in our university courses because they were written at widely varying levels and did not have the pedagogic organization and features, such as review questions, that are so important to students.

We searched for a textbook that offered balanced coverage of both the business and technology elements of electronic commerce, but found that no such textbook existed. *Electronic Commerce* is our attempt to fill that void. The book begins with an explanation of the economic foundations of electronic commerce. This sets the stage for the descriptions of electronic commerce infrastructure that follow. The book then explains the main technologies that are used to implement online business activities. After the reader has gained an understanding of the foundations and technological implementation issues, the book introduces a number of different business strategies that companies are using for electronic commerce. By studying the business strategies, the reader can see how the economic framework and the specific technologies come together in actual business applications. Of course, these business applications must operate in the global environment of business, so the book also includes overview discussions of international, legal, ethical, and tax issues that can arise in the conduct of electronic commerce. The book also explains how project planning and management techniques can help make online business initiatives successful.

ORGANIZATION AND COVERAGE

Electronic Commerce introduces readers to both the theory and practice of doing business over the Internet and World Wide Web. The first two chapters provide an introduction to electronic commerce and the elements of its infrastructure. Chapters 3 through 7 describe the technologies of electronic commerce, including electronic commerce software, electronic commerce security issues, and electronic payment systems. Chapters 8 through 10 present business strategies for electronic commerce, including branding, customer relationship management, purchasing, electronic data interchange, supply-chain management, auction sites, virtual communities, and Web portals. Chapters 11 and 12 conclude the book with a discussion of topics that are important to electronic commerce but are neither business nor technology issues. These topics include international, legal, ethics, and tax issues, and project management issues.

Chapter 1, Introduction to Electronic Commerce, begins with an explanation and overview of commerce, then defines electronic commerce and describes how companies use it to create new products and services and improve many other standard business activities. Chapter 1 also describes the history of the Internet and the Web, provides an overview of the economic structures in which businesses operate, and describes how electronic commerce fits into those structures. Two themes are introduced in this chapter that recur throughout later chapters: that examining a firm's value chain can suggest opportunities for electronic commerce initiatives, and that reductions in transaction costs are important elements of many electronic commerce intitiatives.

Chapter 2, Infrastructure for Electronic Commerce, introduces the Internet infrastructure, packet-switched networks, several Web markup languages, and popular Internet applications, protocols, and utility programs. Chapter 2 also describes the language of the Web, HTTP, explains the similarities and differences of intranets and extranets, and discusses Internet connection options and tradeoffs.

Chapter 3, Web-Based Tools for Electronic Commerce, builds on Chapter 2 with coverage of Web-based tools for electronic commerce, including Web server hardware and software options and their various strengths and scalability. Web site-hosting options and their implications are also discussed. Chapter 3 presents extensive overviews of several Web performance evaluation and tuning tools.

Chapter 4, Electronic Commerce Software, describes the basic electronic commerce software functions, building on the Web server software and hardware presented in previous chapters. Also covered in Chapter 4 are fundamental services, including catalog display, transaction processing, and shopping carts. The chapter discusses a wide range of software commerce choices for a range of business from small stores to enterprise-class firms. Chapter 4 also provides you a chance to sample several free and low-cost storefront software packages and describes the several advertising options that both attracts customers and provides revenue.

Chapter 5, Security Threats to Electronic Commerce, discusses the many internal and external security threats to electronic commerce. Also covered are the roles of copyright and intellectual property security and threats to them, including cybersquatting, domain name stealing, and recent new models for music delivery that challenge copyright laws. Chapter 5 explains the vulnerability of the communication channels carrying information between one location and another as well as that of commerce servers.

Chapter 6, Implementing Security for Electronic Commerce, describes security threat countermeasures, including antivirus software and encryption, to combat the several types of threats presented in Chapter 5. Also covered are special Internet protocols and message authentication codes that provide message protection and message deletion protection. Chapter 6 provides a detailed discussion of digital certificates and certification authorities, the tools used to verify user identification. In addition to these protections, the chapter describes intellectual property threats and some interesting approaches to protect graphics and audio resources for sale on the Internet.

Chapter 7, Electronic Payment Systems, presents a comprehensive discussion of electronic payment systems, covering electronic cash, electronic wallet technologies, stored-value cards, and credit and charge cards. In this chapter you will find discussions both of failed electronic payment systems and related organizations as well as new, promising payment methods. Chapter 7 describes how payment systems operate to both approve transactions and credit merchants for sales.

Chapter 8, Strategies for Marketing, Sales, and Promotion, explains that by understanding how the Web differs from other media, businesses can create an effective Web presence that delivers value to visitors. The chapter describes how firms that understand the nature of communication on the Web can identify and reach the largest possible number of qualified customers. Technology-enabled customer relationship management and rational branding on the Web are compared to their more traditional counterparts. The ideas behind viral marketing and permission marketing are also introduced. The chapter also explains how some businesses on the Web are sharing and transferring brand benefits through affiliate marketing and cooperative efforts among brand owners.

Chapter 9, Strategies for Purchasing and Support Activities: From Electronic Data Interchange to Electronic Commerce, explores the variety of methods that companies are using to improve their purchasing and logistics primary activities with Internet and Web technologies, and are making similar improvements in a wide range of support activities. Chapter 9 explains how the emerging network model of organization, described in Chapter 1, is being used by firms to extend the reach of their enterprise planning and control activities. Chapter 9 provides an overview of EDI and explores how the Internet is now providing an inexpensive EDI communications channel that allows smaller businesses to reap EDI's benefits. Chapter 9 also explains how the Internet and the Web have become an important force driving the adoption of supply chain management techniques in a variety of industries.

Chapter 10, Strategies for Web Auctions, Virtual Communities, and Web Portals, outlines how companies are now using the Web to do things that they have never done before, such as operating auction sites, creating virtual communities, and serving as Web portals. The chapter presents the key characteristics of six major auction types and describes how firms are using Web auction sites to sell goods to their customers and generate advertising revenue. The chapter explains how new companies have formed to take advantage of the Web's ability to bring together people and organizations that share narrow interests but are geographically dispersed. Businesses are creating virtual communities with their customers and suppliers and are using these communities to sell goods and services. The chapter notes that major Web search engine sites have evolved into Web portals and explains how smaller businesses are beginning to use similar Web portal strategies to improve brand awareness, increase sales, keep visitors coming back, and generate advertising revenue.

Chapter 11, The Environment of Electronic Commerce: International, Legal, Ethics, and Tax Issues, discusses the challenges posed to businesses by differing language, culture, laws and infrastructure when they conduct electronic commerce across international borders. Chapter 11 notes that variations and inadequacies in the infrastructure that supports the Internet worldwide can make it challenging to conduct electronic commerce in certain countries. Chapter 11 explains that although companies conducting electronic commerce are subject to the same laws and taxes as other companies, those that engage in electronic commerce face a large number of laws and taxes sooner than traditional companies usually face them. The large number of government units that have jurisdiction and the power to tax makes it essential that companies doing business on the Web understand the potential liabilities of doing business with customers in those jurisdictions.

Chapter 12, Planning for Electronic Business: Resource and Implementation Issues, presents an overview of key elements that are typically included in business plans for electronic commerce implementations. These elements include the setting of objectives and estimated costs and benefits of the project. Chapter 12 describes how companies develop and implement an outsourcing strategy for electronic commerce projects and also covers the use of project management as a formal way to plan and control specific tasks and resources used in electronic commerce projects. The chapter also includes a discussion of staffing strategies and describes the critical staffing areas of business management, application specialists, customer service staff, systems administration, network operations staff, and database administration.

FEATURES

Electronic Commerce is unique in its field because it includes the following features:

- **Business Case Approach** A business case introduces each chapter and provides a unifying theme for the chapter. The case provides a backdrop for the material described in the chapter. Each case has been chosen carefully to illustrate the role and use of electronic commerce.

- **Summaries** Each chapter concludes with a Summary that concisely recaps the most important concepts in the chapter.

- **Online Companion** An Online Companion is a set of Web pages, one for each chapter, maintained by the publisher for readers of this book. The Online Companion complements the book and contains links to hundreds of essential, up-to-date electronic commerce resources that further illustrate or graphically demonstrate the points that the book discusses. The Online Companion is continually monitored for changes in links so that its links remain "alive" and point to relevant Web sites. You can find the Online Companion for this book on Course Technology's Web site at www.course.com/mis, by searching on *Electronic Commerce*.

- **Embedded Online Companion References** Throughout each chapter there are Embedded Online Companion References that indicate the name of a link included in the Online Companion. For example, text set in bold, sans-serif letters ("**Metabot Pro**") indicates that there is a like-named link in the Online Companion. The links in the Online Companion are organized under chapter and subchapter headings that correspond to those in the book. The Online Companion also contains many supplemental links that do not appear in the book.

- **Review Questions and Exercises** Every chapter concludes with meaningful review materials that include both conceptual discussion questions and exercises. The exercises frequently involve scenarios and hands-on experiences that result in a computer output or a typed paper. One of the exercises, for example, asks you to create a small electronic commerce site using the free and easy-to-use tools supplied by a major storefront mall. Others ask you to research an issue using the Web, answer questions, and produce a summary of what you discovered.

- **For Further Study and Research** Every chapter contains a comprehensive list of references to magazine articles, newspaper articles, and journal papers that you can read to learn more about topics contained in the chapter. These readings are drawn from the technical and business literature that exists on electronic commerce as we go to press. The Online Companion will be periodically updated to include references to materials published after the book is in print. In addition to recent articles about electronic commerce, the reference list includes older works that have proved seminal or are foundation articles that help define key elements of electronic commerce.

TEACHING TOOLS

When this book is used in an academic setting, instructors may obtain the following teaching tools from Course Technology:

- **Instructor's Manual** The Instructor's Manual has been carefully prepared and tested to ensure its accuracy and dependability. The Instructor's Manual is available through the Course Technology Faculty Online Companion on the World Wide Web. (Call your customer service representative for the exact URL and to obtain your username and password.)

- **Course Test Manager** Course Test Manager (CTM) is a state-of-the-art, Windows-based testing software program developed by Course Technology for the exclusive use of instructors who adopt this or other Course Technology books. The full-featured program allows you to select questions from any of the book's sections, preview the test, and either export it to a rich text file format or print it directly. CTM can automatically grade the tests students take at the computer and can generate statistical information on individual as well as group performance. A CTM test bank has been written to accompany this book. It contains multiple-choice, true/false, short answer, and essay questions. The test bank is included on the Instructor's CD-ROM.

- **Classroom Presentations** Microsoft PowerPoint presentations are available for each chapter of this book to assist instructors in classroom lectures or to make available to students. The Classroom Presentations are included on the Instructor's CD-ROM.

Creating a quality text is a collaborative effort between author and publisher. We work as a team to provide the highest quality book possible. The authors want to acknowledge the work of the seasoned professionals at Course Technology. We thank Mac Mendelsohn, who first suggested to us the idea of writing an electronic commerce book. In addition, we thank Kristen Duerr, Publisher; Jennifer Locke, Managing Editor; Margarita Donovan, Product Manager; Elena Montillo, Production Editor; Toby Shelton, Marketing Manager; and Janet Aras, Editorial Assistant, for their tireless work and dedication to the project. We also thank Amanda Brodkin, our fabulous Development Editor. Amanda took our often-opaque paragraphs and made them crystal clear. She was always ready with encouragement and fresh ideas when we were running low on them. We want to thank Matt Olson, Yahoo! Store Western Region Sales representative, for providing us with complimentary Yahoo! Store space.

We want to thank the following reviewers for their insightful comments and suggestions at various stages of the book's development: Tina Ashford, Macon State College; Robert Chi, California State University-Long Beach; Perry M. Hidalgo, Gwinnett Technical Institute; Diane Lockwood, Albers School of Business and Economics, Seattle University; Michael P. Martel, Culverhouse School of Accountancy, University of Alabama; Pete Partin, Forethought Financial Services; and William E. McTammany, Florida Community College at Jacksonville. Our special thanks go to reviewer A. Lee Gilbert of Nanyang Technological University in Singapore, who provided extremely detailed comments and many useful suggestions for improving Chapter 12.

We want to acknowledge the role that the University of San Diego had in making this book possible. The University provided research funding that gave us time to work on the first edition of this book and provided us with fellow faculty members who were always happy to discuss and critically evaluate our ideas for the book. We are especially grateful to Tom Buckles, our favorite marketing professor, who provided many useful suggestions, pointed out a number of valuable research resources, and was willing to sit and discuss ideas for this book long after everyone else had left the building. We are equally appreciative of the many hours of consultation provided by Rahul Singh, a fellow professor of information systems, regarding the more technical elements of electronic commerce infrastructure. The University of San Diego School of Business Administration also provided the research assistance of MBA student Anthony Coury, who applied his considerable legal knowledge to reviewing Chapter 11 and suggesting many improvements. We are grateful to Anthony for his assistance and willingness to share his expertise.

Finally, we want to express our deep appreciation for the continuous support and encouragement of our spouses, Cathy Cosby and Nancy Perry. They demonstrated remarkable patience as we worked both ends of the clock to complete this second edition on a tight schedule. Without their support and cooperation, we would not have attempted to revise this book. We also thank our children for tolerating our absences while we were busy writing.

Gary P. Schneider
James T. Perry

DEDICATION

To Cathy, Ben, Annie, and Maggie

Gary P. Schneider

To Stirling

You are a wonderful young man, full of promise and enthusiasm. Mom and I are very proud of you.

James T. Perry

ABOUT THE AUTHORS

Gary Schneider is an Associate Professor of Accounting and Information Systems at the University of San Diego, where he teaches courses in electronic commerce, database design, and management control systems. He has published 20 books and more than 60 research papers on a variety of accounting, information systems, and management topics. Gary's research has been funded by the Irvine Foundation and the U.S. Office of Naval Research. His work has appeared in the *Journal of Information Systems*, *Interfaces*, and the *Information Systems Audit & Control Journal*. He has served as editor of the *Accounting Systems and Technology Reporter*, as associate editor of the *Journal of Global Information Management*, and on the editorial boards of the *Journal of Database Management* and the *Information Systems Audit & Control Journal*. Gary has lectured on electronic commerce topics at universities and businesses in the United States, South America, and Asia. He has provided consulting and training services to a number of major clients, including the GartnerGroup, Gateway, General Electric, LexFusion, and Qualcomm. In 1999, he was named a Fellow of the Gartner Institute. Gary is a licensed C.P.A. in Ohio, where he practiced public accounting for 14 years. He holds a Ph.D. in accounting information systems from the University of Tennessee, an M.B.A. in accounting from Xavier University, and a B.A. in economics from the University of Cincinnati.

Jim Perry is a Professor of Management Information Systems at the University of San Diego School of Business. He is the coauthor of over 36 textbooks and more than a dozen articles on computer security, database management systems, multimedia delivery systems, and chief programmer teams. Jim is a charter member of the Association for Information Systems. He holds a Ph.D. in computer science from the Pennsylvania State University and a B.S. in mathematics from Purdue University. Jim has worked as a computer security consultant to various private and governmental organizations, including the Jet Propulsion Laboratory. He was a consultant on the Strategic Defense Initiative ("Star Wars") project and served as a member of the computer security oversight committee.

INTRODUCTION TO
ELECTRONIC COMMERCE

INTRODUCTION

In 1994, a 29-year-old financial analyst and fund manager named Jeff Bezos became intrigued by the rapid growth of the Internet. Looking for a way to capitalize on this hot new marketing tool, he made a list of 20 products that might sell well on the Internet. After some intense analysis, he determined that books were at the top of that list. Six years later, **Amazon.com**, the company he formed to sell books on the Internet, had annual sales of over $2 billion and a list of more than 20 million customers. Bezos had no experience in the book-selling business, but he realized that books were small-ticket commodity items and were easy and inexpensive to ship. He knew many customers would be willing to buy books without inspecting them in person and that books could be impulse purchase items if properly promoted. Over 4 million book titles are in print at any one time throughout the world; over 1 million of those are in English. However, even the largest bookstore cannot stock more than 200,000 books. Bezos had identified a strategic opportunity for selling online.

As important as this selling opportunity was, the structure of the supply side of the business was equally important to Amazon.com's success. Bezos found that there were a large number of book publishers, none of which held a dominant position in the book-selling marketplace. Thus, it was unlikely that a single supplier would restrict Bezos' supply of books or enter his market as a competitor. He decided to locate his firm in Seattle, close to a large pool of programming talent and near one of the largest book distribution warehouses in the world.

Bezos encouraged early customers to submit reviews of books, which he posted with the publisher's information about the book. This customer participation served as a substitute for the corner bookshop staff's friendly advice and recommendations. Bezos saw the power of the Internet in reaching small, highly focused market segments, but he realized that his comprehensive bookstore could not be all things to all people. Therefore, he created a sales associate program in which Web sites devoted to a particular topic, such as model railroading, could provide links to Amazon.com books that related to that topic. In return, Amazon.com would remit a percentage of referred sales to the referring site.

As it has grown, Amazon.com has continued to identify strategic opportunities. In 1998, it began selling music CDs and videotapes. The Web site's software can track a customer's purchases and recommend similar book, CD, or video titles. Customers can request notification whenever a particular author publishes a new book or a particular recording artist releases a new CD.

Amazon.com has continued to extend its product line offerings, which now include a variety of consumer goods, including electronics, software, art and collectibles, housewares, and toys. By relentlessly paying attention to every process involved in buying, promoting, selling, and shipping consumer goods, and by working to improve each process continuously, Bezos and Amazon.com have become one of the first highly visible success stories in electronic commerce.

LEARNING OBJECTIVES

In this chapter, you will learn about:

- The basic elements of electronic commerce
- Differences between electronic commerce and traditional commerce
- Advantages and disadvantages of using electronic commerce to conduct business activities
- The international nature of electronic commerce
- The ways in which the growth of the Internet and the World Wide Web has stimulated the emergence of electronic commerce
- Economic forces that have created a business environment that fosters electronic commerce
- The ways by which businesses use value chains to identify electronic commerce opportunities

In this section, you will learn about traditional commerce and the activities it includes. Then, you will learn how electronic commerce uses various technologies to implement some or all of those activities.

To many people, the term electronic commerce (sometimes shortened to e-commerce) means shopping on the part of the Internet called the World Wide Web (the Web). Although consumer shopping on the Web is expected to exceed $300 billion by 2004, electronic commerce is much broader and encompasses many more business activities than just Web shopping. In fact, the total volume of all business activities on the Web is expected to exceed $4 trillion by 2004. Some people and businesses use the term **electronic business** (or **e-business**) when they are talking about electronic commerce in this broader sense. However, most people use the terms electronic commerce and electronic business interchangeably. In this book, we will use the term **electronic commerce** in its broadest definition: business activities conducted using electronic data transmission via the Internet and the World Wide Web. Figure 1-1 shows the three main elements of electronic commerce. These elements include:

- Consumer shopping on the Web, called **business-to-consumer** (or **B2C**)
- Transactions conducted between businesses on the Web, called **business-to-business** (or **B2B**)
- The transactions and business processes that support selling and purchasing activities on the Web

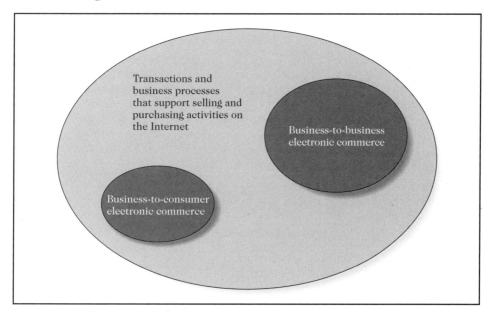

Transactions and business processes that support selling and purchasing activities on the Internet

Business-to-business electronic commerce

Business-to-consumer electronic commerce

Figure 1-1   Elements of electronic commerce

To understand these distinctions better, consider an example of a company that manufactures stereo speakers. The company might sell its finished product to consumers on the Web, which would be B2C electronic commerce. It might also purchase the materials it uses to make the speakers on the Web from other companies, which would be B2B electronic commerce. In addition to buying and selling, however, the company must undertake many activities to convert the purchased materials into speakers. These activities might include, for example, hiring and managing the people who make the speakers, renting or buying the facilities in which the speakers are made and stored, shipping the speakers, maintaining accounting records, purchasing insurance, developing advertising campaigns, and designing new versions of the speakers. An increasing number of these transactions and business processes can be done on the Web.

In terms of dollar volume and number of transactions, B2B electronic commerce is greater than B2C electronic commerce. However, the dollar volume of the transactions and business processes that support B2C and B2B activities is greater than both of them combined.

Although the Web has made online shopping possible for many businesses and individuals, in a broader sense, electronic commerce has existed for many years. For decades, banks have been using **electronic funds transfers** (**EFTs**, also called **wire transfers**), which are electronic transmissions of account exchange information over private communications networks.

Businesses also have been engaging in a form of electronic commerce, known as electronic data interchange, for many years. **Electronic data interchange** (**EDI**) occurs when one business transmits computer-readable data in a standard format to another business. In the 1960s, businesses realized that many of the documents they exchanged related to the shipping of goods—such as invoices, purchase orders, and bills of lading—and included the same set of information for almost every transaction. They also realized that they were spending a good deal of time and money entering these data into their computers, printing paper forms, and then reentering the data on the other side of the transaction. Although the purchase order, invoice, and bill of lading for each transaction contained much of the same information—such as item numbers, descriptions, prices, and quantities—each paper form usually had its own unique format for presenting that information. By creating a set of standard formats for transmitting that information electronically, businesses were able to reduce errors, avoid printing and mailing costs, and eliminate the need to reenter the data.

Businesses that engage in EDI with each other are called **trading partners**. The standard formats used in EDI contain the same information that businesses have always included in their standard paper invoices, purchase orders, and shipping documents. Firms such as **General Electric** and **Wal-Mart** have been pioneers in using EDI to improve their purchasing processes and their relationships with suppliers. Other firms, such as **Sterling**, **Commerce One**, and **Harbinger**, played a key role in facilitating EDI between firms by developing needed software and providing connectivity.

One serious problem that potential adopters of EDI faced was the high cost of implementation. Until quite recently, doing EDI meant buying expensive computer hardware and software, then either establishing direct network connections (using leased telephone lines) to all trading partners or subscribing to a value added network. A **value added network (VAN)** is an independent firm that offers connection and EDI transaction forwarding services to buyers and sellers engaged in EDI. Before the Internet as we know it today existed, VANs provided the connections between most trading partners and were responsible for ensuring the security of the data transmitted. VANs usually charged a fixed monthly fee plus a per-transaction charge, adding to the already significant expense of implementing EDI. Many smaller firms were unable to afford to participate in EDI and lost important customers, who went elsewhere to buy. **Open Market** was one of the first firms to move EDI traffic to the Internet, but many other EDI software development and consulting firms have joined in this trend. Experts estimate that EDI transaction activity on the Internet could exceed $1 trillion by 2003.

A good way to understand the full range of electronic commerce is to learn about the activities that companies undertake when they do any kind of business (that is, when they engage in commerce), and then to learn how these firms might undertake these activities electronically. In the next two sections, you will learn about the elements of traditional commerce and then see how these elements are carried out electronically using electronic commerce.

Traditional Commerce

In addition to buying or selling, firms engage in many other activities that keep them in business. For example, a seller of a product must identify demand, promote its product to potential buyers, accept orders, deliver its product, bill and accept payment for its product, and support its customers' use of its product after the sale. In many cases, sellers will customize or create a product to a customer's specifications. Similarly, product buyers also engage in additional activities. They must examine their needs, identify products that might meet those needs, and evaluate those products. Next, buyers must order the selected product, arrange for delivery, and pay for the product. In many cases, buyers need to maintain contact with the seller for warranty and other maintenance on the product. Of course, buyers and sellers can engage in these transactions not just for products, but for services too. When you think broadly about commerce, you see that it can involve individuals, business firms, not-for-profit organizations, and governmental entities as both buyers and sellers.

The origins of traditional commerce occurred before recorded history, when our ancestors first decided to specialize their everyday activities. Instead of each family unit having to grow crops, hunt for meat, and make tools, families developed skills in one of these areas and traded some of their production for other needs. For example, the tool-making family would exchange tools for grain from the crop-growing family. Services were bought and sold in these primitive economies, too. For example, the local shaman would cast spells or intercede with the deities in exchange for food and tools.

Eventually, bartering gave way to the use of currency, making transactions easier to settle; however, the basic mechanics of trade were the same. One member of society had created something of value that another member of society desired. **Commerce**, or **doing business**, is a negotiated exchange of valuable objects or services between at least two parties and includes all activities that each of the parties undertakes to complete the transaction.

The Buyer

You can examine any commerce transaction from either the buyer's or the seller's viewpoint. The elements of traditional commerce in which a buyer engages appear in Figure 1-2.

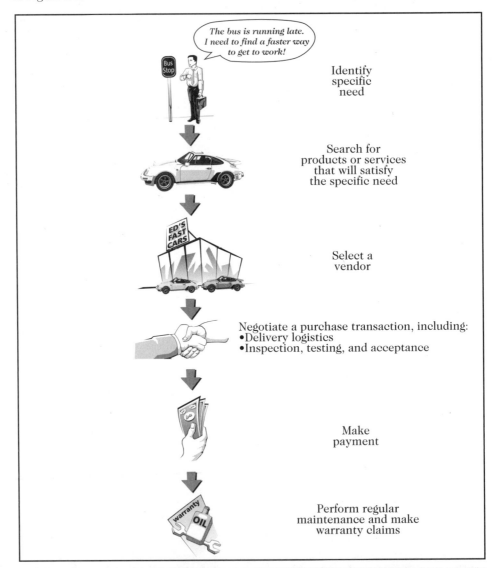

The bus is running late. I need to find a faster way to get to work!

Identify specific need

Search for products or services that will satisfy the specific need

Select a vendor

Negotiate a purchase transaction, including:
•Delivery logistics
•Inspection, testing, and acceptance

Make payment

Perform regular maintenance and make warranty claims

Figure 1-2   *Elements of traditional commerce: the buyer's side*

A buyer begins by identifying a need. This may be a simple need, such as when an individual decides "I'm hungry and would like to find some lunch," or it may be very complex, such as when a city council decides "We must find a way to generate clean power that will meet our city's needs in 25 years." The need-identification process for a hungry individual may require no more than a quick consideration of which fast-food outlets are nearby and open. To identify the specific needs for the power-generation example could require many people working in an organized way for a long period of time. Most need-identification processes fall between these two time- and resource-consumption extremes.

Once buyers have identified their specific needs, they must find products or services that will satisfy those needs. In traditional commerce, buyers use a variety of search techniques. They may consult catalogs, ask friends, read advertisements, or examine directories. The Yellow Pages is a good example of a directory that buyers often use to find products and services. Buyers may consult salespersons to gather information about specific features and capabilities of products they are considering. Companies often have highly structured procedures for finding products and services that satisfy recurring needs of their businesses.

After buyers have selected a product or service that will meet the identified need, they must select a vendor that can supply that product or service. Buyers in traditional commerce contact vendors in a variety of ways, including by telephone, by mail, and by contact at trade shows. After choosing a vendor, the buyer negotiates a purchase transaction. This transaction may have many elements, including delivery date, method of shipment, price, warranty, and payment terms, and will often include detailed specifications to be confirmed by inspection when the product is delivered or the service is performed. When the buyer is a business, the negotiation of a purchase transaction can be very complicated. Imagine, for example, the complex ordering, delivery, and inspection activities that must occur when an airline buys a new airplane from an aircraft manufacturer. Businesses often have entire departments devoted to the function of negotiating purchase transactions with their suppliers. These departments are usually named **supply management** or **procurement**.

When the buyer is satisfied that the purchased product or service has met the terms and conditions agreed to by both buyer and seller, the buyer will pay for the purchase. After the sale is complete, the buyer may have further contact with the seller regarding warranty claims, upgrades, and regular maintenance.

The Seller

Each action taken by a buyer engaging in commerce has a corresponding action that is taken by the seller. Figure 1-3 shows the elements of commerce from a seller's viewpoint.

Sellers often undertake **market research** to identify potential customers' needs. Even businesses that have been selling the same product or service for many years are always looking for ways to improve and expand their offerings. Firms conduct surveys, have salespersons talk with customers, run focus groups, and hire outside consultants to help them in this identification process.

Once customer needs have been identified, sellers then create the products and services that they feel will meet those needs. This creation activity includes design, testing, and production activities.

The next step for sellers is to make potential customers aware that the new product or service exists. Sellers engage in many different kinds of advertising and promotional activities to communicate information about their products and services to existing and potential customers.

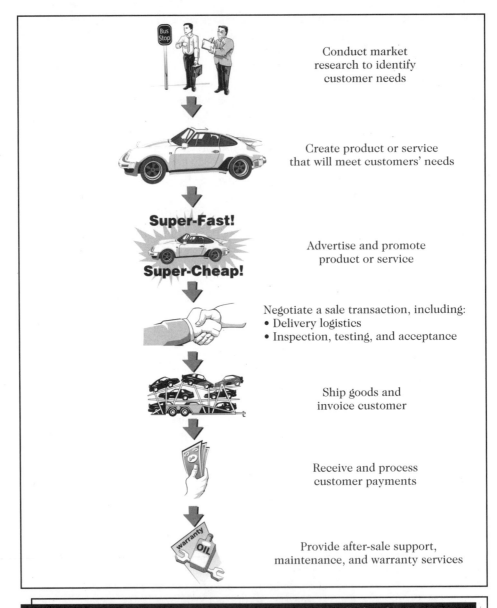

Conduct market research to identify customer needs

Create product or service that will meet customers' needs

Advertise and promote product or service

Negotiate a sale transaction, including:
• Delivery logistics
• Inspection, testing, and acceptance

Ship goods and invoice customer

Receive and process customer payments

Provide after-sale support, maintenance, and warranty services

Figure 1-3   *Elements of traditional commerce: the seller's side*

Once a customer responds to the seller's promotion activities, the two parties must negotiate the details of a purchase transaction. In some cases, this is simple; for example, many retail transactions involve nothing more than a buyer entering a seller's store, selecting and inspecting items to purchase, and paying for them. In other cases, purchase transactions require prolonged negotiations to settle the terms of delivery, inspection, testing, and acceptance.

After the seller and buyer resolve delivery logistics, the seller ships the goods or provides the service and sends an invoice to the buyer. In some businesses, the seller will also provide a monthly billing statement to each customer that summarizes the invoicing and payment activity of that customer. In some cases, the seller will require payment before or at the time of shipment. However, most businesses sell to each other on credit, so the seller must keep a record of the sale and wait for the customer to pay. Most businesses maintain sophisticated systems for receiving and processing customer payments. They want to track the amounts they are owed and ensure that payments they receive are credited to the proper customer and invoice.

Following the conclusion of the sale transaction, the seller will often provide continuing after-sale support for the product or service. In many cases, the seller is bound by contract or statute to guarantee or warrant that the product or service sold will perform satisfactorily for a specific period of time. The seller provides support, maintenance, and warranty work to help ensure that the customer is satisfied and will return to buy again.

Activities as Business Processes

You may have noticed that each element of commerce—whether viewed from the buyer's or the seller's standpoint—can include a number of distinct activities. For example, when buyers arrange for delivery of purchased items, they often will buy shipping services from a company other than the one selling the items. This service purchase transaction is a part of the delivery logistics function, or activity.

Another example occurs when sellers engage in advertising and promotion activities. To do this, a seller might purchase the services of advertising agencies, copywriters, and market research firms. The seller might also purchase products (for example, display racks or cartons) from other companies to use in promotional displays or in the ads themselves. Other sellers might use their own employees to undertake these advertising and promotional activities. For these firms, commerce includes the coordination and management of these employee activities.

The activities in which businesses engage as they accomplish a specific element of commerce are often referred to as **business processes**. Transferring funds, placing orders, sending invoices, and shipping goods to customers are all types of business processes. For example, the business process of shipping goods to customers might include a number of activities, such as inspecting the goods, packing the goods, negotiating with a freight company to deliver the goods, creating and printing the shipping documents, and loading the goods onto the truck.

Electronic Commerce

Over the thousands of years that people have engaged in commerce with one another, they have adopted the tools and technologies that became available. For example, the advent of sailing ships in ancient times opened new avenues of trade to buyers and sellers. More recent innovations, such as the printing press, the steam engine, and the telephone, have each changed the way in which people conducted commerce activities.

For decades, firms have used various electronic communications tools to conduct different kinds of business transactions. Banks have used EFTs to move customers' money around the world, all kinds of businesses have used EDI to place orders and send invoices, and retailers have used television advertising to generate telephone orders from the general public for all kinds of merchandise.

Our definition of electronic commerce, stated earlier in this chapter, mentions the use of electronic data transmission to implement or enhance business processes. Some people use the term **Internet commerce** to mean electronic commerce that specifically uses the Internet or the Web as its data transmission medium. Other terms exist for what we call electronic commerce in this book. Since the field of electronic commerce is so new, people and businesses sometimes use terms in different ways. For example, *IBM* has defined electronic business to be "the transformation of key business processes through the use of Internet technologies."

This book will show you how to use electronic data transmission technologies, primarily those that are part of the Internet, to improve existing business processes and identify new business opportunities. An important aspect of electronic commerce is that it can be used by firms to adapt to change. The business world is changing more rapidly now than ever before. Although much of this book is devoted to explaining technologies, its focus is on the business of electronic commerce. The technologies are only business-process enablers. Figure 1-4 compares a traditional commerce version of a typical business transaction with an electronic commerce version of the same transaction.

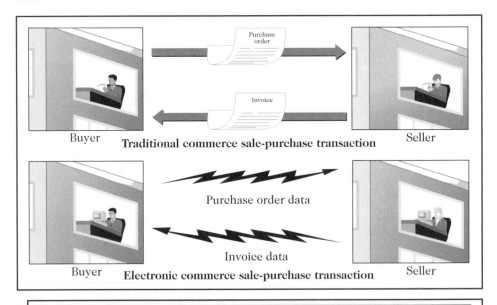

Figure 1-4   *Comparing traditional commerce and electronic commerce*

Business Processes in Commerce

An important function of this book is to help you learn how to identify those business processes that firms can accomplish more effectively by using electronic commerce technologies. In many cases, business processes use traditional commerce activities very effectively, and technology cannot improve upon them. Products that buyers prefer to touch, smell, or examine closely are difficult to sell using electronic commerce. For example, customers might be reluctant to buy high-fashion clothing and perishable food products, such as meat or produce, if they cannot examine the products closely before agreeing to purchase them.

Retail merchants have years of traditional commerce experience in creating store environments that help convince customers to buy. This combination of store design, layout, and product display knowledge is called **merchandising**. In addition, many salespeople have developed skills that allow them to identify customer needs and find products or services that meet those needs. The arts of merchandising and personal selling can be difficult to practice over an electronic link.

Branded merchandise and products, such as books or CDs, can be easily sold using electronic commerce. Since one copy of a new book is identical to other copies, and since the customer is not concerned about fit, freshness, or other qualities, customers are usually willing to order a title without examining the specific copy they will receive. The advantages of electronic commerce, including the ability of one site to offer a wider selection of titles than even the largest physical bookstore, can outweigh the advantages of a traditional bookstore, such as the customer's ability to browse. In later chapters, this book will show you how to evaluate the advantages and disadvantages of using electronic commerce for specific business processes.

Figure 1-5 lists 12 examples of business processes. Four of these are well suited to electronic commerce and four are better suited to traditional commerce. The four business processes listed in the middle column are well suited to a combination of the two approaches. Of course, these classifications depend on the current state of available technologies and thus might change as new tools for implementing electronic commerce emerge. These classifications also depend on the way business organizations are structured. As these business structures change, the suitability of various business processes to electronic or traditional forms of commerce might also change.

Well suited to electronic commerce	Suited to a combination of electronic and traditional commerce strategies	Well suited to traditional commerce
• Sale/purchase of books and CDs and other commodities	• Sale/purchase of automobiles	• Sale/purchase of high-fashion clothing
• Online delivery of software	• Online banking	• Sale/purchase of perishable food products
• Promotion and delivery of travel services	• Roommate-matching services	• Processing of small-denomination transactions
• Online shipment tracking	• Sale/purchase of investment and insurance products	• Sale of high-value jewelry and antiques

Figure 1-5 *Business process suitability to types of commerce*

One business process that is especially well suited to electronic commerce is the selling of commodity items. A **commodity** item is a product or service that has become standardized and well-known. Office supplies, computers, and airline transportation are all examples of commodity products or services, as are the books and CDs sold by Amazon.com. On the other hand, business processes tend to benefit from traditional commerce when personal selling skills are a factor, as in real estate sales, or when the condition of the products is best determined by personal inspection, as in purchases of high-fashion clothing or perishable food products.

Combinations of electronic and traditional commerce strategies work best when the business process includes both commodity and personal inspection elements. For example, many people are finding information on the Web about automobiles, but few people would buy a used car without driving and inspecting it personally. In this case, electronic commerce provides a good way for buyers to obtain information about available models, options, reliability, prices, and dealerships, but the variability of individual cars makes the traditional commerce component of personal inspection a key part of the transaction negotiation. It is still not possible to take a test drive on the Web. The next two sections summarize some advantages and disadvantages of electronic commerce.

Advantages of Electronic Commerce

Firms are interested in electronic commerce because, quite simply, it can help increase profits. All the advantages of electronic commerce for business entities can be summarized in one statement: Electronic commerce can increase sales and decrease costs. Advertising done well on the Web can get even a small firm's promotional message out to potential customers in every country in the world. A firm can use electronic commerce to reach narrow market segments that are geographically scattered. The Web is particularly useful in creating virtual communities that become ideal target markets for specific types of products or services. A virtual community is a gathering of people who share a common interest, but instead of this gathering occurring in the physical world, it takes place on the Internet. Virtual communities will be explained in more detail in Chapter 10.

A business can reduce the costs of handling sales inquiries, providing price quotes, and determining product availability by using electronic commerce in its sales support and order-taking processes. **Cisco Systems** currently sells almost all of its computer equipment through its Web site. Since no customer service representatives are involved in making these sales, Cisco operates very efficiently. In 1998, when 72 percent of its sales were on the Web, Cisco estimated that it avoided handling 500,000 calls per month and saved $500 million in that year alone. Many other businesses have emulated the **Cisco Global Networked Business Model** in recent years.

Just as electronic commerce increases sales opportunities for the seller, it increases purchasing opportunities for the buyer. Businesses can use electronic commerce in their purchasing processes to identify new suppliers and business partners. Negotiating price and delivery terms is easier in electronic commerce, because the Web can provide competitive bid information very efficiently. Electronic commerce increases the speed and accuracy with which businesses can exchange information, which reduces costs on both sides of transactions.

Electronic commerce provides buyers with a wider range of choices than traditional commerce, since they can consider many different products and services from a wider variety of sellers. This wide variety is available for consumers to evaluate 24 hours a day, every day. Some buyers prefer a great deal of information to use in deciding on a purchase; others prefer less. Electronic commerce provides buyers with an easy way to customize the level of detail in the information they obtain about a prospective purchase. Instead of waiting days for the mail to bring a catalog or product specification sheet, or even minutes for a fax transmission, buyers can have instant access to detailed information on the Web. Some products, such as software, audio clips, or images, can even be delivered through the Internet, which reduces the time buyers must wait to begin enjoying their purchases.

The benefits of electronic commerce also extend to the general welfare of society. Electronic payments of tax refunds, public retirement, and welfare support cost less to issue and arrive securely and quickly when transmitted over the Internet. Furthermore, electronic payments can be easier to audit and monitor than payments made by check, which can help protect against fraud and theft losses. To the extent that electronic commerce enables people to work from home, we all benefit from the reduction in commuter-caused traffic and pollution. Electronic commerce can make products and services available in remote areas. For example, distance education is making it possible for people to learn skills and earn degrees no matter where they live or which hours they have available for study.

Disadvantages of Electronic Commerce

Some business processes may never lend themselves to electronic commerce. For example, perishable foods and high-cost items such as jewelry or antiques may be impossible to inspect adequately from a remote location, regardless of any technologies that might be devised in the future. Most of the disadvantages of electronic commerce today, however, stem from the newness and rapidly developing pace of the underlying technologies. These disadvantages will disappear as electronic commerce matures and becomes more available to and accepted by the general population.

Many products and services require that a critical mass of potential buyers be equipped and willing to buy through the Internet. For example, online grocers such as **Peapod** offer their delivery services only in a few cities. As more of Peapod's potential customers become connected to the Internet and begin to feel comfortable with purchasing online, the business should be able to expand into more geographic areas. Peapod is a good example of how difficult it can be to build a business in an industry that requires this kind of critical mass. Although it was one of the first online groceries, it has had a difficult time staying in business. It was even offline for a few weeks in mid-2000 before being acquired by a European firm that was willing to invest additional cash to keep Peapod in operation.

Businesses often calculate return-on-investment numbers before committing to any new technology. This has been difficult to do for investments in electronic commerce, because the costs and benefits have been hard to quantify. Costs, which are a function of technology, can change dramatically during even short-lived electronic commerce implementation projects, because the underlying technologies are changing so rapidly. Many firms have had trouble recruiting and retaining employees with the technological, design, and business process skills needed to create an effective electronic commerce presence. Another problem facing firms that want to do business on the Internet is the difficulty of integrating existing databases and transaction-processing software designed for traditional commerce into the software that enables electronic commerce.

In addition to technology and software issues, many businesses face cultural and legal obstacles to conducting electronic commerce. Some consumers are still somewhat fearful of sending their credit card numbers over the Internet. Other consumers are simply resistant to change and are uncomfortable viewing merchandise on a computer screen rather than in person. The legal environment in which electronic commerce is conducted is full of unclear and conflicting laws. In many cases, government regulators have not kept up with technologies. Laws that govern commerce were written when signed documents were a reasonable expectation in any business transaction. However, as more businesses and individuals find the benefits of electronic commerce to be compelling, many of these technology and culture-related disadvantages will disappear.

International Electronic Commerce

The Internet brings people together from every country in the world. It reduces the distances between people in many ways. About 60 percent of all electronic commerce sites on the Web are in English, although sites in other languages and in multiple languages are appearing with increasing frequency. Once the language barrier is overcome, the technology exists for any business to conduct electronic commerce with any other business or consumer, anywhere in the world.

Unfortunately, the political structures of the world have not kept up with Internet technology, so doing business internationally presents a number of challenges. Currency conversions, tariffs, import and export restrictions, local business customs, and the laws of each country in which a trading partner resides can make international electronic commerce difficult.

Many of the international issues that arise relate to legal, tax, and privacy concerns. Each country has the right to pass laws and levy taxes on businesses that operate within their jurisdiction. European countries, for example, have very strict laws that limit the collection and use of personal information that companies gather in the course of doing business with consumers. Even within the United States, individual states and counties have the power to levy sales and use taxes on goods and services. In other countries, national sales and value-added taxes are imposed on an even broader list of business activities. Later in this book, you will learn more about the international, legal, and tax issues that can arise when conducting electronic commerce.

The main technological development that has allowed electronic commerce to grow beyond its beginnings in bank EFTs and business-to-business EDI is the emergence of the Internet and the Web. The next section presents a short outline of this technology and describes how it came to exist and grow to its present state.

THE INTERNET AND WORLD WIDE WEB

Millions of people use the Internet every day, but only a small percentage of them really understand how it works. The **Internet** is a large system of interconnected computer networks that spans the globe. Using the Internet, you can communicate with other people throughout the world by means of electronic mail; read online versions of newspapers, magazines, academic journals, and books; join discussion groups on almost any conceivable topic; participate in games and simulations; and obtain free computer software. In recent years, the Internet has allowed commercial enterprises to connect with one another and with customers. Today, all kinds of businesses provide information about their products and services on the Internet. Many of these businesses use the Internet to market and sell their products and services. The part of the Internet known as the **World Wide Web**, or, more simply, the **Web**, is a subset of the computers on the Internet that are connected to each other in a specific way that makes those computers and their contents easily accessible to each other. The most important thing about the Web is that it includes an easy-to-use standard interface. This interface makes it possible for people who are not computer experts to use the Web to access a variety of Internet resources.

Origins of the Internet

In the early 1960s, the U.S. Department of Defense became very concerned about the possible effects of nuclear attack on its computing facilities. The Defense Department realized that the weapons of the future would require powerful computers for coordination and control. The powerful computers of that time were all large mainframe computers, so the Defense Department began examining ways to connect these computers to each other and also to weapons installations that were distributed all over the world. The Defense Department agency charged with this task hired many of the best communications technology researchers and for many years funded research at leading universities and institutes to explore the task of creating a worldwide network that could remain operational even if parts of the network were destroyed by enemy military action or sabotage. These researchers worked to devise ways to build networks that could operate independently—that is, networks that would not require a central computer to control network operations.

The world's telephone companies were the early models for networked computers, because early networks of computers used leased telephone company lines for their connections. Telephone company systems of that time established a single connection between sender and receiver for each telephone call, and that connection carried all data along a single path. When a company wanted to connect computers it owned at two different locations, the company placed a telephone call to establish the connection and then connected one computer to each end of that single connection.

The Defense Department was concerned about the inherent risk of this single-channel method for connecting computers, and its researchers developed a different method of sending information through multiple channels. In this method, files and messages are broken into packets that are labeled electronically with codes for their origin and destination. The packets travel from computer to computer along the network until they reach their destination. The destination computer collects the packets and reassembles the original data from the pieces in each packet. Each computer that an individual packet encounters on its trip through the network determines the best way to move the packet forward to its destination.

In 1969, these Defense Department researchers used this network model to connect four computers—one each at the University of California at Los Angeles, SRI International, the University of California at Santa Barbara, and the University of Utah. Over subsequent years, many researchers in the academic community connected to this network and contributed to the technological developments that increased the speed and efficiency with which the network operated. At the same time, researchers at other universities were creating their own networks using similar technologies.

New Uses for the Internet

Although the goals of the Defense Department network were still to control weapons systems and transfer research files, other uses for this vast network began to appear in the early 1970s. In 1972, a researcher wrote a program that could send and receive messages over the network. E-mail was born and became widely used very quickly. The number of network users in the military and education research communities continued to grow. Many of these new participants used the networking technology to transfer files and access computers remotely. The network software included two tools for performing these tasks. **File Transfer Protocol (FTP)** enabled users to transfer files between computers, and **Telnet** let users log on to their computer accounts from remote sites. Both FTP and Telnet are still widely used on the Internet today for file transfers and remote logins, even though more advanced techniques are now available that allow multimedia transmissions such as real-time audio and video clips.

The first e-mail mailing lists also appeared on these networks. A **mailing list** is an e-mail address that forwards any message it receives to any user who has subscribed to the list. In 1979, a group of students and programmers at Duke University and the University of North Carolina started **Usenet**, an abbreviation for **User's News Network**. Usenet allows anyone who connects to the network to read and post articles on a variety of subjects. Usenet survives on the Internet today, with over 1000 different topic areas that are called newsgroups. Other researchers even created game-playing software for use on these interconnected networks.

Although the people using these networks were developing many creative applications, use of the networks was limited to those members of the research and academic communities who had access to the networks. Between 1979 and 1989, these new and interesting network applications were improved and tested by an increasing number of users. The Defense Department's networking software became more widely used as academic and research institutes realized the benefits of having a common communications network. The explosion of personal computer use during that time also helped more people become comfortable with computing. In the late 1980s, these independent academic and research networks merged into what we now call the Internet.

Commercial Use of the Internet

As personal computers became more powerful, affordable, and available during the 1980s, companies increasingly used them to construct their own internal networks. Although these networks included e-mail software that employees could use to send messages to each other, businesses wanted their employees to be able to communicate with people outside their corporate networks. The Defense Department network and most of the other academic networks that had teamed up with it were receiving funding from the **National Science Foundation** (NSF). The NSF prohibited commercial network traffic on its networks, so businesses turned to commercial e-mail service providers to handle their e-mail needs. Larger firms built their own networks that used leased telephone lines to connect field offices to corporate headquarters.

In 1989, the NSF permitted two commercial e-mail services, MCI Mail and CompuServe, to establish limited connections to the Internet for the sole purpose of exchanging e-mail transmissions with users of the Internet. These connections allowed commercial enterprises to send e-mail directly to Internet addresses and allowed members of the research and education communities on the Internet to send e-mail directly to MCI Mail and CompuServe addresses. The NSF justified this limited commercial use of the Internet as a service that would primarily benefit the Internet's noncommercial users. As the 1990s began, people from all walks of life—not just scientists or academic researchers—started thinking of these networks as the global resource that we now know as the Internet. Although this network of networks had grown from four Defense Department computers in 1969 to more than 300,000 computers on many interconnected networks by 1990, the greatest growth in the Internet was yet to come.

Growth of the Internet

In 1991, the NSF further eased its restrictions on Internet commercial activity and began implementing plans to privatize the Internet. The privatization of the Internet was substantially completed in 1995, when the NSF turned over the operation of the main Internet connections to a group of privately owned companies. The new structure of the Internet was based on four **network access points** (**NAPs**), each operated by a separate company. The San Francisco NAP is operated by Pacific Bell, the New York NAP is operated by Sprint, the Chicago NAP is operated by Ameritech, and the Washington, D.C., NAP is operated by MFS Corporation. These companies, which are known as **network access providers**, sell Internet access rights directly to larger customers and indirectly to smaller firms through other companies, called **Internet service providers** (**ISPs**).

The Internet was a phenomenon that truly sneaked up on an unsuspecting world. The researchers who had been so involved in the creation and growth of the Internet just accepted it as part of their working environment. People outside the research community were largely unaware of the potential offered by a large interconnected set of computer networks. Figure 1-6 shows the number of **Internet hosts**, which are computers directly connected to the Internet, for the years 1991 through 2000. As you can see, the growth has been dramatic.

In under 30 years, the Internet became one of the most amazing technological and social accomplishments of the 20th century. Millions of people are using a complex, interconnected network of computers. These computers run thousands of different software packages. The computers are located in almost every country of the world. Every year billions of dollars change hands over the Internet in exchange for all kinds of products and services. All of this activity occurs with no central coordination point or control, which is especially interesting given that the Internet began as a way for the military to maintain control while under attack.

The opening of the Internet to business activity helped increase the Internet's growth dramatically; however, there was another development that worked hand in hand with the commercialization of the Internet to spur its growth. That development was the World Wide Web.

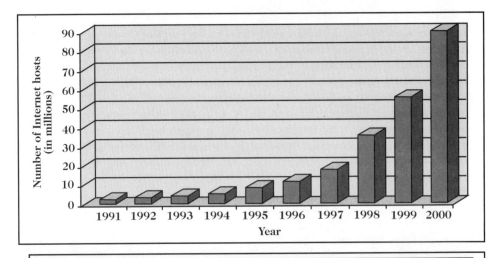

Figure 1-6   *Growth in the number of Internet hosts, 1991–2000*

Emergence of the World Wide Web

The Web is more a way of thinking about and organizing information storage and retrieval than it is a technology. As such, its history goes back many years. Two important innovations played key roles in making the Internet easier to use and more accessible to people who were not research scientists. These two innovations were hypertext and graphical user interfaces.

The Development of Hypertext

In 1945, **Vannevar Bush**, who was director of the U.S. Office of Scientific Research and Development, wrote an *Atlantic Monthly* article about ways that scientists could apply the skills they learned during World War II to peacetime activities. The article included a number of visionary ideas about future uses of technology to organize and facilitate efficient access to information. Bush speculated that engineers eventually would build a machine that he called the Memex, a memory extension device that would store all a person's books, records, letters, and research results on microfilm. Bush's Memex would include mechanical aids, such as microfilm readers and indexes that would help users consult their collected knowledge quickly and flexibly. In the 1960s, Ted Nelson described a similar system in which text on one page links to text on other pages. Nelson called his page-linking system **hypertext**. Douglas Engelbart, who also invented the computer mouse, created the first experimental hypertext system on one of the large computers of the 1960s. In 1987, Nelson published *Literary Machines*, a book in which he outlined project Xanadu, a global system for online hypertext publishing and commerce.

In 1989, Tim Berners-Lee was trying to improve the laboratory research document-handling procedures for his employer, CERN: European Laboratory for Particle Physics. CERN had been connected to the Internet for two years, but its scientists wanted to find better ways to circulate their scientific papers and data among the high-energy physics research community throughout the world. Berners-Lee proposed a hypertext development project intended to provide this data-sharing functionality.

Over the next two years, Berners-Lee developed the code for a hypertext server program and made it available on the Internet. A **hypertext server** is a computer that stores files written in the hypertext markup language and lets other computers connect to it and read those files. Hypertext servers used on the Web today are usually called **Web servers**. **Hypertext markup language** (**HTML**), which Berners-Lee developed from his original hypertext server program, is a language that includes a set of codes (or tags) attached to text. These codes describe the relationships among text elements. For example, HTML includes tags that indicate which text is part of a header element, which text is part of a paragraph element, and which text is part of a numbered list element. One important type of tag is the hypertext link tag. A **hypertext link**, or **hyperlink**, points to another location in the same or another HTML document.

Graphical Interfaces for Hypertext

You can use several different types of software to read HTML documents, but most people use a Web browser such as Netscape Navigator or Microsoft Internet Explorer. A **Web browser** is a software interface that lets users read (or browse) HTML documents and move from one HTML document to another through text formatted with hypertext link tags in each file. If the HTML documents are on computers connected to the Internet, you can use a Web browser to move from an HTML document on one computer to an HTML document on any other computer on the Internet. HTML is based on **Standard Generalized Markup Language** (**SGML**), which organizations have used for many years to manage large document-filing systems. Another markup language derived from SGML is now being used on the Web to facilitate business transaction processing. This **eXtensible Markup Language** (**XML**) allows users to define new meanings for its commands or extend (thus, "extensible") the existing meanings of its commands. XML is similar to HTML, but can be used to do more things on the Web. Figure 1-7 shows the relationships of these markup languages.

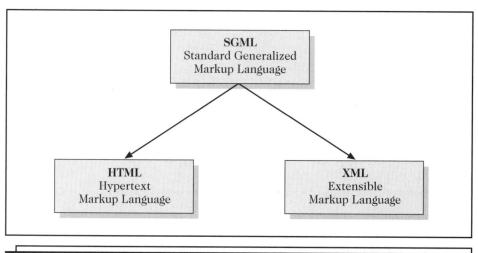

Figure 1-7   *Markup languages*

An HTML document differs from a word processing document in that it does not specify how a particular text element will appear. For example, you might use word processing software to create a document heading by setting the heading text font to Arial, its font size to 14 points, and its position to centered. The document would display and print these exact settings whenever you opened the document in that word processor. In contrast, an HTML document would simply include a heading tag with the heading text. Many different programs can read an HTML document. Each program recognizes the heading tag and displays the text in whatever manner each program normally displays headings. Different programs might display the text differently, but always with the characteristics of a heading.

A Web browser presents an HTML document in an easy-to-read format in the browser's graphical user interface. A **graphical user interface (GUI)** is a way of presenting program control functions and program output to users. It uses pictures, icons, and other graphical elements instead of just displaying text. Almost all personal computers today use a GUI such as Microsoft Windows or the Macintosh user interface.

Berners-Lee called his system of hyperlinked HTML documents the World Wide Web. The Web caught on quickly in the scientific research community, but few people outside that community had software that could read the HTML documents. In 1993, a group of students led by Marc Andreessen at the University of Illinois wrote Mosaic, the first GUI program that could read HTML and use HTML hyperlinks to navigate from page to page on computers anywhere on the Internet. Mosaic was the first Web browser that became widely available for personal computers, and it is still in use today.

Programmers quickly realized that a functional system of pages connected by hypertext links would provide many new Internet users with an easy way to access information on the Internet. Businesses recognized the profit-making potential offered by a worldwide network of easy-to-use computers. In 1994, Andreessen and other members of the University of Illinois Mosaic team joined with James Clark of **Silicon Graphics** to found **Netscape Communications**. Their first product, the Netscape Navigator Web browser program based on Mosaic, was an instant success. Netscape became one of the fastest growing software companies ever. **Microsoft** created its Internet Explorer Web browser and entered the market soon after Netscape's success became apparent. A number of other Web browsers exist, but these two products dominate the market today.

The number of Web sites has grown even more rapidly than the Internet itself. The number of Web sites is currently estimated to be about 16 million, and the number of Web pages is likely over 1 billion. Each Web site might include hundreds or even thousands of individual Web pages. Therefore, nobody really knows how many Web pages exist. For example, researchers at **BrightPlanet** estimated that the number of Web sites could be more than 500 million. Figure 1-8 shows that the growth rate of the Web has increased dramatically beginning in 1996.

As more people gain access to the Web, commercial interest in using the Web to conduct business will increase and the variety of nonbusiness uses will become even greater. Although the Web has already grown very rapidly, many experts believe that it will continue to grow at an increasing rate for the foreseeable future. In Chapter 2, you will learn more about how Internet and Web technologies work to enable electronic commerce. The next section discusses some of the broad economic issues that relate to the emergence and future growth of electronic commerce.

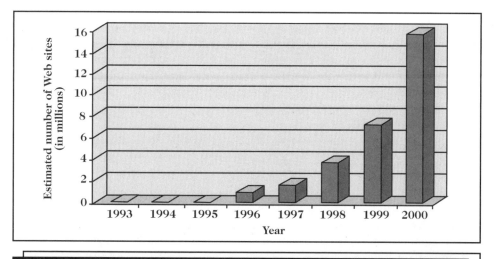

Figure 1-8   *Growth of the World Wide Web*

ECONOMIC FORCES AND ELECTRONIC COMMERCE

Economics is the study of how people allocate scarce resources. One important way that people allocate resources is through commerce (the other major way people allocate resources is through government actions, such as taxes or subsidies). Many economists are interested in how people organize their commerce activities. One important way people do this is to participate in markets. Economists use a formal definition of **market** that includes two conditions: first, that the potential sellers of a good come into contact with potential buyers, and second, that a medium of exchange is available. This medium of exchange can be currency or barter. Most economists agree that markets are strong and effective mechanisms for allocating scarce resources. Thus, one would expect most business transactions to occur within markets. However, much business activity today occurs within large **hierarchical business organizations**, which economists generally refer to as **firms**.

Most hierarchical organizations are headed by a top-level president or chief operating officer. Reporting to the president are a number of vice presidents who, in turn, have a larger number of middle managers who report to them, and so on. An organization can have a relatively flat hierarchy, in which there are only a few levels of management, or it can feature many reporting levels. In either case, the bottom level features the largest number of employees of any type, and is usually made up of production workers or service providers. Thus, the hierarchical organization will always have a pyramid-shaped structure.

These large firms often conduct many different business activities entirely within the organizational structure of the firm and participate in markets only for the purchase of raw materials and for the selling of their completed products. If markets are indeed highly effective mechanisms for allocating scarce resources, these large corporations should participate in markets at every stage of their production and value-generation processes. Nobel laureate Ronald Coase wrote an essay in 1937 in which he questioned why individuals who engaged in commerce often created firms to organize their activities. He was particularly interested in the hierarchical structure of these business organizations. Coase concluded that transaction costs were the main motivation for moving economic activity from markets to hierarchically structured firms.

Transaction Costs

Transaction costs are the total of all costs that a buyer and a seller incur as they gather information and negotiate a purchase-sale transaction. Although brokerage fees and sales commissions can be a part of transaction costs, the cost of information search and acquisition is the greatest element. Another significant component of transaction costs can be the investment a seller makes in equipment or the hiring of skilled employees to supply the product or service to the buyer.

To understand better how transaction costs occur in markets, consider a sweater production example. A sweater dealer could obtain sweaters by engaging in market transactions with a number of independent sweater knitters. Transaction costs incurred by the dealer would include the costs of identifying the independent knitters, visiting them to negotiate the purchase price, arranging for delivery of the sweaters, and inspecting the sweaters when they arrived. The knitters would also incur costs, such as the purchase of knitting tools and yarn. Since individual knitters could not know whether any sweater dealer would ever buy sweaters from them, the investments they must make to enter the sweater knitting business would have an uncertain yield. This risk is a significant transaction cost for the knitters.

After purchasing the sweaters, the dealer takes them to a different market in which sweater dealers meet and do business with the retail shops that sell sweaters to the consumer. The dealers can use these market negotiations to find out which sweater colors and patterns are in demand and can then use that information to negotiate price and other terms in the knitters' market. A diagram of this set of markets appears in Figure 1-9.

Markets and Hierarchies

Coase reasoned that when transaction costs were high, businesspersons would form organizations to replace market-negotiated transactions with a hierarchical form of organization that features a supervision and monitoring system. Instead of negotiating with individuals to purchase sweaters they had manufactured, a hierarchical organization would hire knitters, then supervise and monitor their work activities. Although the costs of creating and maintaining a supervision and monitoring system are high, they can be lower than transaction costs in many instances. In the sweater example, the sweater dealer would hire knitters, supply them with yarn and knitting tools, and would supervise their knitting activities. This supervision could be done mainly by first-line supervisors, who might be drawn from the ranks of the more skilled knitters. The practice of an existing firm replacing one of its supplier markets with its own hierarchical structure for creating the supplied product is called **vertical integration**. Figure 1-10 shows how our wool sweater example would look after the knitters were vertically integrated into the hierarchical structure of the sweater dealer's organization.

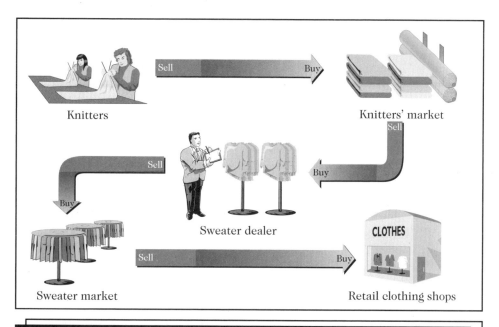

Figure 1-9 *Market forms of economic organization*

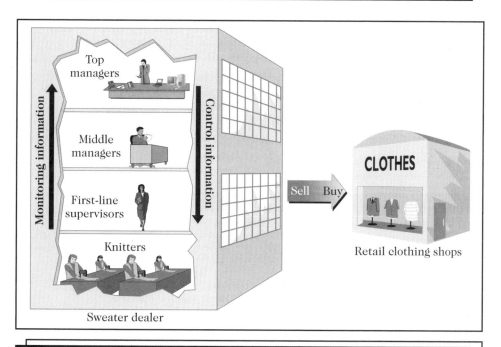

Figure 1-10 *Hierarchically-structured form of economic organization*

Oliver Williamson, an economist who extended Coase's analysis, noted that industries with complex manufacturing and assembly operations tended to include many firms that used hierarchical structures and were substantially vertically integrated.

Many of the manufacturing and administrative innovations that occurred in businesses during the 20th century increased the efficiency and effectiveness of hierarchical monitoring activities. Assembly lines and other mass-production technologies allowed work to be broken down into small, easily supervised procedures. The advent of computers brought tremendous increases in the ability of upper-level managers to monitor and control the detailed activities of their subordinates. Some of these direct measurement techniques are even more effective than the first-line supervisors on the shop floor.

As improvements in monitoring became commonplace, the size and level of vertical integration of firms increased consistently during the years from the Industrial Revolution through the present. In some very large organizations, however, monitoring systems have not kept pace with the organization's increase in size. This creates problems because the whole economic viability of the firm depends on its ability to effectively track the operational activities at the very lowest levels of the firm. These firms have instituted decentralization programs that allow business units to function as separate organizations and negotiate transactions with other business units as if they were operating in a market rather than as part of the same firm. These decentralization approaches are simply a return to the highly effective market mechanisms that worked so well before the firm vertically integrated itself.

Exceptions to the general trend toward hierarchies do exist. Many commodities, such as wheat, sugar, and crude oil, are still traded in markets. The commodity nature of the products traded in these markets reduces transaction costs significantly. There are a large number of potential buyers for an agricultural commodity such as wheat, and the farmer does not make any special investment in customizing or modifying the product for a particular customer. Thus, neither buyers nor sellers in commodity markets experience significant transaction costs.

The Role of Electronic Commerce

Businesses and individuals can use electronic commerce to reduce transaction costs by improving the flow of information and increasing coordination of actions. By reducing the cost of searching for information about potential buyers and sellers and increasing the number of potential market participants, electronic commerce can change the attractiveness of vertical integration for many firms. It is not clear yet whether widespread adoption of electronic commerce will cause hierarchical organization structures to revert to their former market-based structures, but it certainly is a distinct possibility.

To see how electronic commerce can change the level and nature of transaction costs, consider an employment transaction. The agreement to employ a person has high transaction costs for the seller—the employee who sells his or her services. These transaction costs include a commitment to forego other employment and career development opportunities. Individuals make a high investment in learning and adapting to the culture of their employers. If accepting the job involves a move, the new employee can incur very high costs, including actual costs of the move and related costs, such as the loss of a spouse's job. Much of this investment is company-specific; it would not transfer to a new job at the same location with the same company. If a sufficient number of employees throughout the world can telecommute—that is, perform their job tasks from any location using electronic commerce technologies—then many of these transaction costs could be eliminated. Instead of uprooting a spouse and family to move, a worker could accept a new job by simply logging on to a different Internet server!

Some researchers have argued that many companies and strategic business units operate in an economic structure that exists *between* markets and hierarchies. In this **network economic structure**, firms coordinate their strategies, resources, and skill sets by forming a long-term, stable relationship based on a shared purpose. Network organizations are particularly well suited to technology industries that are information-intensive. In our sweater example, the knitters might organize into networks of smaller organizations that specialize in certain styles or designs. Some of the particularly skilled knitters might leave the sweater dealer to organize their own firm to produce custom-knit sweaters. Some of the sweater dealer's marketing employees might form an independent firm that conducts market research on what the retail shops plan to buy in the upcoming months. They could sell their research reports to both the sweater dealer and the custom-knitting firm. As market conditions change, these smaller and more nimble organizations can continually reinvent themselves and take advantage of new opportunities that arise in the sweater markets. An illustration of such a network organization appears in Figure 1-11.

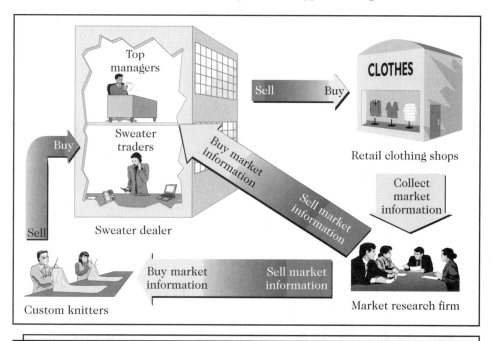

Figure 1-11 *Network form of economic organization*

Electronic commerce can make such networks, which rely extensively on information sharing, much easier to construct and maintain. Some researchers believe that these network forms of organizing commerce will become predominant in the near future. One of these researchers, Manuel Castells, has even predicted that economic networks will become the organizing structure for all social interactions among people. Thomas Petzinger, a columnist for *The Wall Street Journal*, has written extensively about these new patterns of work and commerce in his newspaper columns and in his 1999 book, *The New Pioneers*.

One interesting role for electronic commerce is in the improvement of existing markets and the creation of completely new markets. You will learn more about this aspect of electronic commerce in later chapters of this book. Regardless of how businesses in a particular industry organize themselves—as markets, hierarchies, or networks—you will need a way to identify business processes and evaluate whether electronic commerce is suitable for each process. The next section presents one good structure for examining business processes.

VALUE CHAINS IN ELECTRONIC COMMERCE

Electronic commerce includes so many activities that it can be difficult for managers to decide where and how to use it in their businesses. One way to focus on specific business processes as candidates for electronic commerce is to break the business down into a series of value-adding activities that combine to generate profits and meet other goals of the firm. In this section, you will learn about one popular way to analyze business activities as a sequence of activities that create value for the firm.

Commerce is conducted by firms of all sizes. Smaller firms can focus on one product, distribution channel, or type of customer. Larger firms often sell many different products and services through a variety of distribution channels to several types of customers. In these larger firms, managers organize their work around the activities of strategic business units. A **strategic business unit**, or simply **business unit**, is one particular combination of product, distribution channel, and customer type. Multiple business units owned by a common set of shareholders make up a firm, and multiple firms that sell similar products to similar customers make up an industry.

Strategic Business Unit Value Chains

In his 1985 book, *Competitive Advantage*, Michael Porter introduced the idea of value chains. A **value chain** is a way of organizing the activities that each strategic business unit undertakes to design, produce, promote, market, deliver, and support the products or services it sells. In addition to these **primary activities**, Porter also includes **supporting activities**, such as human resource management and purchasing, in the value chain model. Figure 1-12 shows a value chain for a strategic business unit engaged in manufacturing a product, including both primary and supporting activities.

The left-to-right flow in Figure 1-12 does not imply a strict time sequence for these processes. For example, a business unit may engage in marketing activities before purchasing materials and supplies. For each business unit, the primary activities include:

- **Identify customers:** activities that help the firm find new customers and new ways to serve old customers, including market research and customer satisfaction surveys
- **Design:** activities that take a product from concept to manufacturing, including concept research, engineering, and test marketing
- **Purchase materials and supplies:** procurement activities, including vendor selection, qualification, negotiating long-term supply contracts, and monitoring quality and timeliness of delivery

- **Manufacture:** activities that transform materials and labor into finished products, including fabricating, assembling, finishing, testing, and packaging
- **Market and sell:** activities that give buyers a way to purchase and that provide inducements for them to do so, including advertising, promoting, managing salespersons, pricing, and identifying and monitoring sales and distribution channels
- **Deliver:** activities that store, distribute, and ship the final product, including warehousing, handling materials, consolidating freight, selecting shippers, and monitoring timeliness of delivery
- **Provide after-sale service and support:** activities that promote a continuing relationship with the customer, including installing, testing, maintaining, repairing, fulfilling warranties, and replacing parts

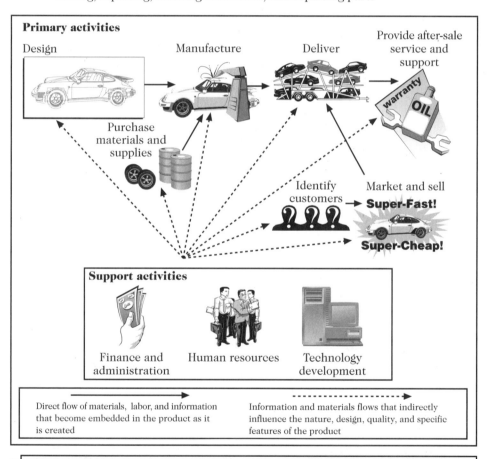

Figure 1-12   *Value chain for a strategic business unit*

The importance of each primary activity depends on the product or service the business unit provides and to which customers it sells. If a strategic business unit provides a service, its value chain would include a *Provide service* activity instead of the *Manufacture* activity shown in Figure 1-12. The other activities in its value chain would be similar to those for a product manufacturing business unit. Each business unit must also undertake support activities that provide the infrastructure for the unit's primary activities. These support activities appear in Figure 1-12 and include:

- **Finance and administration:** activities that provide the firm's basic infrastructure, including accounting, paying bills, borrowing funds, reporting to government regulators, and ensuring compliance with relevant laws
- **Human resources:** activities that coordinate the management of employees, including recruiting, hiring, training, compensation, and providing benefits
- **Technology development:** activities that help improve the product or service that the firm is selling and that help improve the business processes in every primary activity, including basic research, applied research and development, process improvement studies, and field tests of maintenance procedures

Industry Value Chains

Porter's book also identifies the importance of examining where the strategic business unit fits within its industry. Porter uses the term **value system** to describe the larger stream of activities into which a particular business unit's value chain is embedded. However, many subsequent researchers and business consultants have used the term **industry value chain** when referring to value systems. When a business unit delivers a product to its customer, that customer may, for example, use the product as purchased materials in its value chain. By becoming aware of how other business units in the industry value chain conduct their activities, managers can identify new opportunities for cost reduction, product improvement, or channel reconfiguration. An example of an industry value chain appears in Figure 1-13. This value chain is for a wooden chair and traces the life of the product from trees in a forest to its grave in a landfill or at a sawdust recycler.

Each business unit (logger, sawmill, lumberyard, chair factory, retailer, consumer, and recycler) shown in Figure 1-13 has its own value chain. For example, the sawmill purchases logs from the tree harvester and combines them in its manufacturing process with inputs such as labor and saw blades from other sources. Among the sawmill customers are the chair factory shown in Figure 1-13 and other users of cut lumber. Examining this industry value chain could be useful for the sawmill that is considering entering the tree harvesting business or the furniture retailer who is thinking about partnering with a trucking line. The industry value chain identifies opportunities up and down the product's life cycle for increasing the efficiency or quality of the product.

The Role of Electronic Commerce

One opportunity that many businesses are finding as they examine their industry value chains is that electronic commerce can play a role in reducing costs, improving product quality, reaching new customers or suppliers, and creating new ways of selling existing products. For example, a software developer who releases annual updates to programs might consider removing the software retailer from the distribution channel for updates by offering to send the updates through the Internet directly to the consumer. This change would modify the software developer's industry value chain and would be an opportunity for increasing sales revenue (the software developer retains the margin a dealer would have added to the price of the update), but it would not appear as part of the software developer business unit value chain. By examining elements of the value chain outside the individual business unit, managers can identify many business opportunities, including opportunities that can be exploited using electronic communications technologies—electronic commerce.

The value chain concept is a useful way to think about business strategy in general. When firms are considering electronic commerce, the value chain can be an excellent way to organize their examination of business processes within their business units and in other parts of their product's life cycle. Using the value chain reinforces the idea that electronic commerce should be a business solution, not a technology implemented for its own sake.

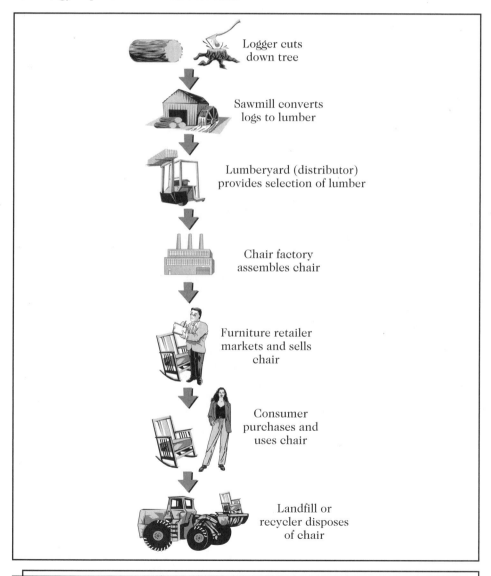

| **Figure 1-13** | *Industry value chain for a wooden chair* |

Summary

In this chapter, you have learned that commerce, the negotiated exchange of goods or services, has been practiced in traditional ways for thousands of years. Electronic commerce is the application of new technologies, particularly Internet and Web technologies, to help individuals, businesses, and other organizations better conduct business. Not all activities lend themselves to improvement with these technologies, but many do. Using electronic commerce, some businesses have been able to create new products and services, and others have improved the promotion, marketing, and delivery of existing offerings. Firms have also found many ways to use electronic commerce to improve purchasing and supply activities, identify new customers, and operate their finance, administration, and human resource management activities more efficiently.

You learned about the history of the Internet and the Web, including how these technologies emerged from research projects and grew to be the supporting infrastructure for electronic commerce today. We also examined an overview of markets, hierarchies, and networks—the economic structures in which businesses operate—and described how electronic commerce fits into those structures. Porter's ideas about value chains at the business unit and industry levels were presented, and you learned how to use value chains as a way to organize business processes and analyze their suitability for electronic commerce implementation.

Key Terms

Business Processes
Business-to-Business (B2B)
Business-to-Consumer (B2C)
Commerce
Commodity
Doing Business
Electronic Business (e-business)
Electronic Commerce (e-commerce)
Electronic Data Interchange (EDI)
Electronic Funds Transfer (EFT)
Extensible Markup Language (XML)
File Transfer Protocol (FTP)
Firm
Graphical User Interface (GUI)
Hierarchical Business Organization
Hypertext
Hypertext Link (hyperlink)
Hypertext Markup Language (HTML)
Hypertext Server
Industry Value Chain
Internet
Internet Commerce
Internet Host
Internet Service Providers (ISPs)
Mailing List

Market
Market Research
Merchandising
Network Access Points (NAPs)
Network Access Providers
Network Economic Structure
Primary Activities
Procurement
Standard Generalized Markup Language (SGML)
Strategic Business Unit (business unit)
Supply Management
Supporting Activities
Telnet
Trading Partners
Transaction Costs
Usenet (User's News Network)
Value Added Network (VAN)
Value Chain
Value System
Vertical Integration
Web Browser
Web Server
Wire Transfer
World Wide Web (Web)

Review Questions

1. What are the key differences between traditional commerce and electronic commerce?

2. What two developments contributed to the emergence of the Internet as an electronic commerce infrastructure?

3. What are transaction costs and why are they important?

4. How might electronic commerce help change an industry's economic structure from a hierarchy to a network?

5. How might a manager use an industry value chain to identify new applications for electronic commerce?

Exercises

1. You have decided to buy a new laser printer for your home office. List specific activities that you must undertake as you gather information about printer capabilities and features. Use the **CompUSA**, **Office Depot**, and **Staples** Web sites to gather information. Write a short summary of the process you undertook so that others who plan to undertake a similar task can use your information.

2. **Germany** and **China** have laws that prohibit the use of certain kinds of political speech and adult images in the conduct of commerce. Each of these countries has recently taken steps to enforce these laws by prosecuting firms and individuals who violate them while conducting electronic commerce. Outline the problems that such enforcement actions can create for firms and their non-German or non-Chinese customers who engage in electronic commerce. Propose a solution for each problem you identify.

3. You are a consultant to a firm that wants to establish an electronic commerce presence selling videos to consumers. You believe that after-sale service and support will be an important way in which this firm can differentiate itself from other firms in this business. You have decided that the online book industry is a good model for the video business, and you want to adapt business processes and practices from the after-sale service and support primary activity of that industry's value chain for your client. Use the Online Companion links for this exercise to examine the Web sites for **Amazon.com**, **Barnes and Noble**, **Books-A-Million**, **Borders**, and **Internet Bookshop**. Identify three electronic commerce business processes that one or more of these booksellers use and that you feel would work for your client. As you work to identify these business processes, remember that you may be able to observe specific Web site features that are a result of such processes. You may also be able to infer the existence of business processes by reading statements that these companies make on their Web sites. Explain how you would adapt each process to meet the needs of the video retailing business.

Anders, G. 1998. "Click and Buy," *The Wall Street Journal*, December 7, R4.

Bergman, M. 2000. *The Deep Web: Surfacing Hidden Value.* Sioux Falls, SD: BrightPlanet.com. Available online at: (http://www.completeplanet.com/Tutorials/DeepWeb/).

Berry, J. 2000. "E-Business Still an Elusive Goal for Most Companies," *Internetweek*, April 24, 41–42.

Biggs, M. 1998. "E-Commerce Is Hot Today, But E-Business Is the Gift That Keeps on Giving All Year Long," *InfoWorld*, 20(51), December 21, 82.

Blackmon, D. 2000. "Where the Money Is," *The Wall Street Journal*, April 17, R30.

Brache, A. and J. Webb. 2000. "The Eight Deadly Assumptions of E-Business," *Journal of Business Strategy*, 21(3), May–June, 13–17.

Business Week. 2000. "Jeff Bezos," May 15, EB28.

Castells, M. 1996. *The Rise of the Network Society.* Cambridge, MA: Blackwell.

Champy, J. 2000. "Re-Engineering Redux," *Computerworld*, 34(17), April 24, 47.

Coase, R. 1937. "The Nature of the Firm," *Economica*, 4(4), November, 386–405.

Corcoran, C. 2000. "Who's on Top, and in Trouble, in E-Tail's Various Categories," *The Wall Street Journal Interactive Edition*, January 4. Available online at (http://interactive.wsj.com/archive/retrieve.cgi?id=SB946927753582315731.djm).

Dvorak, J. 1999. "E-Biz Busters," *PC Computing*, 12(2), February, 101.

Ernst & Young, LLP. 1999. *The Second Annual Ernst & Young Internet Shopping Study: The Digital Channel Continues to Gather Steam.* New York: Ernst & Young, LLP.

Fry, J. 1998. "Person to Person," *The Wall Street Journal*, December 7, R12.

Garner, R. 1999. "The E-Commerce Connection," *Sales & Marketing Management*, 151(1), January, 40–44.

Gosh, S. 1998. "Making Business Sense of the Internet," *Harvard Business Review*, 76(2), March–April, 126–135.

Greenemeier, L. 2000. "Andersen Plans for E-Business Growth," *Informationweek*, April 24, 40.

Hamel, G. and J. Sampler. 1998. "The E-Corporation," *Fortune*, 138(11), December 7, 80–87.

Hammer, M. and J. Champy. 1993. *Reengineering the Corporation: A Manifesto for Business Revolution.* New York: HarperBusiness.

Hannon, N. 1998. *The Business of the Internet.* Cambridge, MA: Course Technology.

Hertzberg, R. 1999. "Four Key Rules of Commerce," *Internet World*, January 11, 8.

Hof, R. 1998. "A New Chapter for Amazon.com," *Business Week*, August 17, 39.

Hofman, M. 2000. "Brave New Companies: The Metamorphosis," *Inc*, 22(4), March 14, 52–60.

Kalakota, R. and A. Whinston. 1996. *Readings in Electronic Commerce.* Reading, MA: Addison-Wesley.

LaMonica, M. 2000 "Buying an E-Business Strategy on the Cheap," *InfoWorld*, 22(17), April 24, 5.

Nelson, T. H. 1987. *Literary Machines.* Swarthmore, PA: Nelson.

Mandel, M. 1999. "The Internet Economy," *Business Week*, February 22, 30–32.

Miller, M. 1999. "The Net Changes Everything," *PC Magazine*, 18(3), February 9, 4.

Neuborne, E. 2000. "E-Tail: Gleaming Storefronts with Nothing Inside," *Business Week*, May 1, 94–95.

Petzinger, T. 1999. *The New Pioneers: The Men and Women Who Are Transforming the Workplace and Marketplace.* New York: Simon & Schuster.

Porter, M. 1985. *Competitive Advantage.* New York: Free Press.

Porter, M. 1998. "Clusters and the New Economics of Competition," *Harvard Business Review*, 76(6), November–December, 77–90.

Powell, W. 1990. "Neither Market nor Hierarchy: Network Forms of Organization," *Research in Organizational Behavior*, 12(3), 295–336.

Ring, R. and A. Van de Ven. 1992. "Structuring Cooperative Relationships Between Organizations," *Strategic Management Journal*, 13(4), 483–498.

Shapiro, A. 1999. *The Control Revolution: How the Internet Is Putting Individuals in Charge and Changing the World We Know*. New York: The Century Foundation.

Shapiro, C. and H. Varian. 1999. *Information Rules: A Strategic Guide to the Network Economy*. Boston: Harvard Business School Press.

Schonfeld, E. 1999. "The Exchange Economy," *Fortune*, 139(3), February 15, 67–68.

Schulz, J. 2000. "The World's Big Four," *Journal of Commerce*, May 8, 29.

Schwartz, N. 2000. "Playing the Internet's Next Gold Rush," *Fortune*, 141(10), May 15, 178–183.

Shapiro, C. and H. Varian. 1998. *Information Rules: A Strategic Guide to the Network Economy*. Cambridge, MA: Harvard Business School Press.

Wilder, C. 1998. "The New E-Business Paradigm," *InformationWeek*, September 14, 18.

Williamson, O. 1975. *Markets and Hierarchies: Analysis and Antitrust Implications*. New York: Free Press.

Williamson, O. 1985. *The Economic Institutions of Capitalism*. New York: Free Press.

Willis, C. and S. Donahue. 1998. "Does Amazon.com Really Matter?" *Forbes*, 161(7), April 6, 55–58.

INFRASTRUCTURE FOR ELECTRONIC COMMERCE

Dell Computer Corporation is one of the most successful personal computer retailers in the history of PC sales. One of the world's largest computer manufacturers, Dell recently exceeded $25 billion in annual sales. Dell's customers include large and small corporations, governmental agencies, educational institutes, and individuals.

Through the early 1990s, Dell sold directly to the consumer through its toll-free telephone line. A few years ago, Dell expanded its sales to the Internet and has increasingly logged a significant percentage of its overall sales from the Internet. Thousands of customers now visit and place orders through the **Dell Computer Corporation Electronic Commerce Site**. Dell records millions of dollars in online sales each day. In addition to increasing business, using the Web to sell has lowered overhead for the company. Sales made directly through the Dell Web site mean that fewer humans are involved in the transactions. Technical support, including answers to the most frequently asked questions, is available online as well. The Web site is a significant part of Dell's strategy for continued growth in the 21st century. Company officials predict that within the next few years more than half of Dell's sales will be from the Web.

Supporting such burgeoning online sales is a robust infrastructure of communication devices and networks, Dell servers, and electronic commerce software from Microsoft. Both the infrastructure and software have been chosen for their capabilities and their ability to scale up as the customer base increases. Dell's state-of-the-art Web site will help it maintain its status as an extremely successful PC retailer.

LEARNING OBJECTIVES

In this chapter, you will learn about:

- The general structure of the network of networks supporting the Internet and electronic commerce
- Protocols that move commerce and communications across the Internet and that send and receive electronic mail
- Internet utility programs to trace, locate, and verify the status of Internet host sites
- Popular Internet applications, including electronic mail, Telnet, and File Transfer Protocol (FTP)
- The history and use of Web markup languages, including SGML, HTML, and XML
- HTML tags and links
- Web client and server architectures and the messages they send to each other
- The differences and similarities between internets, intranets, and extranets
- Options for connecting to the Internet and their cost and bandwidth trade-offs

TECHNOLOGY OVERVIEW

Several technologies must be in place for electronic commerce to exist. The most obvious one is the Internet. Beyond that system of interconnected networks, many other sophisticated software and hardware components are needed to provide the required support structure, including database software, network switches and hubs, encryption hardware and software, multimedia support, and, of course, the World Wide Web. All elements that support electronic commerce, including hardware and software, are in a constant state of change. Any business that engages in electronic commerce and that hopes to compete in the future must quickly adapt to new Internet technologies as they become available. The frenetic pace of the Web and technology requires that businesses be nimble and accept inevitable changes in the way they conduct business on the Web. Firms that aren't flexible will quickly lose their Web-based business to those that are. Online consumers say poor Web site performance (slow response time, for instance) drives them to abandon some electronic commerce sites in favor of those with faster response times. The anticipated electronic commerce overload will require companies to find faster and more efficient ways to deal with the ever-increasing crush of online shoppers and the increasing traffic between businesses. However, those companies that strategically plan and execute their electronic commerce strategies can experience huge payoffs—with online business volumes sometimes doubling in less than a year.

This chapter introduces to you the technologies that support electronic commerce. First, you will look at data routing options and the underlying protocols that move information on the Internet and deliver that information to its destination. Then, you will look at the other building blocks that support the Internet, the Web, and electronic commerce.

PACKET-SWITCHED NETWORKS

The early models (dating back to the 1950s) for networked computers were the local and long distance telephone companies. Most early computer networks used leased telephone company lines for their connections. In those days, a telephone call established a single connection between the caller and the receiver. Once a connection was established, data traveled along that single path. Telephone company switching equipment (both mechanical and computerized) selected specific telephone lines, or **circuits**, that were connected to create the single path between caller and receiver. This centrally controlled, single-connection model is known as **circuit switching**.

While circuit switching works well for telephone calls, using the same technique for sending data across a large network or a network of networks does not work well. Establishing point-to-point connections for each pair of sender/receivers is both expensive and difficult to manage. Instead, the Internet uses packet switching, a less expensive and more easily managed technique to move data between two points. In a **packet-switched** network, files and messages are broken down into packets that are labeled electronically with codes that indicate both their origin and destination. Packets travel from computer to computer along the network until they reach their destination. The destination computer collects the packets and reassembles the original data from the pieces in each packet. In packet switching, the best route to move the packet forward to its destination is determined by each computer that an individual packet encounters on its trip from its origin to its destination. Figure 2-1 illustrates a packet-switched network. Computers that make these determinations are often called **routers**, and the programs that determine the best path to follow are called **routing algorithms**.

Packet switching is the method used to move data on the Internet. There are several benefits to using packet switching. One is that long streams of data can be broken down into small, manageable data chunks, allowing the small packets to be distributed over a large number of possible paths to balance the traffic across the network. Another advantage is that it is relatively inexpensive to replace damaged data packets after they arrive. If a data packet is altered in transit, only a single packet is retransmitted.

The earliest packet-switched network, called ARPANET, connected only a few universities and research centers. This experimental wide area network (WAN) grew over the next few years and used the **Network Control Protocol (NCP)**. A **protocol** is a collection of rules for formatting, ordering, and error-checking data sent across a network. Protocols determine how the sending device indicates that it has finished sending a message and how the receiving device indicates that it has received (or not received) the message. The **open architecture** philosophy developed for the evolving ARPANET, which later became the Internet, included four key points that have contributed to the success of the Internet:

- Independent networks should not require any internal changes in order to be connected to the network.
- Packets that do not arrive at their destinations must be retransmitted from their source network.
- The router computers do not retain information about the packets that they handle.
- No global control exists over the network.

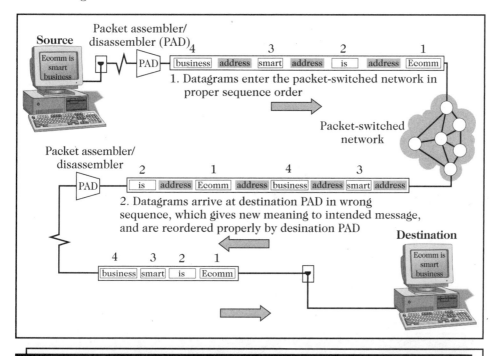

Figure 2-1   *Packet-switched network and message packets*

The TCP/IP Internet Protocol

The two protocols that support the basic operation of the Internet are the **Transmission Control Protocol (TCP)** and the **Internet Protocol (IP)**. Developed by Internet pioneers Vincent Cerf and Robert Kahn, these protocols establish fundamental rules about how data are moved across networks and how network connections are established and broken. The common acronym **TCP/IP** refers to the two protocols.

TCP/IP is a two-layered program. It includes rules that computers on a network use to establish and break connections. TCP controls the assembly of a message into smaller packets before it is transmitted over the Internet, and it controls the reassembly of packets once they reach their destination. The IP protocol includes rules for routing individual data packets from their source to their destination. IP handles all the addressing details for each packet, ensuring that each is labeled with the correct destination address. Together, these two protocols are technically superior to the NCP that ARPANET originally used.

Figure 2-2 illustrates TCP/IP's architecture, which is divided into duty-based functional layers. The five layers function as one unit when delivering information from one location on the Internet to another. The lowest, most fundamental layer is the hardware layer, which handles the individual pieces of equipment attached to the Internet. The highest layer is the application layer, where various Internet-serving applications run. Each layer provides services for the layer above it. While the exact details of the TCP/IP layers are beyond the scope of this text, it is important to note where some components lie within the architecture. The TCP protocol, for example, operates in the transport layer, and the Internet layer contains, among others, the IP protocol.

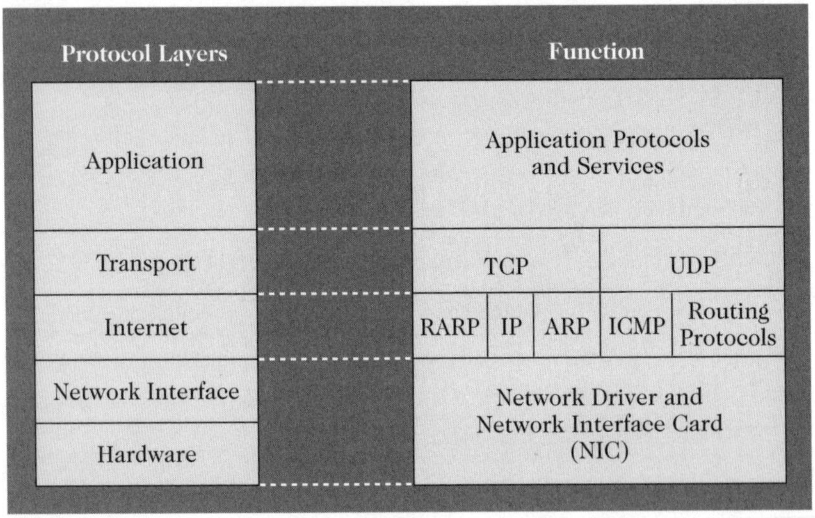

Protocol Layers		Function				
Application		Application Protocols and Services				
Transport		TCP			UDP	
Internet		RARP	IP	ARP	ICMP	Routing Protocols
Network Interface		Network Driver and Network Interface Card (NIC)				
Hardware						

Figure 2-2   TCP/IP architecture

In addition to its Internet function, TCP/IP continues to be used in **local area networks (LANs)**. A LAN connects workstations and personal computers in a single network, usually within a single physical location. Each individual computer in a LAN has its own CPU (central processing unit) to execute programs, but it can access data and devices on any other computer attached to the network. In this way, many users can share devices, such as laser printers and scanners. The TCP/IP protocol is standard in all Windows 95/98, Windows NT, and Windows 2000 computers.

The term *Internet* was first used in a 1974 article about the TCP protocol written by Cerf and Kahn. The development of the TCP/IP protocol in the history of the Internet is so significant that many people consider Vincent Cerf to be the father of the Internet.

IP Address and Domain Names

Internet addresses are represented in several ways, but all the formats are translated to a 32-bit number called an **IP address**. These 32-bit numbers eventually will be phased out, because there are not enough numbers to meet the current and future demand for addresses. The replacement IP address is a 128-bit number—a value that significantly expands the number of available addresses.

When the IP protocol assembles a message packet prior to sending it to its destination on the Internet, the message contains both the source IP address and the destination IP address. IP numbers (addresses) appear as a series of up to four separate numbers delineated by a period. Called a **dotted quad**, an address such as 126.204.89.56 uniquely identifies a computer connected to the Internet. Each of the four numbers can range from zero to 255, so the possible IP addresses begin with 0.0.0.0 and go up to 255.255.255.255. Generally, the first of the four possible numbers identifies a computer's network. The remaining numbers usually identify a particular computer (node) on the network.

Internet users find the dotted quad notation difficult to remember and error-prone. In its place, most people use the naming convention called the **Uniform Resource Locator (URL)**. A URL consists of names and abbreviations that are much easier to remember than numbers. A URL consists of at least two, and as many as four, parts. A simple two-part URL contains the protocol used to access a resource on the Internet followed by the location of the resource. For example, a URL such as http://www.adobe.com reveals that the HTTP is the protocol being used to access the resource, which is a corporate computer called www.adobe.com.

One of several Internet protocols, HTTP is the hypertext transfer protocol. It is the World Wide Web (the Web) access protocol that delivers and displays Web pages. The HTTP protocol, like other Internet protocols, defines how an Internet resource is accessed. Regardless of the method used to designate a resource in a URL, the resource's address must be converted to a 32- or 128-bit IP address. HTTP will be covered in more detail later in this chapter.

An address such as www.microsoft.com or www.course.com is called a **domain name**. Domain names can contain two or more word groups separated by periods. The leftmost part of a domain name (excluding www) is the most specific portion of the name. Each part of a domain name becomes more general as you move to the right, and the rightmost portion of a domain name is the most general. For example, the domain name www.breezy.compsci.nebraska.edu contains five parts separated by periods. Beginning at the left, *www* indicates a Web address. Next, *breezy* indicates the name of one of possibly several computers owned by the computer science department, which the next name, *compsci*, indicates. The institute (the University of Nebraska) is identified by *nebraska*. Finally, the name *edu* stands for educational institute and is one of the top-level domain names. Some domain names include a country designation. If the country designation is missing, then the country is understood to be the United States. Country names are also top-level domain names. Examples of top-level domain names are shown in Figure 2-3.

Name	Description
.com	U.S., Commercial
.edu	U.S., Educational
.gov	U.S., Government
.mil	U.S., Military
.net	U.S., General
.org	U.S., Nonprofit
.au	Australia
.ca	Canada
.de	Germany
.fr	France
.jp	Japan
.md	Moldova
.pt	Portugal
.uk	United Kingdom
.us	United States

Figure 2-3   *Top-level domain names*

Other Internet Protocols

As illustrated in Figure 2-2, TCP/IP incorporates a wide variety of application layer protocols that provide services to users. Sometimes known as application services, these include Web page display, network management facilities, remote login, file copying, electronic mail, and directory services. Some TCP/IP application services are used very frequently and others are used only occasionally. Several of the more popular protocols are outlined next.

HTTP

HTTP, short for **Hypertext Transfer Protocol**, is the Internet protocol responsible for transferring and displaying Web pages. The original HTTP protocol developed in 1991 was rather simple, but it has been continually evolving. HTTP runs in the application layer of the TCP/IP model (see Figure 2-2). Like other Internet protocols, HTTP employs the client/server model in which a user's (the client's) Web browser opens an HTTP session and sends a request for a Web page to a remote server. In response, the server creates an HTTP response message that is sent back to the client's (requester's) Web browser. (A **client/server model** is a network in which each computer on the network is either a client or a server. A **client** is a PC or workstation on which users run applications. A **server** is a powerful computer dedicated to managing disk drives, printers, or network traffic.) The response contains the page displayed by the client browser. After the client determines that the message it received is correct, the TCP/IP connection is closed and the HTTP session ends.

If a Web page contains objects such as movies, sound, or graphics, the client makes a request for each object. Thus, a Web page containing a background sound and three graphics requires five separate server request messages to retrieve the four objects—the background sound and three graphics—and the page that references these objects. The number of HTTP users has grown tremendously in recent years, and consequently HTTP has had an enormous impact on the Internet.

SMTP, POP, and IMAP

Electronic mail, or **e-mail**, that is sent across the Internet is managed and stored by programs and hardware collectively known as **mail servers**. Unlike your personal computer, these mail servers and their programs must be up and running 24 hours a day. Otherwise, mail cannot be received or sent. SMTP and POP are two protocols responsible for sending and retrieving e-mail using the client/server model. The e-mail client program running on a user's computer requests mail delivery from a mail server using the standard **Simple Mail Transfer Protocol (SMTP)**. A wide variety of electronic mail messaging facilities, such as Eudora, UNIX mail, and PINE, must adhere to the published SMTP standard to use the protocol to send mail to a mail server. SMTP specifies the exact format of a mail message and describes how mail is to be administered. SMTP provides application-level services to users connected to a local area network.

POP, short for **Post Office Protocol**, is responsible for retrieving e-mail from a mail server. It requests that the mail server take one of the following actions: retrieve mail from the mail server and subsequently delete it from the server; retrieve mail from the server but not delete it; or not retrieve mail, but simply ask whether new mail has arrived. The POP protocol provides support for **Multipurpose Internet Mail Extensions (MIME)**, which allow the user to attach binary file messages, such as formatted word processing documents or spreadsheets, to e-mail. When you read your mail, POP dictates that the entire mail message is immediately downloaded to your computer and that the message is no longer maintained on the server.

IMAP, the **Interactive Mail Access Protocol**, is a newer protocol that has advantages over POP and that will likely, one day, replace POP. IMAP, like POP, can download e-mail, delete mail from the server, or simply ask if there is new mail. IMAP, however, improves upon some of the shortcomings of POP. For instance, it defines how a client program asks a mail server to present available mail. IMAP can request that the server download only selected e-mail messages rather than all messages. Your e-mail client can view just the heading and the sender of the e-mail message and then decide whether to download the message. IMAP, through your client e-mail program, allows you to create and manipulate mail folders or mailboxes on the server, delete messages, and search for certain parts of a note or an entire note—all without downloading mail from the server to your PC.

FTP

The **File Transfer Protocol (FTP)** is the part of TCP/IP that implements a mechanism to transfer files between TCP/IP-connected computers. FTP also uses the client/server model and permits files to be transferred in both directions—from the client to the server or vice versa. FTP transfers both binary data and ASCII text, and you can select either of the two transfer modes. Binary data include files containing word-processed documents, worksheets, graphics, and other data. ASCII text refers to files containing only characters available through the keyboard but no formatting—the same types of files that the Windows program Notepad can create. Some versions of FTP allow you to transfer EBCDIC (extended binary coded decimal interchange code) files. **EBCDIC,** pronounced *eb-sih-dik*, is an eight-bit code system invented by IBM to represent characters as numbers. FTP can transfer files one at a time, or it can transfer many files at once. FTP also provides useful ancillary services, such as displaying remote and local computers' directories, changing the current client's or server's active directory, and creating and removing local and remote directories. FTP uses the TCP protocol and its built-in error controls to copy files reliably with complete accuracy from one computer to another.

Accessing a remote computer with FTP requires you to log on to the remote computer. If you have an account on the remote computer, you can supply FTP with your username and password. FTP then establishes contact with the remote computer and logs you on to your account on that computer. This **full privilege FTP** access allows you to send files to the remote computer and download files from it. Another way to access a remote computer is called anonymous FTP. **Anonymous FTP** allows you to log on as a guest. By entering the username *anonymous* and a password that is, by custom, your e-mail address, you can access limited parts of the remote computer. FTP applications will be discussed later in this chapter.

Internet Utility Programs

TCP/IP supports a wide variety of utility programs or tools. Many of the utility programs allow users to use the Internet more efficiently. Some of the more popular utility programs include Finger, ping, Talk, Tracert, and VisualRoute. In this section, you will learn about several of these useful programs and see examples of how they work. Each utility program is available from many different software producers. We urge you to search the Internet and download and try several different developers' offerings for programs that interest you. Sites such as **TUCOWS** or **Download.com** provide convenient access to thousands of programs from one Web site.

Finger

Finger is an Internet utility program that runs on UNIX computers and allows a user to obtain limited information about other network users. You can issue a Finger command to determine which users are logged on to a given network or to get some details about a particular user on a network. For example, you can ascertain the last time a user logged on to the network and reveal other information that the user has entered into the system.

Many organizations do not allow the Finger command to access information about their network from outside their organization, for privacy and security reasons. For example, if you issue a Finger command for Microsoft Corporation (finger www.Microsoft.com), you will receive no response. Many e-mail programs have the Finger program built into them, so you can send the command while reading your e-mail.

Figure 2-4 shows an edited example of the Finger command and its output. Finger typically produces five columns. The first column (Login) contains the user's login name. The second column (Name) contains a description of the user's department or supervisory unit. Column three (TTY) contains an internal machine address of each user's computer. The fourth column (Idle) indicates how long the user has been logged on to the computer. The fifth column (When) indicates the day and time when the user first logged on. (Some people log on and remain connected for days, weeks, and even months at a time.)

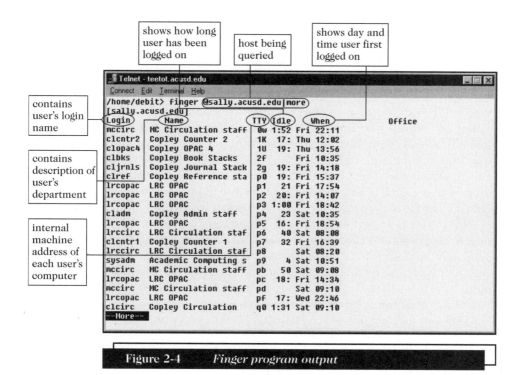

shows how long user has been logged on

host being queried

shows day and time user first logged on

contains user's login name

contains description of user's department

internal machine address of each user's computer

Figure 2-4   *Finger program output*

Ping

Ping, short for Packet Internet Groper, tests the connectivity between two Internet hosts and determines if a host is active on the network. It works by sending two packets (short messages) to the specified address and waiting for a reply. Ping is used primarily to troubleshoot Internet connections. Many freeware and shareware ping utilities are available on the Internet. It is a fast and helpful utility that is useful, for example, to determine if an electronic commerce site is online. If you suspect that the site www.amazingstuff.com may be offline, you can send out a ping and see if it echoes back by typing

```
ping www.amazingstuff.com
```

in a DOS window and waiting for the three-line reply. If you receive no response within a second or two, then it is likely that the commerce site is offline.

Ping provides performance data about the connection between Internet computers, such as the number of hosts (**hops**) between them and the time it takes to send a simple message from one computer to another. Ping is most commonly executed from the MS-DOS prompt, although several Windows-based ping programs are available. To run ping, you simply type *ping* followed by the IP address or the domain name of the host you are trying to reach. For example, to determine if the U.S. Census Bureau's Web site is online, you can type the following at the MS-DOS prompt (or use a ping client):

```
ping www.census.gov
```

or

```
ping 148.129.129.31
```

Experiment with ping by trying one of the ping client programs available from Internet download sites such as **TUCOWS** or **Download.com**. (The Online Companion has a list of links to several of the programs described in this chapter.)

Tracert and Other Route-Tracing Programs

Tracert (TRACE RouTe) sends data packets to every computer on the path (Internet) between your computer and another computer and clocks the packets' round-trip times. This provides an indication of the time it takes a message to travel from your computer to a remote one and back, ensures that the remote computer is online, and pinpoints any data "traffic" jams. Similar in function to ping, route tracing programs calculate and display the number of hops between computers and the time it takes to traverse the entire one-way path between machines. Route tracing programs such as Tracert work by sending a series of packets to a particular destination. Each computer (called a router) along the Internet path between your computer and the destination computer sends back to your machine the IP address of each router and the round-trip time it took to reach it. After the trace program completes its execution, you will see how many hops and how much time it took to reach each node and to travel the entire path.

Besides providing travel time statistics, graphical user interface trace route programs provide a trace of the route taken from the source to the destination, plotted on a map. You can use route-tracing programs to determine where the biggest delays are. Even if you aren't interested in determining where delays occur on the Internet, it is interesting because it illustrates (somewhat) how the Internet functions. **VisualRoute** is typical of the several Windows GUI route-tracing programs available for download, trial, and purchase on the Internet. Figure 2-5 shows an example of a route traced between a West coast (San Diego) computer facility and an East coast computer facility—Harvard University—using the VisualRoute program.

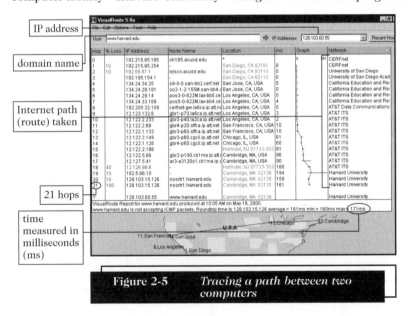

Figure 2-5   *Tracing a path between two computers*

Internet Applications

Three representative Internet applications persist from the very early days of the ARPANET: electronic mail, Telnet, and FTP. These very popular programs preceded the most popular use of the Internet: the World Wide Web. Using them, you can access the Internet and check business information using any Internet-capable computer with a dial-up network that simulates the TCP/IP protocol over dialed lines. There is practically no place on earth that is out of range of the Internet.

Electronic Mail

Electronic mail can be traced back to the 1970s and the ARPANET. Although the goals of the ARPANET were to control weapons systems and transfer research files, more general communications uses for this vast network became obvious in the early 1970s. In 1972, Ray Tomlinson, an ARPANET researcher, wrote a program to send and receive messages over the network. Today, e-mail is the most popular form of business communication—individually surpassing the telephone, conventional mail, and facsimile in volume. In some countries, phone service is so bad that more than a few people have resorted to the more reliable e-mail system to communicate with one another.

Not only was e-mail one of the first Internet applications, it is the reason that so many people are attracted to the Internet. E-mail conveys messages from one destination to another in a few seconds. Messages can contain simple ASCII text, or they can contain character formatting similar to word processing programs.

One very attractive feature of e-mail is that you can send documents, pictures, movies, worksheets, or other important pieces of information along with the message itself. These **attachments** are frequently the most important part of the message. A business-to-business e-mail message attachment might contain an invoice, the latest 200-page wholesale catalog, or a complete and compressed set of Web pages describing a company's products being sold online.

Usually, people use one of several popular e-mail client programs such as Eudora, Netscape Messenger, Microsoft Outlook Express, or one of the several Web-based e-mail systems such as Excite email, HotMail, or Yahoo! Mail. Netscape Messenger and Microsoft Outlook Express are especially popular because they are packed with the Netscape and Internet Explorer Web browsers, respectively. Figure 2-6 shows a typical e-mail client program (Microsoft's Outlook Express) with an attached file ready to be sent over the Internet.

Many electronic commerce sites use e-mail to confirm the receipt of customer orders and to confirm shipment or delivery of items ordered over the Internet. Software purchasing and delivery over the Internet also use e-mail to send important information about the purchase to the buyer. Suppose you decide to download a trial version (also called **limited edition** or **limited use**) of software—which can be used for free for a limited time period or a limited number of uses. After using it for a while, you decide the product is just what you need and you decide to purchase it. Since the software is already downloaded, you merely need to purchase the software license with a credit card through the vendor's Web site. Normally, the vendor then sends a long, unique code to your e-mail address. This code is a key that unlocks the software for unlimited use or eliminates the "nag" screen that the limited edition displays. As you can see, e-mail facilitates electronic commerce transactions, such as purchasing software and goods.

attachments

e-mail message body

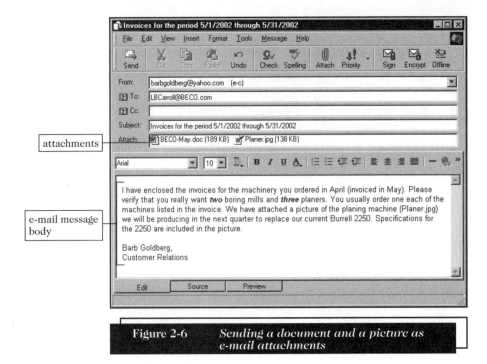

Figure 2-6   *Sending a document and a picture as e-mail attachments*

E-mail has a downside. Perhaps the single, largest annoyance is spam. **Spam** is electronic junk mail. You receive this in the form of solicitations ("Want to make big $$$ working on your computer?"), advertisements, or postings from newsgroups. Besides wasting your time and allotted e-mail disk space, spam, which is also known as **bulk mail**, consumes a lot of Internet capability. If someone sends a useless e-mail to 100,000 people, then for a few moments, the unsolicited mail consumes resources that could otherwise be used for more beneficial purposes. Spam also robs the Internet of bandwidth. **Bandwidth** is the amount of data that can be transmitted in a fixed amount of time. It is usually expressed as the number of bits per second (bps). Many grass-roots and corporate organizations have decided to fight spam aggressively. AOL, for example, has taken an active role in limiting spam (also known as spamming) through legal channels. The most effective deterrent is the threat to a spammer (one who sends spam) of losing his or her account on his or her Internet service provider (ISP). Without an ISP, there is no vehicle to transmit the unwanted e-mail. ISPs are discussed later in this chapter. (There is some debate about the origin of the term *spam*. Most believe it comes from the Monty Python song dedicated to the mysterious meat that comes in a tin: "Spam spam spam spam, spam spam spam spam, lovely spam, wonderful spam…" Just like the song, spam is a tiresome and worthless repetition of worthless text.)

Telnet

Telnet is an application that allows you to log on to a remote computer that is attached to the Internet. Several Telnet clients are available on the Internet, but you may be satisfied with the one supplied with Windows systems: Telnet.exe, which is limited but adequate. It permits you to alter the shape of the cursor, change background and foreground colors, and change the character font on the local (client) computer. Unlike e-mail, Telnet exposes your computer to the commands and programs of the remote host. Performing the functions of a **terminal emulation** program, Telnet allows you to type commands and other character strings that are passed directly to the remote host computer. That computer acts upon the commands that you enter through Telnet. Telnet emulates a few terminals, such as the industry standards of VT-52 and VT-100, but oddly, it does not emulate the very popular IBM 3270 terminal. You can log on to the United States Library of Congress network via Telnet using any Telnet client and entering the address locis.loc.gov. You can also use your Web browser as a Telnet client by entering the URL telnet://locis.loc.gov and pressing Enter.

Figure 2-7 shows an example of a Telnet session using the Windows-supplied Telnet program. The first screen of the Library of Congress site is displayed. You typically type `quit` or follow the instructions on the host computer to end the Telnet session. For UNIX-based systems, the command you enter to quit is usually `logout` or `logoff`.

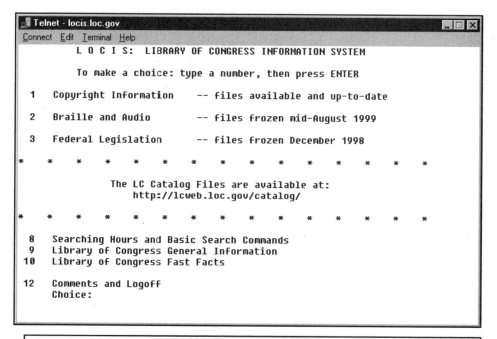

Figure 2-7   *Example of a Telnet session*

Salespeople on the road or even in a foreign country can use the Internet and a Telnet client to log on to their company's computer to check the status of orders and inventory, or to access other business information.

FTP

FTP is the fastest way to deliver digital business information from one computer to another. Perhaps the most popular use of FTP is for the sale and delivery of software packages and updates. Microsoft Corporation, for instance, offers a large number of software updates, online and free of charge. Whether you use a Web browser or an

FTP client program, you can download books, manuals, or complete software suites. Using anonymous FTP, business clients and students alike can access commerce sites to download files they want to purchase. With an account consisting of a username and password, you can access a corporate or university computer in a more privileged way than with an anonymous account, with the ability to send files, documents, and any other digital information both to the host computer (**upload**) and to your personal computer (**download**).

Figure 2-8 illustrates an FTP session. The host site URL is ftp.mcafee.com, which is the computer from which you can download the latest McAfee antivirus data files. In the left panel are the folders and files found on the local computer's disk. The right panel displays folders and files on the remote computer's disk (McAfee's site, in this example). Buttons in both panels allow you to switch directories, create new directories, and perform other directory maintenance tasks independently on the local and remote computers.

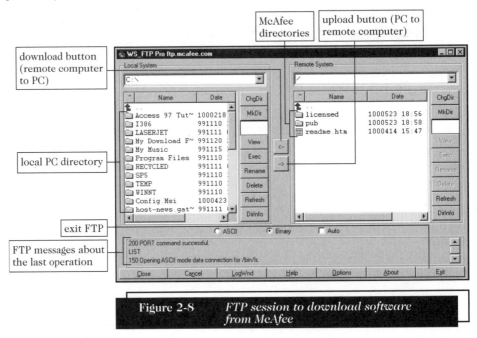

Figure 2-8   *FTP session to download software from McAfee*

MARKUP LANGUAGES AND THE WEB

Of course, the most popular use of the Internet is the World Wide Web. Web pages number in the millions, and this section discusses how Web pages are constructed.

Historically, the term *markup* has described annotations and handwritten notes found on manuscript pages that tell a compositor or typist how a particular page should be laid out or typeset. There is a universal set of copyediting symbols for marking up paper manuscripts. Similarly, electronic pages are marked with tags to govern the display and formatting of text elements. Three markup languages are presented in this chapter: SGML, the grandfather of the markup languages; HTML, a derivative of SGML; and XML, a newer derivative of SGML.

Overview of SGML, HTML, and XML

SGML, HTML, and XML are the three most important markup languages. As you learned in Chapter 1, SGML is the parent language from which both HTML and XML were derived. Each language has a unique purpose. SGML is a rich meta language that is useful for defining an almost endless supply of markup languages. (A **meta language** is a language—a set of language elements—that can be used to define other languages—HTML, for example.) HTML is particularly useful for displaying Web pages. XML—which is not well known among the Web-browsing public—defines data structures important for a wide range of data exchange activities, including electronic commerce. Each is briefly described next.

Standard Generalized Markup Language

In the 1960s, scientists were working on the definition of a Generalized Markup Language (GML) for describing electronic documents and how they are formatted. In 1986, the International Standards Organization (ISO) adopted a particular version of the standard called Standard Generalized Markup Language (SGML). SGML offers a system of marking up documents that is independent of any software application. It also contains an international standard that defines device-independent and machine-independent methods for representing electronic documents. SGML is powerful and well suited for organizations that need exacting standards. Many organizations—especially those with special or complex requirements for document management, such as the U.S. Department of Defense, the Association of American Publishers, Hewlett-Packard, and Kodak—use SGML.

SGML has the following advantages:

- It has long-term viability as a carefully regulated ISO standard (since 1986).
- It is nonproprietary and platform-independent and will outlive most current applications.
- It supports user-defined tags and architecture to complement the required richness of documents.

Although it is a complete specification, SGML is not well suited for quick development of Web pages. While sophisticated, it is lacking in other areas, mainly:

- It is costly and complicated to set up, requiring real technical expertise beyond that of most Web designers.
- SGML's tools are relatively expensive compared to those of HTML.
- Creating document type definitions with SGML can be expensive—especially in human labor.
- SGML has a steep learning curve.

Hypertext Markup Language

Recall from Chapter 1 that Tim Berners-Lee invented Hypertext Markup Language (HTML), a document production language that includes a set of tags that define the format and style of a document. The codes describe the relationships between the document's text elements. The *hypertext* part of the name has its roots in the 1960s— *Literary Machines* author Ted Nelson coined it. Nelson envisioned a page-linking system that would interconnect related pages, regardless of where they were stored. The *markup language* portion of the name is rooted in conventional publishing.

HTML is based on SGML. HTML is an instance of one particular SGML document type—Document Type Definition (DTD)—that is both easier to learn and use than SGML. The HTML DTD, for example, is used by all the documents on the Web. In the early days of HTML—in the early 1990s—the then-current version of HTML worked fine for creating text-based documents with headings, title bar titles, bullets, lines, and ordered lists. But users raised the bar on Web page elements, requesting exacting graphic positioning, tables, and frames, and Web designers requested new features weekly. In addition, software vendors continued to ask for enhancements to HTML. In response to these demands, Microsoft added features to HTML that work only in its Internet Explorer browser, and Netscape added features that work only in its flagship product, Netscape Navigator (a part of the Netscape Communicator program suite). The HTML that Berners-Lee produced was, in effect, a trimmed-down version of SGML. He eliminated SGML features that are rarely needed and added, for example, hyperlinks to link Web documents. SGML, on the other hand, is fully extensible, providing the capability to develop a different DTD that provides for different and possibly more capable document markup elements. New versions of HTML are simply changes made to the original SGML-based HTML DTD. Browsers such as Navigator or Internet Explorer cannot read SGML—they can only read HTML. So, SGML is relatively unknown to most Web designers.

You will learn more about HTML in a later section of this chapter. One of the newest SGML derivatives is XML, or eXtensible Markup Language, which is described next.

Extensible Markup Language

EXtensible Markup Language (XML), like HTML, is a descendant of SGML. XML is a relatively new language. It represents an industry wide effort to define *which* data are displayed on a Web page, whereas HTML determines *how* a page is displayed. XML, like HTML, consists of both data and markup elements. XML allows designers to easily describe and deliver structured data from any application in a standard, consistent way.

Most Web experts agree that XML is quickly becoming the Web's preferred method to convey data and structure. While HTML provides a rich variety of tags that describe a page's layout, HTML cannot describe a page's actual *content*—it cannot interpret the meaning of the data on a page. XML, on the other hand, describes a page's contents. In addition, XML's data-tracking capability is propelling it ahead, because it is altering the way businesses share data with one another and the way Web-based applications search databases and files. HTML is analogous to Microsoft Word's character and paragraph formatting—it affects the *formatting* of words on a page. Word's styles are analogous to XML. When you include the heading styles such as Heading1 in a document, you establish a structure—an outline—to a document.

XML defines a document's structure by marking the start and end of each of its elements. XML users can check that each part of a document occurs where it should in the data stream being sent from one location to another on the Internet.

Among its other benefits, XML:

- Provides metadata (data about information) that helps people find information as well as helps information consumers and information purveyors find each other

- Makes processing data possible with inexpensive software
- Simplifies the exchange of data between businesses and fosters creation of platform-independent protocols that fuel data of electronic commerce
- Delivers information that can aid automatic processing by electronic agents working on behalf of businesses or people

Using paired start and stop tags, XML marks data in much the same way as you would define a record structure in a database system. For example, suppose your company sells products on the Web. Your Web pages contain descriptions and beautiful graphics featuring the various products you sell. The pages are written in HTML. Unlike the product descriptions, the product data—information about individual products, including their price, identification number, and quantity on hand—are formatted with XML and exist as data. That is, XML is embedded within an HTML document. Figure 2-9 shows an example of one XML-formatted product with four elements and two records. The first record contains details about the coffee called Kenya AA, while the second record contains details about Yemen Mocha. Each of these coffee records contains four elements called ItemName, CountryOfOrigin, WholesaleCost, and OnHand.

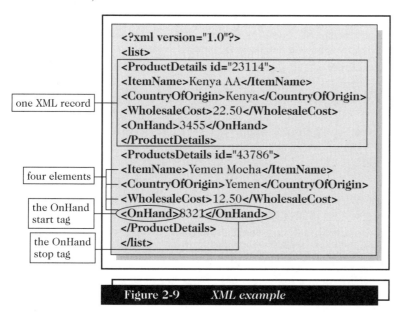

Figure 2-9   *XML example*

All tags, except for the first one, consist of properly nested start/stop pairs. Though the tags resemble those used in HTML (which you will learn about in the next section), there is one big difference between the code example in Figure 2-9 and typical HTML code: In the XML example, there is no information about how the data are to be displayed or formatted (in terms of size, position, color, and so on). Formatting information, if required, would come from elsewhere—another Web page, an application receiving the XML file, or another document. The idea behind XML is analogous to address entries in a personal digital assistant (PDA), such as a Palm Pilot, or a database such as Oracle. You enter records, one by one, about people and their phone numbers into individual fields of a database. When you want to produce a hard copy of your PDA address book, it is a simple matter to produce an attractively formatted report using Microsoft Word. Records are merged from the database into the Word document. Similarly, XML handles the database structure details. When it comes time to format that data, HTML handles the task. Unlike many other markup languages, XML allows you to create your own customized markup language—your own tags. (Thus, it is an *extensible* markup language.) There are several widely accepted XML-based languages in everyday use on the Web.

You can always call on HTML to handle the presentation and formatting details of the data presented by the XML file. This formatting capability is especially important when you want to produce a response to a Web site query. For example, assume that your Web site sells coffee, and you receive a query requesting that you list all coffee products that you have on hand from South Africa or Kenya. XML contains structure information that permits the search request (finding the appropriate data), and HTML has powerful presentation and formatting capabilities to display the retrieved product information.

Despite the obvious strengths of XML, HTML is still the biggest influence on the Web today. However, many believe that XML soon will displace HTML as the most popular and widely used markup language—especially in the very large business-to-business market in which partner businesses exchange important, structured business documents. Evidence of the booming interest in XML is everywhere. XML books fill entire bookstore shelves, whereas there were no books on the subject just a few years ago. XML experts believe that XML's benefits include smarter search engines and faster responses from Web servers.

It is important to understand more about HTML before moving on. The next section provides an overview of HTML and illustrates how pages can be formatted. The formatting differences between representative Web browsers are also discussed.

More About HTML

The World Wide Web organizes interlinked pages of information residing on sites around the world. Hypertext Markup Language is the language of choice to display millions of those pages, and the HTTP protocol delivers pages between servers and applications. Hyperlinks within document pages form a "web" of document pages. You can traverse the interwoven pages by clicking a hyperlink to move from one page to another. If you so choose, you can read the pages of a document in serial order, or you can read them in whatever order you prefer by following hyperlinks. Figure 2-10 shows how topics can be related by hyperlinks in a nonlinear fashion, permitting you to read either serially or nonserially.

HTML Tags

An HTML document contains both document content and tags. The document content consists of all the information actually appearing on the computer screen, including text, graphics, and video. The **tags** are the HTML codes inserted in a document that specify how a complete document or a portion of it should be formatted and arranged onscreen. (Tags are used by all format specifications—including SGML, XML, and HTML—that store documents as text files.) HTML tags often are used in pairs and have a simple structure. The general form is:

`<tagname properties>` Displayed information affected by tag `</tagname>`

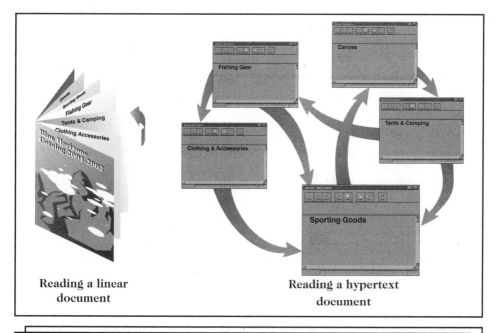

Figure 2-10   *Traditional versus hyperlinked document pages*

A specific example of a tag pair is a boldface character formatting tag that causes the word *best* to appear in boldface type:

The <B>**best**</B> way to view this document is with a Web browser.

Tags are easily distinguished in an HTML document, because each tag is enclosed in brackets (<>). You can write tags in either lowercase or uppercase letters; the tag <b> has the exact same meaning as the tag <B>. Though most tags are **two-sided**—requiring both an opening and a closing tag—some are not. The latter are known as **one-sided tags**. The **opening tag** comes first, followed by text that the tag affects. The **closing tag** is identified by the slash (/) that precedes the tag's name. If you were to omit the closing bold tab in the preceding example, the entire sentence, and more, would be bolded. Sometimes an opening tag contains one or more property modifiers that further refine how the tag operates. A tag's property may modify a text display, or it may designate where to find a graphic element. For example, the paragraph tag, which is a one-sided tag, defines the beginning of a text paragraph. One of its optional properties indicates a paragraph's alignment. For instance, the following is an HTML segment that displays and right-aligns a paragraph in the browser window.

```
<P align="right"> This will right-align the paragraph, based
on the width of the user's browser and computer screen, so
that the end of each line (except the last) lines up with the
right border of the screen. The left ends of each line will
not be aligned. This is known as ragged left alignment.
```

Figure 2-11 illustrates the appearance of the previous paragraph when Microsoft's Internet Explorer Web browser displays it. Because the window is not maximized, each line of text automatically wraps to a new line to accommodate the browser's window size. The paragraph automatically readjusts its margins if you change the browser's window size.

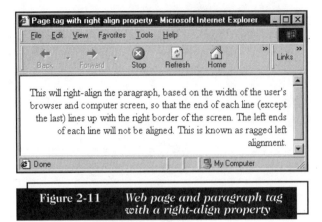

Figure 2-11 *Web page and paragraph tag with a right-align property*

Many tags are defined for HTML. Several sources and textbooks are available describing the HTML tags and their use, and you may wish to consult them for an in-depth look at HTML. Here, you will examine a few of the tags to gain a broad understanding of HTML document structures without becoming overly concerned with HTML code details.

HTML code defines the structure and formatting of a Web page, but a page may look different in two different browsers. Look at the underlying HTML page in Figure 2-12 and then compare it to the displayed page shown in Figure 2-13.

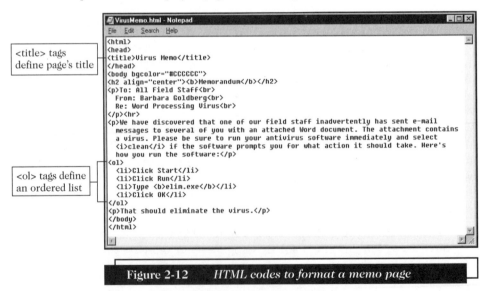

Figure 2-12 *HTML codes to format a memo page*

When you study Figure 2-12 and then examine the Web browser result in Figure 2-13, you can probably see how each tag pair formats the page. The `<title>` tag, for example, creates the title that appears in the browser's title bar. (The text of a Web page's title is one of the key elements that is indexed by several of the search engines when they scan Web pages for entries on a regular basis.) Notice that the opening `<body>` tag contains the modifier *bgcolor*. You probably guessed that it stands for background color. The value of the color code is in hexadecimal, a base 16 numbering scheme that computers use. It indicates how much red (first), green (second), and blue (third) to mix together to get the desired color. (*CC* translates to 204 base 10.) Near the bottom is a numbered list delimited by the `<ol>` `</ol>` opening and closing tags. Between those tags are three list items (`<li>` tags) that are automatically numbered sequentially, beginning at 1. You may find it interesting to examine the underlying HTML code of other pages. Examine the code of any Web page you are viewing in a browser by executing View, Source (Internet Explorer) or View, Document Source (Netscape Navigator) on the menu bar.

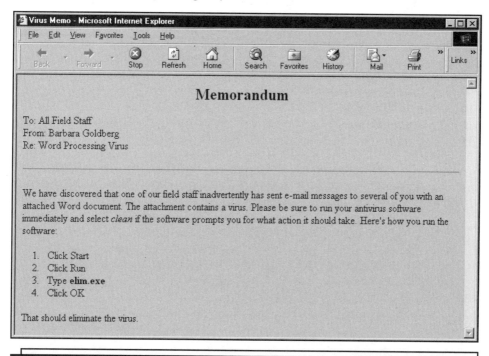

Figure 2-13   *Internet Explorer display of the memo page*

HTML Links

What makes the Web *really* interesting—something more than simply an electronic storage facility for a collection of independent documents—is the HTML hyperlink. Hyperlinks are bits of text that connect the current document to another location in the same document, to another document on the same host machine, or to another document somewhere else on the Internet. Linking documents to one another through hyperlinks creates the web of documents comprising the World Wide Web. Hyperlinks are created using the HTML **anchor tag**. Whether you are linking to text within the same document or to a document on a distant computer, the anchor tag has the same basic form:

```
<A HREF="address">Visible link text</A>
```

Like other tags, anchor tags have opening and closing tags. The opening tag also specifies the HREF property, which specifies the remote or local document's address. Clicking the text following the opening link transfers control to the HREF address—wherever that happens to be. Suppose you are creating an electronic, Web-based résumé with your university's name and address under the Education heading. Instead of presenting a "flat" university name in the résumé, you can create a hyperlink and connect it to the university name in your document. Anyone viewing your résumé can click the link, which will lead the reader to your university's home page. The following example shows the HTML code to create a hyperlink to another Web server:

```
<A HREF="http://www.purdue.edu">Purdue University</A>
```

Similarly, you could create a local link to another part of the same document—perhaps page 3 of your résumé—with the following link and HTML code:

```
<A HREF="#references">References are found here</A>
```

In both preceding examples, the text *between* the anchors appears on the Web page as a hyperlink. Most browsers display the link in blue and underline it. But no matter how the link appears, whenever you move your mouse over a hyperlink, the mouse pointer changes from an arrow to a pointing hand.

An electronic commerce application, like any other Web application, uses links to direct customers to various pages on the company's server and to other secure servers. The way links lead customers through pages can affect the user-friendliness rating of the site, and that can impact the customer's impression of the company. You can use different link structures on your Web site. Experience and customer response will help you decide which is best for your organization. Two popular link structures are linear and hierarchical. A **linear hyperlink structure** resembles conventional paper documents in that the reader begins on the first page and clicks a "next" button to move to the next page in a serial fashion. Few other paths are provided. This sort of structure works well when customers fill out forms prior to a purchase or other agreement. In this case, the customer reads and responds to page one, and then moves on to the next page. This process continues until the entire form is completed. The only Web page navigation choices the user typically has are "back" and "continue."

Another very popular link arrangement is called a hierarchical structure. In a **hierarchical hyperlink structure**, the Web user opens an introductory or home page. That page usually contains one or more links to other pages, and those pages, in turn, link to other pages. This hierarchical arrangement resembles an upside-down tree in which the root is at the top and the branches are below it. Hierarchical structures are excellent means of leading customers from general topics or products to specific product models and quantities. A company's home page might contain links to help, company history, company officers, order processing, frequently asked questions, and product catalogs. Figure 2-14 illustrates the linear and hierarchical structures. Of course, pages combining linear page with hierarchical structures are possible.

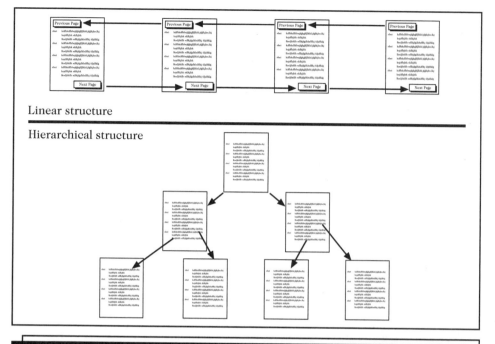

Linear structure

Hierarchical structure

Figure 2-14   *Hyperlink structures*

HTML Version History

The Hypertext Markup Language has undergone several modifications since it was first introduced as version 1.0 in the summer of 1991. At that time, Berners-Lee developed the specifications for version 1.0 and posted them on the Internet. Seeking to make the HTML language platform-independent, he first suggested and developed a language compilation program—a (Web) **browser**. Shortly after that, new browsers began to show up everywhere. Each browser seemed to spawn variants on the Berners-Lee specification, and resulted in the creation of many proprietary HTML versions. That diverged from the original purpose of HTML—a platform-independent Web page viewer. Not until the World Wide Web Consortium released the next version of the HTML specification, version 2.0, did an industry standard develop.

HTML 2.0 was released in September 1995. It was the first of several revisions to move toward a standard. At that same time, two new browsers were released: Microsoft's Internet Explorer 2.0 and Netscape's Navigator 2.0. Netscape gave away copies of its browser software at COMDEX, an industry trade show, to help spur acceptance of the browser—a very successful strategy, as history has shown. Version 2.0 of HTML is the first version to support in-line graphics and fill-out forms. HTML 2.0 marked the start and explosion of the widespread use of the Internet generally and the World Wide Web particularly.

HTML 3.0 was introduced a short while after version 2.0. In 1997, version 3.2 was introduced. Built upon version 3.0, version 3.2 provided support for tables, complex numbers, and text flow around images. Because proprietary versions of HTML (versions that only work on particular browsers) were being developed, the World Wide Web

Consortium created a committee to develop HTML version 3.2 in an attempt to further standardize the language. The **World Wide Web Consortium (W3C)** was founded in 1994. Its purpose is to serve as a leader in maintaining Web standards and common protocols and to promote both the Web's evolution and interoperability. Though version 3.0 was missing important features that were already being used by other Web browsers, existing major browsers eventually supported HTML 3.2.

HTML version 4.0 was released by W3C in December 1997. Version 4.0 included support for the OBJECT tag and Cascading Style Sheets—features that previously were not supported by HTML. With the OBJECT tag, Web page designers can embed scripting language code directly in HTML pages. **Scripting language code** allows a downloaded Web page to execute programs on the user's computer. **Cascading Style Sheets (CSS)** provide Web developers with more control over the format of displayed pages. Similar to predefined document styles in word processing programs, CSS lets designers define formatting styles that can be reapplied whenever they are needed. *Cascading* simply means that designers can apply many style sheets to the same Web page.

In addition to providing OBJECT tags and CSS, version 4.0 introduced internationalization to the language. For example, internationalization features included the ability to render text right to left, the way several languages are read. HTML 4.0 provided accessibility features as well. Accessibility features in HTML accommodate Web users who have disabilities. An example of an accessibility feature is providing a text alternative for images for those with limited vision. Text content can be provided to a reader as synthesized text or Braille. (The World Wide Web Consortium provides **Web Content Accessibility Guidelines** on its Web site.)

HTML version 4.01 was released in December 1999. Version 4.01 fixed errors in HTML 4.0 and defined new semantics and data types for HTML.

XHTML 1.0 became a W3C recommendation in January 2000. It is a reformulation of HTML 4.01 in XML, which brings the rigor of XML to HTML. Browsers can immediately make use of XML by following a simple set of guidelines provided by the W3C organization.

XHTML Basic was placed in *last call* status recently, meaning that those contributing to the language's design had only a few weeks remaining to comment on the language specification before finalizing the specification. XHTML Basic defines a minimal necessary set of XHTML features that support mobile equipment such as Palm and Handspring devices that can link to the Internet through wireless connections. Handheld devices offer reduced Web browsing capabilities compared with full-featured PCs, so they have different XHTML requirements.

HTML Editors

Before discussing Web servers and Web clients, it is important to discuss HTML editors briefly. HTML editors have varying levels of sophistication. On the low end of the sophistication scale are editors that display HTML code on the screen and allow you to insert HTML tag pairs by clicking selected buttons. These types of editors are more helpful than Notepad for building Web pages, but they provide no drag and drop graphics capability and usually require you to load a page into a browser in order to view it. There are dozens of freeware, shareware, and commercial HTML editors available on the Internet (see the Online Companion for links to free or low-cost HTML editors).

On the high end of the sophistication scale are HTML editors that are, in fact, Web site builder programs. With these, you can create full-scale commercial-grade Web sites, complete with database access, graphics, and fill-in forms. These editors provide a rich environment that displays the Web page, not the HTML code. You can drag and drop objects such as graphics, buttons, or lines onto the page. Software generates the HTML code for the pages. When you are done creating a set of pages for your site, the Web site builder software uploads the pages to your Internet Web server from the PC on which you developed them. Examples of Web site builders include **Microsoft FrontPage** and **Macromedia Dreamweaver**. There are many fine programs in this category also. They are listed in the Online Companion for Chapter 2. Figure 2-15 shows an example of a Dreamweaver site display.

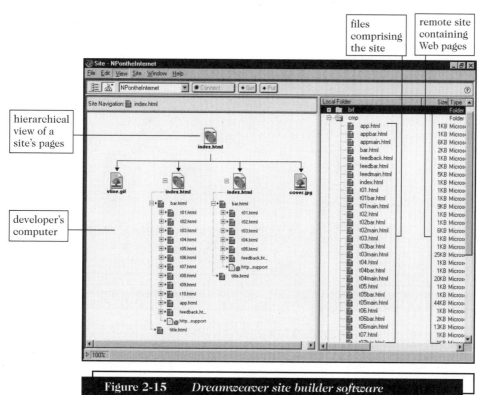

Figure 2-15 *Dreamweaver site builder software*

WEB CLIENTS AND SERVERS

When you use your Internet connection to become part of the Web, your computer becomes a **Web client** in a worldwide client/server network. Your Web browser software—Internet Explorer or Netscape Navigator, for example—is the software that makes your computer work as a Web client. The Internet connects many different types of computers running different operating system software. Because Web software is platform-neutral, it lets your computer communicate with all of these different types of computers easily and effectively. This platform neutrality is a critical ingredient in the rapid spread and widespread acceptance of the Web.

Interlinked Documents

Computers that are connected to the Internet and that contain documents that their owners have made publicly available through their Internet connections are called **Web servers**. Figure 2-16 shows how this client/server structure uses the Internet to provide multiple interconnections among a wide variety of client and server computers.

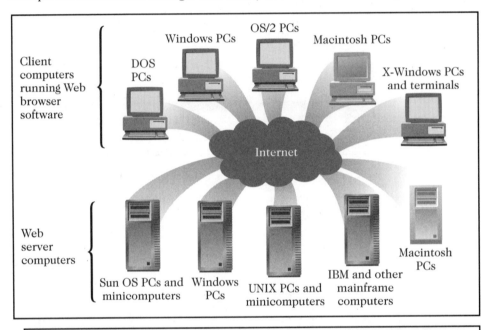

Figure 2-16   *Client/server structure of the World Wide Web*

Documents publicly available on the Web—Web pages—consist of text and HTML code. HTML hyperlinks between documents located on the same computer and on widely dispersed computers form the web of interconnected documents. A document stored on a computer in Kansas City, Missouri, can contain a hyperlink to a document stored on a computer in Athens, Greece. The document in Athens, in turn, can have links to documents at other locations in the world. Web pages are truly international. But, Web pages would be static (devoid of links to other pages, customized client data, or features such as animation) if it weren't for the client/server architecture and the Internet infrastructure on which they travel. When someone using a Web browser clicks a hyperlink, another page from the target URL quickly appears on the user's screen. How does that happen? What interaction is there between the browser and the target computer? The next section describes the nature of the interaction between a Web client and a server.

Web Client/Server Architecture

Client/server architecture may be used on LANs, WANs, and the Web. The main characteristic that these three somewhat diverse uses share is a division of the workload between the server and the client. In each case, the client computers typically request services, including printing, information retrieval, and database access. The partner in these activities is the server, which is responsible for processing the clients' requests. Nearly always, the client does very little work.

While the client's workload is light, the server's workload is not. Besides receiving and interpreting requests from the client, the server must locate information, reprocess it, and request initialization of resources supplied by other applications running on dedicated computers under the server's control. That workload-sharing arrangement is why servers generally must be beefy, expensive computers with lots of disk capacity, fault-tolerant processors, and ample memory.

In contrast to the server, clients require no more capability than is found on any ordinary personal computer. In fact, older-generation PCs are perfectly suitable as client machines—as long as the required software can run on them. The term **thin client** is a popular description of a client's relatively low workload, compared with that of a server. Thin clients, which are diskless, are usually found on local area networks connected to the Internet. For electronic commerce applications, this translates to low-cost computers (client machines) for people wanting to purchase goods and services from a Web-hosted business. In this case, a Web business must shoulder larger costs to purchase and run robust computers and software (servers) to serve a potentially large customer base.

How and what does a client communicate to a server? When the server goes about its work, what sort of information transformation goes on at the server and what does it return? How does the client/server interaction work when used for electronic commerce? The next section answers these questions and more.

Web Client/Server Communication

The division of labor between Web clients and Web servers is quite distinct. The Web client—your computer at the office or at home—requests information from a particular Web server on a distant computer. Using the Internet as the transportation medium, the request is formulated into an HTTP request and sent to the target computer—the server. A moment later, when the target server receives the request, it retrieves the page or other information that the server requested, formulates it as an HTML-formatted page, and sends it back to the requester client via the Internet. When the requested information—an HTML page in this instance— arrives at the client computer, the Web browser software determines that the information is an HTML page. It displays the page on the client machine according to the directions defined in the page's HTML code. Repeatedly, this same general scenario is carried out as the client requests, the server responds, and the client displays the result. Sometimes, a simple client request results in dozens or even hundreds of separate server responses to locate and deliver information.

A Web page containing many graphics and other objects can be slow to appear, because each element (for example, each graphic, each header, and so on) requires a separate request and response. This division of labor between client and server is fixed and well established. Neither the client nor the server can deviate from its assigned responsibilities. However, the exact way a server or a client carries out its respective duties can vary, as discussed in the following sections.

Two-Tier Client/Server

A two-tier model involves only a client and server. All communication takes place between the client on the Internet and the target server. Of course, other computers are involved in the process of transporting packets of information across the Internet. Details of packet transmission are part of the transportation facility that is handled by TCP/IP and are not part of this discussion. The conversation that occurs between a Web browser and a Web server is similar to any conversation between clients and servers generally.

What interactions occur between a browser and a Web server? To understand the processes that make up a two-tier Web transmission, envision the following scenario: You have launched your favorite Web browser and are conducting research on microcomputers. You want to purchase a new one and you'd like to compare prices. You start by looking at Gateway's computers. You begin by typing www.gateway.com, Gateway's URL, and pressing Enter. Your browser then forms the URL into a message and creates a proper HTTP request that it sends on to the Internet. Meanwhile, Gateway.com, the target server, waits for requests to arrive at the server. HTTP-type requests arrive at a particular **port**, which is an endpoint to a TCP/IP connection devoted to a particular type of Internet traffic. Port 80 is monitored for HTTP incoming messages. When the Gateway.com server recognizes your incoming HTTP request, the server establishes a temporary connection with the client computer.

Deciphering the request, the server determines that the client is requesting the home page of Gateway. Once the server locates the Gateway home page, it constructs a proper HTTP message containing the Web page within the message and sends the complete message back to the client using the client-supplied return address URL. When the server's message arrives back at the requesting client—your PC—the browser recognizes that the page is written in HTML—something it can interpret. The browser displays the page on your screen. If the page contains objects requiring browser **plug-ins** (helper applications that are not part of the browser), then the required plug-ins execute. Plug-ins might play an embedded video, animate page elements, or play music of an audio object. Once the server sends the message back to the client, the connection between them is broken. When other objects such as graphics are on a page, then the client must request their delivery from the server with a separate request. Of course, there may be many such requests for pages containing many non-HTML elements. Figure 2-17 represents a high-level view of the communication that occurs between a Web client and a Web server over the Internet.

The preceding scenario illustrates a two-tier Web client and server interaction. A **two-tier architecture** is one in which only a client (tier 1) and a server (tier 2) are involved in the requests and the responses that flow between them over the Internet. In other words, the Web server alone accommodates all Web activity and does not require the services of a backend processor to handle the requests for that site.

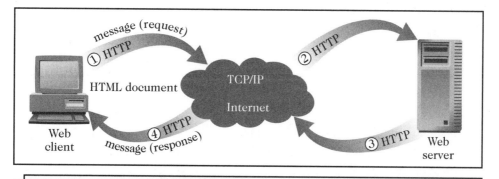

Figure 2-17 *Message flow between a Web client and server*

A typical request message from a client to a server consists of three major parts:

- A request line
- Optional request headers
- An optional entity body

The **request line** contains a command, the name of the target resource server (without the protocol or domain name), and the protocol name and version. Following the request line are name/value pairs that comprise the request header, if it exists. The **request header** contains additional information about the client and more information about the request. Finally, the optional **entity body** is sometimes used to pass bulk information to the server. A blank line separates the entity body from the request header information. This is an example of a request message containing a request line and two request headers:

```
GET /whatsnew/rfc/rfc1939.html HTTP/1.1      request line
Accept: text/html                            request header 1
Accept: audio/x                              request header 2
```

The preceding example of a client request contains the GET command, which asks the server to retrieve a file. Following the command is the path and filename of the needed file. The domain name is absent because it is used by TCP/IP and is not needed in the server request. Last on the request line is `HTTP/1.1`, which indicates that the client is using version 1.1 of the hypertext transfer protocol. Message lines two and three indicate that the client accepts text in HTML format and accepts a specific audio format. TCP/IP is responsible for transporting the message safely and intact to the target server.

Once the server receives the request, it executes the command (for example, it sends back to the client a particular Web page), retrieves a particular Web page from its trove of pages, and then formulates a properly formatted response to send back to the requesting client. A server's response consists of three parts that are identical in structure to a request message: a response header line, one or more response header fields, and an optional entity body. In the response, though, each part has a slightly different function than it does in the request. The **response header line** indicates the HTTP version used by the server, the status of the response (whether the server found what you wanted, didn't find it at all, etc.), and an explanation of the status information. Response header fields follow the response header line. A **response header field** returns information describing the server's attributes. The entity body returns the HTML page requested by the client machine. Though the entity body is optional, it is almost always present. Figure 2-18 shows an example of a response message.

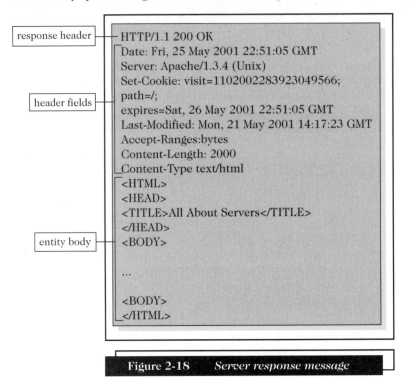

response header	HTTP/1.1 200 OK
header fields	Date: Fri, 25 May 2001 22:51:05 GMT Server: Apache/1.3.4 (Unix) Set-Cookie: visit=1102002283923049566; path=/; expires=Sat, 26 May 2001 22:51:05 GMT Last-Modified: Mon, 21 May 2001 14:17:23 GMT Accept-Ranges:bytes Content-Length: 2000 Content-Type text/html

entity body	`<HTML>`
	`<HEAD>`
	`<TITLE>`All About Servers`</TITLE>`
	`</HEAD>`
	`<BODY>`
	
	...
	
	`<BODY>`
	`</HTML>`

Figure 2-18   Server response message

Three-Tier Client/Server

A **three-tier architecture** builds on the traditional two-tier approach. The first tier is the client, the second tier is the Web server, and the third tier consists of applications and their associated databases that supply non-HTML information to the Web server on request. From a software perspective, the three tiers are client processes (tier 1), Web services (tier 2), and data services (tier 3). Interactions between client and server operate the same way as they do in a two-tier architecture. The third tier provides comprehensive data services, including database operations supported by database software, enterprise resource planning software services, and other services needed to support a robust electronic commerce server.

A typical example of services supported by a database is a catalog with search, update, and display functions. Suppose that a user requests a display of your company's exotic fruit selections. The client request is formulated into an HTTP message, sent over the Internet to the server, and examined by the server. Analysis of the request reveals that the request requires the help of the server's database. The server sends a request to the database to search for, retrieve, and return data about all exotic fruit in the catalog. Database information flows back to the server, and the server formats the response into properly formed HTML and sends a message back to the client over the Internet.

Though the preceding is a very simplified explanation of the three-tier architecture, you can see why electronic commerce sites will require the third layer of hardware and software. The additional software and support hardware are used to track customer purchases stored in shopping carts, look up sales tax rates, keep track of customer preferences, query inventory databases, and keep the company catalog current. Figure 2-19 shows an overview of information flows in a three-tier architecture. Numbers on the flow arrows indicate the order in which the messages flow over the indicated paths.

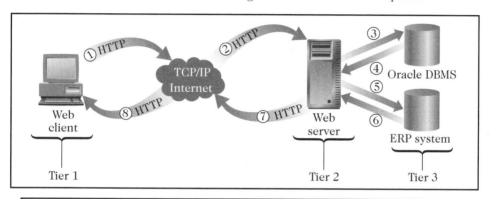

Figure 2-19   *Message flow in a three-tier client/server architecture*

The details of how a Web server communicates with backend machines in tier three are not important here; however, you should have an understanding of the process. A client program can request the services of a backend processor connected to a server through a mechanism called the **common gateway interface (CGI)**. A CGI, which is a protocol, is a common way for Web servers to interact dynamically with clients (users). Web pages that contain forms filled with text boxes, option buttons, and list boxes supply information that CGI programs can use to manipulate databases, store information, or retrieve data. CGI is also known as a server-side solution because all the processing occurs on the Web server, not the client computer. CGI is a standard way of interfacing backend applications with Web servers. CGI applications execute when requested and produce output. Then they formulate and return HTML documents to the Web server.

Electronic commerce is enabled by the ability of CGI technology to update databases in tier three of the architecture—the so-called backend. The backend servers provide programs that dynamically transform data into HTML so that Web browsers can display the results. (HTML version 3.2 was the first version supporting forms, which enabled electronic commerce finally to take off.) CGI hidden fields provide the ability to maintain customer transaction information across a series of Web messages that pass back and forth between a client and a server during a transaction.

INTERNETS, INTRANETS, AND EXTRANETS

It is usually not very long after a company establishes an Internet Web presence that the notion of creating an intranet occurs to management. And, not long after management gives the nod to producing an intranet, someone comes up with the idea of extending the intranet to an extranet. An **intranet** is a Web-based private network that hosts Internet applications on a local area network. Often an intranet facilitates dissemination of company information within the confines of the company and its suborganizations. An **extranet** extends the intranet concept to provide a network that connects a company's network to the networks of its business partners, selected customers, or suppliers. While the facsimile, phone, electronic mail, and express carriers represent the way business has been conducted until now, the extranet is a likely candidate to displace these slow and expensive techniques. Both intranets and extranets have security issues that companies need to consider when choosing their network systems. Security is discussed in detail in Chapters 5 and 6.

Intranets

Technically, intranets are not much different from the Internet, except that only selected individuals are allowed to access an intranet. Based on the client/server model, internal requests for files, documents, or schematic drawings proceed the same way as on the Internet. For example, a regional sales manager, using a Web browser, can inquire about the status of his region's year-to-date sales. The Web client sends an HTTP message to the target company's server using TCP/IP and the company intranet. Authorization checks verify that the requester (the regional manager) has need-to-know access rights to the files. The files are sent back to the requesting client computer on the intranet. The manager is able to look at the latest sales figures for his region and compare them to his projected sales figures.

Intranets are an extremely popular and low-cost way to distribute corporate information. An intranet uses Web browsers and Internet-based protocols, including TCP/IP, FTP, Telnet, HTML, and HTTP. Because corporate intranets are compatible with the Internet, selected information from intranets can be readily shared with external consumers. Another benefit of using an intranet is that different departments of a company having different computer hardware can interact with one another on an intranet. This is because intranet software and protocols are hardware-neutral—they run equally well on a Macintosh, a PC, or a UNIX-based machine.

An intranet server can collect and group information that then can be passed to the Internet for external dissemination. For example, a customer can inquire about the pricing and availability of a particular product or products. An intranet locates the requested information from internal databases, including inventory and work-in-process information, formats the information, and passes it from the intranet to the Internet and on to the customer.

Intranets are reasonably priced because their infrastructure requirements are usually already in place if a company's PCs are on a LAN that connects to the Internet. The intranet infrastructure includes a TCP/IP network, Web authoring software, Web server hardware and software, Web clients, and a firewall server. (Firewalls provide security between the outside world and the private corporate intranet. Firewalls are described in Chapters 5 and 6.) Because intranets use client/server two- and three-tier infrastructures, hardware and software that operates on an intranet works also on the Internet. In addition, using the standard TCP/IP protocol ensures that the intranet capability is available to any company currently using the Internet. In other words, there are not multiple standards to support simultaneously—only one. A three-tier infrastructure is typical of most intranets, because intranets support functions not normally available in the traditional two-tier Internet model.

Intranets can save a company both time and money. In most organizations, intranets are almost always the best way to distribute a wide variety of internal corporate information because producing and distributing paper is usually slower and more expensive than using Web-based communications. For instance, human resources departments have embraced intranets. It saves money and time to post on an intranet the employee handbooks, company policies, and governmental regulations governing the workplace. Anyone who has ever dealt with a voluminous and frequently changing corporate policy manual can appreciate not having to handle printing and distributing monthly and annual policy manual updates. Other information that intranets handle can include job postings, internal performance and product production information, white papers and technical reports, corporate phone books, electronic mail, software manuals, and governmental forms. Training is another area that benefits from an intranet, in the form of defrayed expenses and enhanced convenience. Using an intranet, employees can take online training at any time and any place. Intranet-hosted training is far less expensive than traditional in-person training, because the company avoids the costs of transportation and lodging to send an employee to a training center.

Intranets expedite software distribution and updating. It is expensive to manage and maintain corporate PC software. Intranets can lower the **total cost of ownership (TCO)** by reducing software maintenance and update costs. Computing staff can place software updates and patches on the intranet and then provide a script to update employee workstations automatically the next time they log on. Compare this approach with the time-consuming (and thus expensive) task of individually updating each affected employee's computer. It is no contest—the intranet wins and the companies win. Intranet benefits include:

- Increased, less expensive, environmentally friendly ("green") internal communication
- Low acquisition and deployment costs
- Low maintenance costs
- Increased information accessibility
- Timely, current information availability
- Easy information publication, distribution, and training

Intranets do have some drawbacks, however. They are not free—a company must weigh the benefits against the costs. Currently, it is difficult to calculate precisely an intranet's payback (a figure that accountants and chief financial officers want to know). Some of the intranet tools are immature and not quite ready for wide-scale deployment. Intranets have a way of growing out of hand as different departments and divisions independently add their own information to an intranet server. Distinct organizational intranets require careful monitoring to ensure that they can work seamlessly when needed.

Extranets

Extranets connect companies with suppliers or other business partners. An extranet can be any of the following types: a public network, a secure (private) network, or a virtual private network (VPN). Each has the same capability of sharing information between companies. Information on extranets is secure to prevent security breaches from unauthorized users. Authorized users connect transparently to another company's network through the extranet. Extranets provide the private infrastructure for companies to coordinate their purchases, exchange business documents through EDI, and communicate with one another. In fact, an extranet may be set up through the Internet, but frequently an extranet is a separate network connecting businesses to one another. Extranets can use the Internet for communicating among themselves using traditional Internet protocols, including TCP/IP. Even private networks—separate from the Internet—use Internet protocols and technology to drive communications.

Some extranets start out as intranets, serving employees internally for a number of years. Later, management opens formerly restricted intranet data to Internet users to reduce the workload of its own employees. A good example of this, and one with which you are probably familiar, is FedEx. For years, customers could track their packages by calling a FedEx toll-free number and then giving the operator a tracking number. Package tracking information is displayed on the FedEx operator's console after he or she enters the tracking number. Then, the operator tells the customer about the package's status. All the information about packages is internal to FedEx. A few years ago, FedEx distributed package-tracking software free of charge to anyone who wanted it. Once it was installed on the customer's computer, the software dialed the FedEx computer using a modem, queried the status of the customer's package, and displayed the results on the customer's computer. With the Web in full bloom, FedEx has now eliminated client-machine software and made package tracking available on its Web site. This latest system is called **FedEx Ship** and it provides Web access for package tracking, online airbill creation, shipment logging, and FedEx supply shipment. It lets customers enter their account and airbill numbers, then tap into FedEx's operational system to track the progress of their shipments. Critical information, such as a package's status, is kept on the FedEx extranet. High-volume corporate customers can handle their shipments automatically. The system integrates a customer's order and warehousing data systems with FedEx's intake, billing, and package-tracking software. Figure 2-20 shows the FedEx Ship page. (You can download the client portion of the software from this page.)

Public Network

A **public network** extranet exists when an organization allows the public to access its intranet from any public network, such as the Internet, or when two or more companies agree to link their intranets using a public network. Security is an issue in this configuration, because a public network does not provide any security protection. To secure transactions between cooperating companies, each company must provide protection for outgoing information before that information passes from each intranet onto the public network. Normally a firewall checks inbound information packets coming from the Internet (which is a public network, of course), but firewalls are not 100 percent secure. This is why a public network extranet is rare in practice. There are simply too many risks. Both private networks and virtual private networks provide the added security that most companies require for business-to-business exchanges.

Private Network

A **private network** is a private, leased-line connection between two companies that physically connects their intranets to one another. A **leased line** is a permanent telephone connection between two points; unlike a normal dial-up connection, a leased line is always active. The single advantage of this arrangement is security. The leased line is a dedicated, always-available connection. No party, outside the two who are legitimately attached to the private network, have access to the connection. Thus, the private network inherently provides secrecy and integrity for messages flowing on it.

The single largest drawback to a private network is cost. Leased lines are expensive. Every pair of companies wanting a private network between them requires a separate private (phone) line connecting them. For instance, if a company wants to set up an extranet connection over a private network with seven other companies, the company must pay the cost of seven leased lines, or pair-wise connections. Vendors refer to this as a *scaling* problem—increasing the number of private networks is difficult, costly, and time-consuming. So what are companies to do in order to have a close, private relationship between their intranets? The answer may be an extranet designed from a virtual private network.

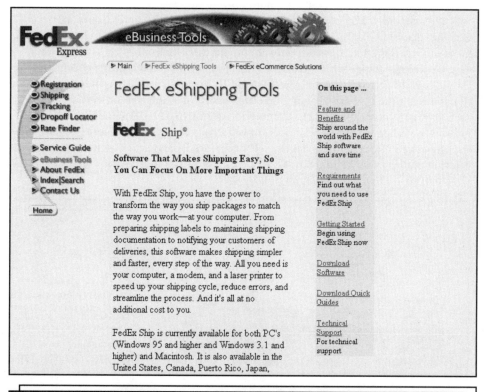

Figure 2-20 *FedEx Ship page*

Virtual Private Network (VPN)

A **virtual private network (VPN)** extranet is a network that uses public networks and their protocols to send sensitive data to partners, customers, suppliers, and employees using a system called "tunneling" or "encapsulation." Tunnels are private passageways through the public Internet that provide secure transmission from one extranet partner to another. A VPN provides security shells, with the most sensitive data under the tightest control. The VPN is like a separate, covered commuter lane on a superhighway (the

Internet) in which the passengers on the commuter lane are protected from being "seen" by the vehicles traveling outside the commuter lane. Company employees in remote locations can send sensitive information to company computers using the VPN private tunnels established on the Internet. This protected tube arrangement between cooperating extranet partners scales very well and is quite inexpensive. Most extranets are implemented either as LAN-to-LAN extranets or client/server extranets. Older systems such as **EDI** typify LAN-to-LAN style extranets. Client/server extranets are popular today. (See the link **VPN Insider** in the Online Companion for more information from a leading VPN source information provider.)

When a company wishes to establish a closer relationship with a supplier or trading partner, a VPN serves to connect them. Establishing VPNs does not require a leased line. The only infrastructure required outside each company's intranets is the Internet. Companies such as **Aventail**—a company providing extranet services—are making VPNs simpler to install and maintain.

Extranets are often confused with VPNs. While a VPN is an extranet, not every extranet is a VPN. (VPNs are described in more detail in Chapter 6.) Virtual private networks are designed to save money, although their main purpose is to create a competitive advantage with the alliance formed between cooperating companies. Unlike private networks using leased lines, VPNs establish short-term logical connections in real time that are broken once the communication session ends. The "virtual" part of VPN means that the connection seems to be a permanent, internal network connection, but the connection is actually short and temporary. Each transaction between two intranets using a VPN is created, carries out its work over the Internet, and is then terminated. Short private bursts of interactions characterize a VPN. Figure 2-21 shows a diagram of a VPN.

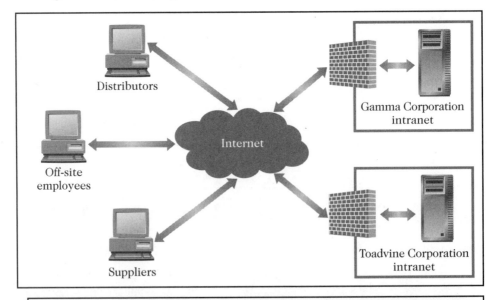

Distributors

Off-site employees

Internet

Suppliers

Gamma Corporation intranet

Toadvine Corporation intranet

Figure 2-21 *Secure VPN extranet*

INTERNET CONNECTION OPTIONS AND TRADE-OFFS

The Internet is a set of interconnected networks. A corporation or individual cannot become part of the Internet without a telephone connection or a connection to a LAN, or an intranet. Larger firms that provide Internet access to other businesses, called **Internet access providers (IAPs)** or **Internet service providers (ISPs)**, usually offer several connection options. This section briefly covers several connection choices and presents their advantages and drawbacks.

Connectivity Overview

ISPs offer several ways to connect to the Internet. One of the major differences between the various ISPs and the services they offer is the connection bandwidth available for each service provider. Recall that bandwidth is the amount of data that can travel through a communication line per unit of time. The higher the bandwidth, the faster data files travel and the faster Web pages appear on your screen. Of course, traffic on the Internet and at your local service provider greatly affects **net bandwidth**—the actual speed that information travels, taking into account traffic on the communication channel at any particular time. When few people are competing for service from an ISP, net bandwidth will approach the carrier's upper limit. On the other hand, you will experience slowdowns during high-traffic periods. Depending on the type of connection you have to the Internet, upstream and downstream bandwidth can be quite different. **Upstream**, sometimes called upload, occurs when you send information from your connection to your ISP. **Downstream**, also called download or downlink, occurs when information flows to your computer from your ISP (for example, when a Web page is sent to your machine). Upstream bandwidth differs from downstream bandwidth for satellite and cable connections. For example, cable modems typically provide a maximum transmission speed of approximately 1 megabit per second from the client to the server. The downstream transmission rate can be as high as 10 megabits per second.

The most common way to connect to an ISP is through the telephone provider. **POTS**, or **plain old telephone service**, uses existing telephone lines and an analog modem to provide a bandwidth of about 56 Kbps (56,000 bits per second). Some telephone companies offer a higher grade of service called **Digital Subscriber Line** or **Digital Subscriber Loop (DSL)** protocol. **Integrated Services Digital Network (ISDN)** was developed to use the DSL protocol suite and has been available in the United States since 1984. ISDN is more expensive than regular telephone service and offers bandwidths of up to 128 Kbps. One of the newest technologies that uses the DSL protocol is **Asymmetric Digital Subscriber Line** (ADSL, also abbreviated **DSL**). It provides transmission bandwidths from 16 to 640 Kbps upstream and from 1.5 to 9 Mbps (million bits per second) downstream.

Cable modems connect to the same broadband coaxial cable that serves a television. (See the Online Companion for links to cable Web pages.) In the United States alone, more than 100 million homes and organizations have broadband cable service available and over 70 million homes subscribe to cable television. The latest estimates indicate that there are more than 2.7 million cable modem subscribers in North America—2 million in the United States.

Connected to a PC through a twisted pair wire and an Ethernet card, cable modems provide both a cost-effective and relatively high-bandwidth connection to an ISP (which, in this case, would be a cable television company such as Cox Cable or MediaOne). Cable connections provide upstream bandwidths of up to 768 Kbps and downstream bandwidths of up to 10 Mbps. (Theoretical speeds are higher both upstream and downstream—but they are rarely obtained.) Unlike ADSL, cable bandwidths vary with the number of other subscribers competing for the shared resource. (ADSL is a private line with no competing traffic.) Transmission speeds can decrease dramatically in heavily subscribed neighborhoods at prime times—in neighborhoods where many people are using cable modems simultaneously. While cable modem sales installations currently exceed DSL modems, DSL modem sales will overtake cable modem sales by 2002, according to a 1999 *PC Magazine* article.

Large firms with high Internet traffic connect to an ISP using higher-bandwidth telephone company connections called **T1** and **T3**. T1 lines operate at 1.544 Mbps, and T3 lines operate at 44.736 Mbps. These connections are much more expensive than POTS or ISDN connections. However, large organizations that need to connect hundreds or thousands of individual users to the Internet require very high bandwidth. **Network Access Providers (NAPs)**, organizations that sell Internet access to Internet service providers, use T1 and T3 lines and also newer **Asynchronous Transfer Mode (ATM)** connections with bandwidths of up to 622 Gbps (gigabytes per second—a gigabyte is 1 billion bits). The NAPs are working with a small group of universities and the National Science Foundation (NSF) to deploy a newly developed successor to the Internet called **Internet 2**. Internet 2 has bandwidths that exceed 1 Gbps.

Figure 2-22 compares seven ways to connect to the Internet. The ATM row refers to the Internet backbone. The "sweet spot" (optimal price and performance) is the cable modem.

Device	Upstream Speed (Kbps)	Downstream Speed (Kbps)	Estimated Startup Fees($)	Estimated Monthly Charges($)
Dial-Up Modem	56	56	20	20
ISDN	128	128	400	80
Cable Modem	768	10,000	200	50
T1 Leased Line	1,544	1,544	3,000	1,100
ADSL	640	9,000	3,000	1,000
T3 Leased Line	44,700	44,700	7,500	8,000
ATM	622,000	620,000		

Figure 2-22   *Internet connection choices*

Pros and Cons

With each available connection there are several advantages and, sometimes, one or two disadvantages. Cost is one of the common disadvantages of the faster connections, but each of the connection types has unique pros and cons.

Dial-Up Connection

Dial-up is the most prevalent and most popular way to connect to the Internet. The main advantage is the relatively low cost to connect to the Internet and, in the United States, the widespread availability of telephone connections. The main disadvantage of a dial-up connection is the slow speed that is particularly evident when downloading a large file, viewing a graphics-intense Web page, or viewing streaming media. In addition, when in use, an Internet dial-up connection ties up a telephone line.

ISDN Connection

An ISDN connection provides higher-speed access than a dial-up connection, and it is available in most urban areas of the United States. On the other hand, ISDN costs more per month than a dial-up connection, and the initial installation fee is relatively expensive.

DSL Connection

A DSL connection is faster than ISDN and uses existing phone lines to provide high-speed access. Although it uses a phone line, DSL does not preclude the phone line for simultaneous telephone use. DSL is faster and more reliable than both dial-up and ISDN, and it is available in most urban areas of the United States. However, DSL is more expensive and its availability is somewhat limited due to a line distance restriction between a consumer and a switch station.

Cable Connection

A cable connection to the Internet is one of the fastest growing of the available connection types—particularly among cable television subscribers. Advantages include the relatively low monthly cost, very high-speed download and reasonably fast upload rates, and the "always on" connection. (Cable companies usually greatly reduce the upload speed to prevent individual users from using their cable modem connections for operating large-scale servers.) The latter advantage is also a disadvantage. The probability of intruders accessing a cable-connected machine is very high compared with other connection methods because a connection is live as long as the PC connected to it is on. Cable connections are not available everywhere, although this problem will diminish rapidly over time. Installation can be a bit expensive, and most users choose to rent a cable modem—an added expense.

T1 Connection

T1 Internet connections are lightning-fast and permit both high-speed FTP and Web browsing. The main disadvantages—at least for individual users—are the high installation costs and the very expensive monthly payments.

T3 Connection

A T3 connection has the same advantages and disadvantages as a T1 connection: A T3 connection is incredibly fast and incredibly expensive. Larger institutions are able to justify the speed need and the very high costs.

Summary

In this chapter, you learned about the protocols, programs, languages, and architectures that support the Internet and the World Wide Web. TCP/IP is the protocol pair that constructs and transports information packets across the Internet. IP addresses, consisting of four numbers, identify computers on the Internet. Domain names such as www.amazon.com identify hosts, but the names are translated into IP addresses for use on the Internet. HTTP, or Hypertext Transfer Protocol, is the protocol that transfers Web pages on the Internet. POP and SMTP are two Internet protocols that receive and send mail. IMAP is poised to be the replacement protocol for POP due to its versatility. The file transfer protocol, FTP, transports files from one computer to another. Telnet provides remote computer access when you supply a username and password. Internet-related utility programs such as Finger and ping determine if another computer is connected to the Internet and the path to it from the computer on which the utility is installed. Electronic mail is a popular Internet application that formats, sends to, and receives mail from an e-mail server.

Hypertext Markup Language, or HTML, derives from the more generic meta language SGML. HTML defines the structure and content of Web pages using special language symbols called tags. Over time, HTML has evolved from its simple roots to version 4.01, which contains a large number of tags that accommodate a rich variety of elements, including graphics, cascading style sheets, and frames. HTML 4.01 also defines semantics and datatypes for HTML. Hyperlinks are HTML tags that contain a URL. The URL can be a local or remote computer. HTTP is responsible for requesting and delivering the page. HTML displays the page on the requesting client machine. The better HTML editors facilitate Web page construction with a plethora of tools and drag-and-drop capabilities.

The Web uses a client/server architecture in which the client computer requests a Web page and a server computer that is hosting the requested page locates and sends a page back to the client. For simple HTTP requests, a two-tier architecture works well. Tier 1 is the client computer and tier 2 is the server. For more complicated Web interactions, such as electronic commerce, requiring services of a database program or other applications, a three-tier architecture is required. Tier 3 adds a backend processor of application programs that compute and deliver information to the Web server in a form that the Web server understands. A backend application could search an inventory database and return catalog entries to a server. The server formats the page and sends it back to the requesting client computer.

Intranets are company-private networks that use the same protocols as the Internet. Organizations host a rich variety of information in Web format. Employees can access the intranet and view or print information they need. Training applications, replete with sound and videos, reside on intranets. When companies want to gain a competitive advantage and collaborate with suppliers and customers, they can connect their intranets to form an extranet. There are three types of extranets: public network, private network, and virtual private network. VPNs, or virtual private networks, provide security at a low cost, whereas public network extranets have no security at all. Private network extranets are dedicated, pair-wise connections and provide security because the network is not shared; however, they are expensive and not easily scalable.

Internet service providers offer several different connections to the Internet. Basic telephone connections are the most economical and easiest to install. They use existing telephone equipment and analog modems to complete the connection. Broadband coaxial cable provides Internet access through the television cable at relatively high speeds. T1 and T3 connections are on the high end of the bandwidth and price spectrum. They provide data transfer rates up to 45 million bits per second. High-volume businesses requiring fast Internet access use T1 and T3 lines. Internet 2 can transmit information at 1 billion bits per second.

Key Terms

Anchor Tag
Anonymous FTP
Asymmetric Digital Subscriber Line (ADSL)
Asynchronous Transfer Mode (ATM)
Attachment
Bandwidth
Browser
Bulk Mail
Cascading Style Sheets (CSS)
Circuit
Circuit Switching
Client
Client/Server Model
Closing Tag
Common Gateway Interface (CGI)
Digital Subscriber Line
Digital Subscriber Loop (DSL)
Domain Name
Dotted Quad
Download
Downstream
EBCDIC
Electronic Mail (e-mail)
Entity Body
Extranet
File Transfer Protocol (FTP)
Full Privilege FTP
Hierarchical Hyperlink Structure
Hops
Hypertext Transfer Protocol (HTTP)
Integrated Services Digital Network (ISDN)
Interactive Mail Access Protocol (IMAP)
Internet 2
Internet Access Provider (IAP)
Internet Protocol (IP)
Internet Service Provider (ISP)
Intranet
IP Address
Leased Line
Limited Edition
Limited Use
Linear Hyperlink Structure
Local Area Network (LAN)
Mail Server

Meta Language
Multipurpose Internet Mail Extensions (MIME)
Net Bandwidth
Network Access Provider (NAP)
Network Control Protocol (NCP)
One-Sided Tag
Open Architecture
Opening Tag
Packet-Switched
Plain Old Telephone Service (POTS)
Plug-In
Port
Post Office Protocol (POP)
Private Network
Protocol
Public Network
Request Header
Request Line
Response Header Field
Response Header Line
Router
Routing Algorithm
Scripting Language Code
Server
Simple Mail Transfer Protocol (SMTP)
Spam
T1
T3
Tags
TCP/IP
Terminal Emulation
Thin Client
Three-Tier Architecture
Total Cost of Ownership (TCO)
Transmission Control Protocol (TCP)
Two-Sided Tag
Two-Tier Architecture
Uniform Resource Locator (URL)
Upload
Upstream
Virtual Private Network (VPN)
Web Client
Web Server

Review Questions

1. Describe in one or two paragraphs the origins of HTML. Be sure to mention and define tags and mention one or more persons involved with HTML's development.

2. Name the protocols that an electronic mail program might use to send and receive mail.

3. What is the role of the World Wide Web Consortium? Use your Web browser to locate the W3C site and conduct your research.

4. Compare the two- and three-tier Web client/server architectures and indicate the role of each. Which architecture is the most likely candidate for an electronic commerce site?

5. Use your Web browser and the Online Companion to search for more information about the pros and cons of ADSL compared with cable connections. Form an opinion in favor of one of the two methods, and write an argument to support your position.

Exercises

1. Radon is a small corporation located in Setubal, Portugal. It is in the process of setting up a Web site to describe Portuguese port wines. First, Radon wants to build a simple series of Web pages to advertise its presence. In stage two, it will contact wineries in Portugal to represent Radon's products on their sites. Radon has hired you to help it investigate HTML before it plunges into full-scale Web page design and deployment. In particular, Radon wants you first to research (briefly) the history of HTML, determine what important new features were introduced in each version, and report back with a one-page description (about 250 words). Next, Radon's senior information systems officer would like you to learn which HTML editors might be good choices for developing pages. For this second report, simply create a table listing five or six editors, the names of the companies that developed the editors, and the approximate retail price of each one. Use the Online Companion links to help you search the Web for information.

2. Bridgewater Engineering Company (BECO), a privately held machine shop, makes industrial-quality, heavy-duty machinery for assembly lines in other factories. It sells its horn presses, grinders, and milling equipment, using a few inside salespersons and telephones. This traditional approach worked well in the startup years, but BECO is getting a lot of competition from abroad. Because you worked for the company during the summers of your college years, BECO's president, Leonard Carroll, knows you and realizes that you are Web-savvy. He wants to form a close relationship with his suppliers—steel companies and small parts manufacturers—so that he can tap into their ordering systems and request supplies when he needs them. Mr. Carroll wants you to describe to him how he can use the Internet to set up such an electronic relationship. What are the alternatives? Are there any companies that could help you and him acquire the software and hardware needed to set up some sort of network? Use the Web and the links in the Online Companion to locate information about extranets and VPN networks. Write a short report comparing various network options and indicate, in writing, the names of at least two companies (from the Web) that could help develop the system for Mr. Carroll. Limit your report to 500 words or less.

3. Baseline High School in Arlington, Texas, wants to connect its network of computers to the Internet. You have been hired to consult with the school and determine the best way to do this. Mrs. Bruton, the principal of Baseline, requests that you produce a report

describing the advantages and drawbacks of various Internet connection options. Write a 500-word report beginning with POTS and proceeding through T3. Use the Web to locate leasing costs. Put together a table comparing the upstream and downstream transmission speeds. Show in the table your estimate of the time that it would take to download a 2 MB plain text file using each of the connection methods.

For Further Study and Research

Barnett, R. 1997. "Connectionless ATM," *Electronics & Communications Engineering Journal*, IEE, Great Britain, October.

Bosak, J. and T. Bray. 1999. "How XML Will Fix the Web: Tags Categorizing Facts, Not Formats, Speed Up Transactions," *Scientific American*, 280(5), May, 89.

Goldfarb, Charles F. 1981. "A Generalized Approach to Document Markup," *ACM Sigplan Notices*, (16)6, June, 68–73.

Gupta, P., S. Lin, and N. McKeown. 1998. "Routing Lookups in Hardware at Memory Access Speeds," *Proceedings of IEEE Infocom*, San Francisco, April.

Holland, R. 2000. "XML helps partners connect," *eWeek*, 17(21), May 22, 44.

Kay, E. 2000. "From EDI to XML," *Computerworld*, 34(25), June 19, 84–85.

Lin, S. and N. McKeown. 1997. "A Simulation Study of IP Switching," *Proceedings of ACM Sigcomm*, September.

Newman, P., T. Lyon, and G. Minshall. 1997. "IP Switching: ATM Under IP," *IEEE/ACM Transactions on Networking*.

PC Magazine. 1999. "High Speed Takeover," 18(6), March 23, 10.

Rendleman, J. 2000. "DSL's Technical Foul," *eWeek*, 17(24), June 12, 28.

Shand, D. 2000. "Simple Sites, Perplexing Problems," *Computerworld*, 34(19), May 8, 80–81.

WEB-BASED TOOLS FOR ELECTRONIC COMMERCE

By just about anyone's analysis, the business-to-business segment of electronic commerce is setting records monthly. Analysts do not agree on the exact numbers, but they predict B2B sales will exceed 1 trillion dollars by 2003. Some argue B2B sales could exceed 3 trillion dollars by 2003.

Fueling a portion of current and predicted sales are small to medium-sized businesses minimizing their costs by purchasing online everything from supplies to raw materials for production of goods. Even the smallest businesses recognize that it is both convenient and cost-effective to order office supplies such as paper clips, copy paper, telephones, and pencils online. The cost savings compared to conventional ordering methods is significant.

Ariba a provider of B2B solutions and services to leading companies around the world, has long recognized the need for enterprise resource planning (ERP) software to facilitate office supply ordering, shipment, and invoicing. It recognized that supplies are something every company must have and that supply expenditures account for over

30 percent of corporate spending. By eliminating paper orders, telephone calls, catalogs, and much wasted time, real cost savings can easily accrue.

While not the most technically interesting use of the Internet, electronic commerce-enabled supplies procurement can provide a significant return for every dollar spent. Procurement experts estimate that savings near 12 percent are possible. For a company spending $3 million on supplies a year (a small amount by most standards), this equals an annual savings approaching $360,000. This chapter explains how to determine what hardware and software are required to implement a system like Ariba's.

LEARNING OBJECTIVES

In this chapter, you will learn about:

- Computers that support Web servers
- Hardware requirements of typical Web server software packages
- Fundamental duties of a Web server
- Other ancillary Web server functions
- Specific Web server software, including Apache, Microsoft Internet Information Server, and Netscape Enterprise Server
- Advanced Web server tools

This chapter provides background on the technical requirements of a system to support electronic commerce. Specific electronic commerce features and functions are covered in later chapters.

WEB SERVER HARDWARE AND PERFORMANCE EVALUATION

A half-dozen years ago, corporate Web sites were simply curiosities that were far less important to businesses than selecting a good accounting package or a human resources management (HRM) application suite. Today, a Web site is frequently the first place consumers go to conduct business. Corporate Web sites have gone beyond the status of "ancillary," moved through "mission critical," and have become the main business focus for many organizations.

Providing an effective Web presence and strategy requires careful planning and good choices. The two main ingredients in a Web server are its hardware—or computer—and its Web server software. In this chapter, you will learn about how much computer "firepower" you need to host Web software. Of course, a popular site (one with many visitors, such as Microsoft's) must have a much larger Web server than a smaller site, such as that of a community hospital. To decide on the level of firepower that you need, you must first determine which Web server software you want to run on the server. Then, you can decide on the hardware to install—since software requirements limit computer hardware choices. We will examine software features later in the chapter. Finally, you must know what server performance measurements and metrics are relevant to your hardware and software combination.

Making Shared Host, Dedicated Host, and Self-Host Decisions

A key question facing most companies wanting to serve up Web pages and, now or later, provide electronic commerce services is whether to host their own Web site or pay someone else to host a Web site for them. An important first step in planning a Web server (and, later, an electronic commerce server) is to find out what the management and sales staff wants to accomplish with the server. Whom do they want to reach? Will the server have to run transactions? To what degree is the company staff involved with the server on a daily basis? Web sites, and the reasons for creating them, can range from a simple development side, an intranet, a marketing ("brochureware") site, a business-to-business portal, a storefront site, or a content delivery site. Each has a different purpose, requires different computer hardware and software, and incurs varying costs. An overview of how the Web site hardware and software requirements vary among these groups follows.

Development Sites

The simplest Web site—and least costly to implement and maintain—is a development site. Companies wanting to experiment with different Web designs and evaluate them can do so with little initial investment. This "try before you buy" approach allows corporate insiders to experiment with the form and fit of the site, make suggestions about content and organization, and ignore security concerns. A development site can reside on an existing PC and be developed with low-cost tools such as FrontPage or Dreamweaver. Testers can access the site through their PCs on the existing LAN or directly through the workstation housing the Web site.

Intranets

Corporate intranets, a step up from a development site, house internal-use memos, corporate handbooks, expense worksheets, newsletters, and various corporate documents. Intranets employ redundant computers, which ensures that corporate intranet servers are available all the time. Because intranets are shielded from the Internet, they don't require special security software.

B2B and B2C Commerce Sites

Commerce sites, both business-to-business and business-to-consumer, must be available 24 hours a day, seven days a week (that is, they must offer high reliability). These commerce sites should use the most reliable and robust servers that the business can afford, and they should have spare servers for handling periodic high traffic volumes that occur (that is, they must offer high availability). In addition to requiring fast and reliable hardware, commerce sites must run Web and commerce software that is very efficient and easily upgraded when increased loads occur. Security software is important whenever a site serves the Internet community outside the confines of corporate firewalls (see Chapter 6 for a discussion of firewalls). Most larger commerce sites maintain an extensive database of user profiles, products, and other information. Unlike B2C sites, B2B-only sites also require certificate servers to issue and analyze electronic authentication information (these security issues are discussed in Chapter 6).

Content Delivery Site

Content delivery sites such as USA Today, The New York Times, and ZDNet sell and deliver content—news, histories, summaries, and any digital information that may be useful in its own right. Delivery must be very speedy. Hardware requirements match those of business-to-business and business-to-commerce sites. In addition, users must be able to locate important articles rapidly with a fast and precise search engine. Databases for content delivery sites must be especially well tuned for quick information delivery.

Other Considerations

When the company's needs go beyond Web service—to building electronic commerce components (which require a specific electronic commerce server)—the company must decide again whether to run servers in-house—to self-host their commerce hardware and software. The company may decide that a third-party Web and electronic commerce provider is a better initial choice for its Web/commerce needs. As you will see in Chapter 4, many small Web stores typically use a third-party host provider for both Web services and electronic commerce—particularly when the company's Web site is relatively small or the company offers a limited number of products for sale. Hosting arrangements can range from shared hosting to dedicated hosting. Shared hosting means that a corporate Web site is on a server that hosts other Web sites simultaneously and is controlled by a third-party service provider. Dedicated hosting occurs when you get your own server that is owned and administered by a service provider that dictates terms of usage. When making Web server hosting decisions, you should ask whether the hardware platform and software combination can be quickly and easily scaled up when Web site traffic grows. Of course, a company's Web server requirements are directly related to its electronic commerce transaction volume and Web site traffic. The most successful electronic commerce solutions are **scalable**, meaning they can be adapted to meet changing requirements. It is costly to be stuck with a Web server that cannot be enhanced to meet increased requirements.

Using a third-party or ISP host provider can be an ideal solution for several reasons. With either type of provider, Web and electronic commerce novices do not need to provide hardware or software to launch the Web site. In addition, third-party hosting services fill the often time-consuming staffing requirements that would otherwise be the company's direct responsibility. Using an ISP means that a small company does not have to establish a direct Internet connection to its business. Smaller companies also can take advantage of third-party electronic commerce software, which provides all the fundamental promotional features needed to sell products online and serve Web pages.

If you are uncertain or cautious about selecting a third-party Web host, turn to the **Web Host Guild** (WHG) for guidance. Formed in 1998, the WHG's goal is to set an industry standard that benefits all hosting companies and protects consumers. The WHG hopes to make Web host certification a part of doing business on the Internet—a sort of Better Business Bureau of the Internet. Though it is a new organization, it has deep roots and a strong board of directors.

Deciding whether to host a Web site in-house has an impact on availability and the number of simultaneous site visitors that a Web site can accommodate without seriously degrading service. A site that can handle many visitors simultaneously is said to have high bandwidth, while sites that have difficulty accommodating large numbers of visitors have low bandwidth. If the number of visitors to a company's Web site is small, a company's slow Internet connection and low bandwidth may be sufficient. On the other hand, a popular Web site with tens of thousands of visitors per hour must either invest in a high-speed Web infrastructure or outsource Web hosting to a third party who provides a very high-speed Internet connection. Nationwide ISPs, such as Sprint and MCI WorldCom, have high-speed backbone Internet access. When you contract with them, bandwidth is not a problem. The **backbone** is the main network of connections that comprises the Internet.

A company shouldn't automatically choose a nationwide provider to initially host a site. A local third-party ISP can be a good option, but it is important to know how the local ISP interacts with the national Internet backbone providers. Even a small Web site can grow quickly and might soon require higher bandwidth. When your local third-party ISP has a direct and good working relationship with national ISPs, then scaling up a Web site to handle increased traffic loads is less complicated. Using smaller hosting companies with connections to larger service providers is important to facilitate fast connections and high bandwidth. Though ISP host services may offer T1-level service, it is important to know *how many* T1 lines they have. Many ISPs will put multiple customers on a single T1 line, which can lower bandwidth. A useful metric to know is the ISP's ratio of customers to T1 lines; a lower ratio is more desirable because the available bandwidth will be greater.

Using an outside Web host instead of using an in-house server means that the staffing burden shifts from your company to the Web host. **EZ Webhost** and **Interland** are examples of Web host companies. Figure 3-1 shows a page from the Interland Web host service Web site. **HostPro** is a Web service provider that provides dedicated Web hosting. HostPro clients can choose either unmanaged or managed hosting. With **managed hosting** (or **managed server hosting**), the service provider manages the operation and oversight of all servers and assigns a dedicated service manager. **Unmanaged hosting** (or **unmanaged server hosting**) shifts the responsibility of maintaining and staffing all servers to the customer—away from the Web service provider. **HostIndex** provides a convenient collection of Web pages that compare Web hosts with one another. Click **HostIndex Web server comparisons** to compare Web host features side by side. **TopHosts.com** provides one of the most comprehensive link collections available to aid in researching Web hosting alternatives and services.

Typically, ISPs have the Web hosting expertise that a small or medium-sized company may not have. ISPs sell both Internet access and their Web and Internet expertise. Creating and maintaining a Web site within your own company and using an existing network can be difficult; any problems that arise are yours to solve—not problems that are the responsibility of the ISP. With the exception of large companies with large Web sites and in-house computer experts, it almost always is cheaper to use outside Web hosting services. While extranets (see Chapter 2) are best run from within the company, Web hosting third-party vendors are a good choice for running and maintaining Web services and running electronic commerce for smaller organizations.

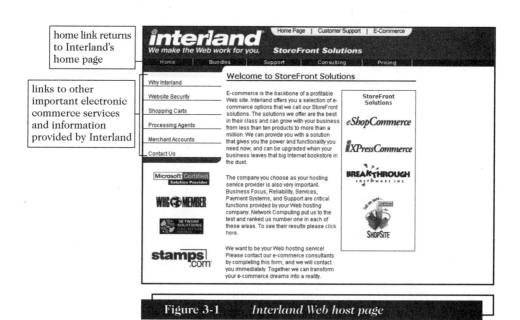

Figure 3-1   *Interland Web host page*

Web Platform Choices

Companies must be forward-thinking when making Web computer hardware choices. Obviously, a fast server is always better than a slower one. But you must think about which server is a good choice for the present, while your business is small, and for the future, when your business has grown. Another important factor in server hardware choices is the internal (intranet) and external traffic or transactions that are likely to occur on the server. For a small organization just starting out, the expected traffic might be a few thousand visits, or hits, per hour initially. Larger, more well-known organizations employing a server for the first time might expect tens of thousands of "eyeballs" at the Web site each hour. Here, planning and testing are your best allies. Because some visitors to your site might be using expensive client machines on high-speed connections and others might be using PC clients on low-speed dial-up connections, you might want more than one machine to service the two different client types.

Hardware decisions go hand in hand with operating system and application server software choices. All three are tied together and collectively determine the performance of your Web system. One of the most important variables in the server decision is whether the server hardware is scalable—that is, whether you can upgrade the server or even connect additional servers together seamlessly. At some point when the server traffic is sufficiently high, you will need to add more computing muscle to your site. Running a large, enterprise-class application server (see the next section) on a personal computer is not feasible. Similarly, purchasing and installing a $50,000 application server for a small site is excessive. Running Microsoft SQL or Oracle database servers on the same computer as the Web or electronic commerce server is not a good idea, because database products have large processing and memory requirements and can slow down Web server response times. Figure 3-2 shows an example of the architecture of a large, scalable, three-tier Web and electronic commerce system.

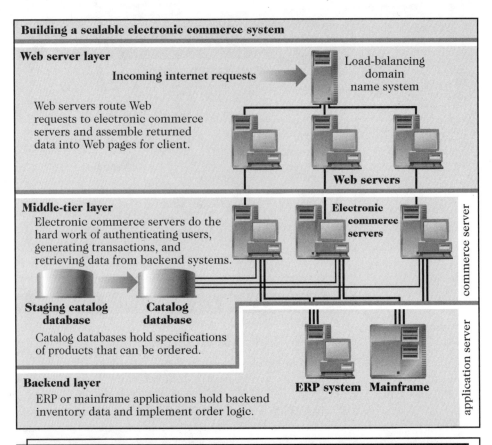

Building a scalable electronic commerce system

Web server layer

Incoming internet requests

Load-balancing domain name system

Web servers route Web requests to electronic commerce servers and assemble returned data into Web pages for client.

Web servers

Middle-tier layer

Electronic commerce servers do the hard work of authenticating users, generating transactions, and retrieving data from backend systems.

Electronic commerce servers

Staging catalog database

Catalog database

Catalog databases hold specifications of products that can be ordered.

Backend layer

ERP or mainframe applications hold backend inventory data and implement order logic.

ERP system   Mainframe

commerce server

application server

Figure 3-2   *A scalable, three-tier Web and electronic commerce architecture*

An **application server** is a middle-tier software and hardware combination that lies between the Internet and a corporate backend server. It contains databases that it serves to users on front-end client computers. One way to classify application servers is through the operating systems that they support. An **operating system** is software that performs basic tasks for the computer on which it resides. These tasks include running programs, allocating computer resources such as memory and disk space to programs, and providing input and output services to devices connected to the computer, including the keyboard and monitor. A computer must have an operating system to run programs. For large systems, the operating system has even more responsibilities, including keeping track of multiple users logged on to the system and ensuring that they do not interfere with one another.

Most application servers run on Windows NT/2000 computers, Linux computers, or UNIX-based operating systems. Each operating system and its supporting machines has distinct advantages and disadvantages. For example, Windows NT or 2000 is simpler to learn and use than the somewhat arcane computer lingo required to set up a UNIX-based application server. However, UNIX-based machines are more popular, and many claim that they are more robust machines on which to run an enterprise-class application server. Advantages of **Linux** are that it is a free operating system, easy to install, and very fast and efficient. According to *PC Magazine*, over half of the worldwide ISPs use Sun Microsystems hardware. Furthermore, three-quarters of all ISPs use Solaris, Sun's proprietary operating system. (Sun is a leader in enterprise network computing and produces powerful workstations and servers, and the operating systems to run them. Click the **Sun Microsystems** link on the Online Companion to learn more.)

Setting up a Web server doesn't necessarily require a lot of money for big UNIX machines or fancy Windows NT or 2000 computers. You can build a small Web server on an inexpensive PC (less than $1500) running Windows 98 or 2000. And in some cases, Web server software is free. Microsoft, for instance, bundles Microsoft Personal Web Server with its Windows 2000 operating systems at no extra cost. You can quickly build a respectable Web server on an inexpensive computer running free Web server software. Of course, this computer and operating system combination won't support Web traffic of 20,000 accesses per second, but it will get you started. With a small investment, you can prototype your Web ideas before making a commitment to purchase a more robust, enterprise-class system.

The best way to choose a server is to run tests on the various hardware and operating system combinations, remembering to consider the system's scalability if needed. **Mindcraft**, one of a handful of companies that is an independent test lab, tests software, hardware systems, and network products for users and says it has "...developed a rigorous quality system that meets international standards... ." Its site (see the Online Companion link **Mindcraft Web server performance reports**) contains a plethora of reports and statistics comparing combinations of application server platforms, operating systems, and Web server software products. Figure 3-3 (on the next page) shows a sample Web page with examples of reports that Mindcraft produces.

Web Server Performance Evaluation

Web servers are an essential component of doing business electronically. Benchmarking Web server hardware and software combinations can help you make informed decisions for your small business or enterprise-class system. **Benchmarking** is testing used to compare the performance of hardware and software. Because the technology is continually changing, any suggestion in this book about particular Web server hardware or software could become quickly obsolete. But guidelines to help make good choices are helpful, so we will offer some basic rules here.

The role of a Web server goes beyond building a Web presence. A Web server facilitates business-to-business and business-to-consumer transactions, hosts company and enterprise applications, and is part of the corporate communications infrastructure. It is important to know what affects Web server performance (and what does not), what activities to measure, how to collect performance information, which software products collect Web server performance data, and what server software and hardware combinations perform well.

When evaluating Web server performance, it is important to know exactly what factors are being measured and ensure that these are important factors relative to the expected use of the Web server. Hardware and operating systems are obviously key areas for benchmarking. A PC with a midrange CPU, small hard drive, and 32 megabytes of memory will perform poorly compared with a high-end workstation or a powerful UNIX-based computer. Operating systems, including Linux, Solaris, and Windows NT, turn in different performance values under various Web benchmark tests. (*PC Magazine* periodically tests Web servers and reports the results in the magazine. Not long ago it tested and compared Windows 2000, Solaris/iPlanet, NetWare, and Linux/Apache. Click **PC Magazine Web server tests** in the Online Companion for the latest results.)

list of Mindcraft services

Mindcraft reports

directory server reports

various white papers comparing servers

click a link to display a summary of the report

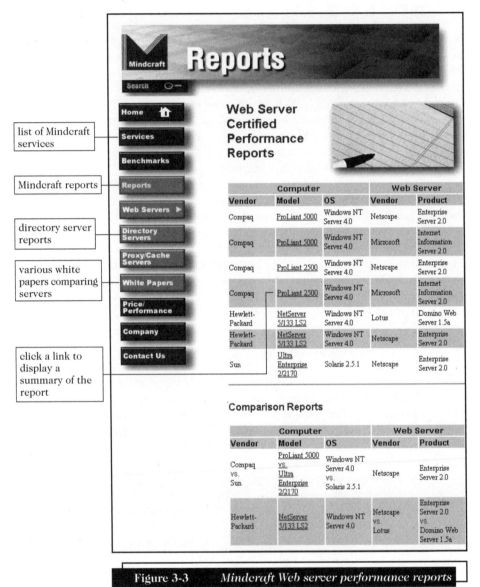

	Computer			Web Server	
Vendor	**Model**	**OS**	**Vendor**	**Product**	
Compaq	ProLiant 5000	Windows NT Server 4.0	Netscape	Enterprise Server 2.0	
Compaq	ProLiant 5000	Windows NT Server 4.0	Microsoft	Internet Information Server 2.0	
Compaq	ProLiant 2500	Windows NT Server 4.0	Netscape	Enterprise Server 2.0	
Compaq	ProLiant 2500	Windows NT Server 4.0	Microsoft	Internet Information Server 2.0	
Hewlett-Packard	NetServer 5/133 LS2	Windows NT Server 4.0	Lotus	Domino Web Server 1.5a	
Hewlett-Packard	NetServer 5/133 LS2	Windows NT Server 4.0	Netscape	Enterprise Server 2.0	
Sun	Ultra Enterprise 2/2170	Solaris 2.5.1	Netscape	Enterprise Server 2.0	

Comparison Reports

	Computer			Web Server	
Vendor	**Model**	**OS**	**Vendor**	**Product**	
Compaq vs. Sun	ProLiant 5000 vs. Ultra Enterprise 2/2170	Windows NT Server 4.0 vs. Solaris 2.5.1	Netscape	Enterprise Server 2.0	
Hewlett-Packard	NetServer 5/133 LS2	Windows NT Server 4.0	Netscape vs. Lotus	Enterprise Server 2.0 vs. Domino Web Server 1.5a	

Figure 3-3   *Mindcraft Web server performance reports*

Another factor that can affect a Web server's performance is the speed of its connection. A server on a T3 connection can deliver Web pages to clients much faster than it could on a T1 connection. (The client's connection is irrelevant when measuring raw server performance.)

Still another factor to consider is how many users the server can handle. User capacity can be difficult to measure, because it depends on the server's line speed, the clients' line speeds, and the typical size of the Web pages that are delivered. What does matter in measuring a server's Web page delivery capability are throughput and response time. **Throughput** is the number of HTTP requests that a particular hardware and software combination can process in a unit of time. **Response time** is the amount of time a server requires to process one request. These values should be well within the anticipated loads a server can experience—even during peak load times.

Finally, the mix and type of Web pages a system is likely to deliver in response to client requests greatly affect performance. A **dynamic page** is a Web page whose content is shaped by a program in response to user requests, whereas a **static page** is an unchanging page retrieved from disk. A server delivering mostly static Web pages will perform better than the same server delivering dynamic Web pages, because static page delivery requires far less computing power than dynamic page delivery. The largest performance differences between competing Web server products appear when servers deliver dynamic pages.

Several Web server benchmarking programs are available. They generate a rich variety of metrics that are crucial in deciding which server to employ. Some benchmarking programs are free and others are available for a few hundred dollars. Figure 3-4 lists several popular Web benchmark programs and their sources. Consult the Online Companion for links to the software and companies that produce and sell these programs. Notable among the Web benchmark vendors is Ziff-Davis, which has written and provides free of charge several Web server benchmark programs, including **NetBench ServerBench**, and **WebBench**. Check for the latest version of these programs at the **Ziff-Davis benchmark programs** link.

Web server benchmark software	Publisher
NetBench	Ziff-Davis
ServerBench	Ziff-Davis
SPEC SFS97	Standard Performance Evaluation Corporation
SPECweb99	Standard Performance Evaluation Corporation
WCAT	Microsoft
WebBench	Ziff-Davis
WebStone	Mindcraft

Figure 3-4   *Web server benchmark software*

It is worthwhile to take a close look at some of the more prevalent benchmarking programs. **WebStone** is the original and is still a very popular Web server benchmark program. It is a typical Web server performance evaluation program, and the information WebStone collects is similar to information collected by other Web benchmark programs. WebStone was developed by Silicon Graphics, and Mindcraft, Inc., has acquired the rights to WebStone. WebStone works by measuring the response of Web servers to a workload it creates. The workload simulates multiple Web clients (users connecting to a Web site with their clients) accessing the Web server. WebStone can simulate over 100 Web clients on a single computer. Webmaster, a program that controls all the testing done by WebStone, runs on one of the client computers and distributes the Web client software and test files to client computers.

After Webmaster starts the execution of a benchmark, it waits for client computers to report the performance that each client measured. When all client performance information is available, Webmaster consolidates the information into a summary report. The files used by the Web client computers determine the performance measured by WebStone. WebStone supplies a standard set of files so administrators can compare the evaluation results fairly for different Web servers. Because of the way that WebStone benchmark tests are structured, evaluation results measure the performance of the combination of the Web server's operating system, Web server software, network connection speed, and CPU speed.

WebStone uses three tests to measure performance: HTML, CGI, and API. The HTML test measures server performance when the client asks the server to fetch and send it an HTML-encoded file (a static Web page). The CGI test causes the Web server to run another program, using the **Common Gateway Interface (CGI)** protocol to pass information to the called program. (Information in Web-based forms uses the CGI protocol to transfer data to and from backend programs.) The third test, called API, tests the Web server's ability to pass information from a Web client to the server's **Application Program Interface (API)**, which is a set of protocols, routines, and tools for building application code blocks. The API request launches another program that locates information for the Web server and passes it back. An example of an API request is a Web client request for information found in a database on another computer.

The WebBench benchmark software, developed by Ziff-Davis Benchmark Operation, produces two main values, or scores. One score reports the number of requests per second to the server, and the other reports throughput (in characters per second). Conveniently, WebBench produces its output in Excel format so that the results can be easily graphed and manipulated. Available free of charge, WebBench runs only on Windows computers.

SPECweb99 is a benchmark program from the Standard Performance Evaluation Corporation, a nonprofit standards organization. SPECweb99 provides system workloads that stress-test Web servers. These workloads originate from representative Internet sites and comprise Web files from 1 KB to 1000 KB. Though not free, the software is inexpensive.

Anyone contemplating purchasing a server that will have heavy traffic should compare standard benchmarks for a variety of hardware and software configurations. If you anticipate making modifications to an existing server but aren't sure what element to tune up, then customized benchmarks that allow you to modify file sizes, cache sizes, and other parameters will give you the greatest flexibility and the most meaningful results. Web managers should run benchmarks regularly for large, enterprise-class Web servers. However, benchmarks are not as meaningful for small Web sites with much smaller numbers of daily visitors. In the latter case, you should concentrate on Web design and site navigation to maximize clients' satisfaction.

Ziff-Davis conducted a number of benchmark tests of several popular Web server software packages on a small number of hardware platforms. The company did features testing and performance testing. Using WebBench, the tests measured the performance of both static HTML requests and dynamic CGI Web server requests. The test workload consisted of over 6000 files ranging in size from a few hundred bytes to files over 500 KB. In total, the files occupied 63 MB of disk space. By modifying the Web server cache size to allow the entire collection of files to be stored in the cache, the testing staff eliminated disk performance differences from the performance equation. That allowed for a fair comparison of servers. (A Web server **cache** is a high-speed memory area set aside to store Web pages. The cache can save time by filling client Web page requests from high-speed memory whenever possible, rather than fetching pages from slower disk storage.) Sixty client computers were used to submit Web requests to the servers being tested. The major conclusions were:

- Every Web server handled static Web pages quickly and efficiently.
- Significant performance differences occurred when Ziff-Davis performed the test suite containing CGI requests—requests for dynamic Web pages.
- Some server software continued to perform well, while other software packages hit maximum requests per second.

It is not important which server software did best, because the results will change as the different suites are developed and newer hardware becomes available. What is important is that the benchmark tests showed that there are observable differences in performance that a Web site manager must consider before purchasing or leasing a Web server.

Besides testing Web servers' raw performance, it is important to test server software features for efficiency and usability. These tests will reveal whether a particular feature is easy to use and whether it performs well. Web server features are described in the next section.

WEB SERVER SOFTWARE FEATURE SETS

Web servers are located publicly on the Internet or privately on organizational intranets, usually behind firewalls. The duties and features of Web servers differ somewhat depending on whether they are publicly accessible. Intranet Web servers often authenticate users by requesting login and password information before permitting users to access selected information. Normally, server software maintains an information log containing time, date, and URL information for every access to the server from inside and outside the organization. The primary duty of a Web or HTTP server is to respond to requests from client programs. Electronic commerce support is implemented by the server, as are backend programs and databases invoked by the server. Backend program and database responses are formatted and passed back to the server, which sends the formatted Web page to the requesting client.

Web server software program features can range from basic to extensive, depending on the software package being used. Web server software features fall into natural groups based on their purpose. All Web server programs provide a certain core feature set, without which the program would not be a Web server at all. O'Reilly Software, producer of WebSite Professional, classifies its Web server's features into these categories: core capabilities, site management, application construction, dynamic content, and electronic commerce. Site development is also a key function of Web server software. While not all Web server programs' features fit precisely and exclusively into one of these classes, the feature classes are a convenient and organized way to discuss Web server software features. Thus, they are used in this text. First, we will examine the core capabilities any Web server program should have.

Essential Capabilities

Recall that the most fundamental duty of a Web server is to process and respond to Web client requests that are sent using the HTTP protocol. For a client request for a Web page, the server program locates and fetches the page, creates an HTTP header, and appends the HTML document to it. For dynamic pages, the server invokes other programs, receives the results from the backend process, formats the response, and sends the pages and other objects to the requesting client program. (Chapter 2 presents some details of this process.) IP-sharing, or a virtual server, is a feature that allows different groups to share a single Web server's Internet Protocol (IP) address. A **virtual server** or **virtual host** is a feature that maintains more than one server on one machine. This means that different groups can each have their own domain name, but all domain names refer to the same physical Web server. For example, ExoticFruitInc Corporation's marketing department could have the domain name www.marketing.ExoticFruitInc.com, while sales could have the domain name www.sales.ExoticFruitInc.com—both names referring to the same ExoticFruitInc Corporation server.

A Web server translates a logical Uniform Resource Locator (URL) into a physical file address. The translated address points directly to a page requested by a client browser and is returned to it. For example, the Web server might translate the URL www.twidleydee.com/infosheet.html into the filename C:\Home\WebserverBase\ Information.html. With the filename available, the operating system can fetch the file and process it.

Security

Security and authentication services are essential for Web server implementations and can authenticate employees accessing the server from the Internet. Security services include validation of usernames and passwords, as well as processing certificates and private/public key pairs (see Chapter 6). Access controls provide or deny access to files based on the username or URL. Servers support **Secure Sockets Layer** (**SSL**), which is a protocol developed by Netscape for transmitting private information securely over the public Internet. Web sites use SSL to accept confidential information, such as credit card numbers, from Web client users.

FTP

Web servers provide File Transfer Protocol (FTP) services whereby users can transfer files to or from the server. Anonymous FTP occurs when a user logs on to a server using the universal password *anonymous*. By convention, anonymous FTP users enter their e-mail addresses for their passwords. Some Web servers disallow anonymous FTP altogether, whereas others permit users to download information from the server to the client, but not the reverse. Additionally, most servers permit Gopher access to their sites. Created at the University of Minnesota, **Gopher** is a system that predates the Web and displays a nongraphical, hierarchically-structured list of files on both Web and Gopher servers. While Gopher has all but disappeared, Web servers still support the system. Figure 3-5 shows an example of a Gopher display of Microsoft's site.

links are indicated by numbers in brackets

single-letter commands to print, quit, obtain help, and more

Figure 3-5 *Using Gopher to access Microsoft.com*

Searching

Search engines and indexing programs are standard in Web servers. Search engines or search tools search either the existing site or the entire Web for requested documents. An indexing program provides full-text indexing whereby indexes are generated for documents stored on the server. When a browser requests a Web site search, the search engine compares the index terms to the requester's search term to see which documents contain matches for the requested term or terms. For example, Microsoft's Web server software, Internet Information Server, contains Indexing Server bundled as part of the Web server package. Indexing software can index several document forms. Normally, the search engine enforces security by returning only documents that the user has permission to view.

Data Analysis

Web servers can capture visitor information, including who is visiting a Web site (the visitor's URL), how long the visitor's Web browser viewed the site, the date and time of each visit, and which pages were displayed. These data are placed into a Web **log file**. As you can imagine, the file grows very quickly—especially for popular sites with thousands of visitors each day. Careful analysis of the log file can be fruitful and reveal many interesting facts about your visitors and what they like (or don't like). To make sense of a log file, it is necessary to run third-party Web log file analysis programs. These programs summarize log file information by querying the log file and either returning gross summary information or accumulating details that reveal how many visitors came to the site per day, hour, or minute, or which hours of the day were peak loading times. Two of the most popular Web log file analysis programs are the **Analog Web server log file analyzer** and the **WebTrends Web server log file analyzer**. Marketwave also publishes several log file analyzers (see **Marketwave Web server log file analyzer** on the Online Companion). Figure 3-6 shows part of one of the reports on Marketwave's Web site. Click the Online Companion entry **Space telescope Web report** to see an example of another type of log file analysis.

Web-Based Tools for Electronic Commerce

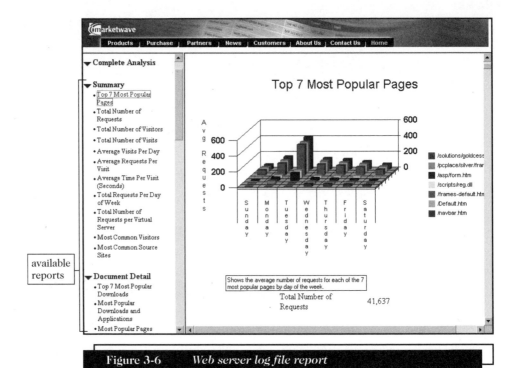

available
reports

Figure 3-6   *Web server log file report*

Site Management

Good Web site management tools offer more than authoring tools such as Microsoft's FrontPage. While FrontPage is an excellent authoring program and has some site management capabilities, it does not offer all the capabilities that a complete and dedicated site management tool provides. For instance, **Allaire** Corporation's **HomeSite** validates graphics, computes page-download times for modem connections, validates links, and validates HTML code.

Dedicated site management tools include a standard set of features, starting with link checking. A link checker examines each page on the site and reports on any URLs that are broken, that seem to be broken, or that are in some way incorrect. It can also identify orphan files—files on the Web site that are not linked to any page. Other important site management features include script checking and HTML validation. Web server log file analysis, described previously in the "Essential Capabilities" section, is also a function of a site management tool. Generally, site management tools scan large sites and create order out of chaos. They can sweep through a site, locate error-prone pages and code, and list broken links. Site management tools can e-mail maintenance results to any address on the Web.

On the company Web site, it is important to regularly check links that point to pages both within and outside the corporate Web site. Some Web software does contain link-checking software, and there are several programs available from third-party vendors. Maintaining a site that is free from bad links is vital, because too many bad links, or dead links, on a site will cause people to jump to another one. A **dead link**, when clicked, displays an error message rather than a Web page. Web-browsing customers are just a click away from going to a competitor's site if they become annoyed at an errant Web link.

You can launch some of the free link-checking and Web site validation programs by entering the address of a Web site's home page and checking a few boxes. Then, the results of the link checker are either displayed automatically or e-mailed to you. Besides checking links, Web site validation programs sometimes check spelling and other structural components of Web pages. The Online Companion contains links to several link checkers. Other checkers that run on your own Web site are also available. While not free, they frequently produce more comprehensive results and more detailed analyses. Examples of these include **Linkbot Pro** and **Big Brother**. Linkbot Pro is an example of a complete Web site analysis tool that scans a site for broken links and more than 50 other potential problems. Linkbot Pro generates graphical reports detailing any errors it finds. Figure 3-7 (on the next page) shows a report produced by Linkbot Pro. **LinxCop**, one of several reverse link checkers available, ensures that designated Web sites have a link to your home page. LinxCop provides a convenient way to automatically check that advertising links driving customers to your Web site are actually in place at other sites.

With remote server administration, a Web site administrator can control the company's Web site from any Internet connection. While all Web sites provide administrative controls—most through a local console or through a Web browser—it is convenient for an administrator to be able to tweak and fix the server from wherever he or she happens to be (at home, on the road, etc.). For example, an administrator can install **Web Site Garage** on any Internet-connected Windows or NT machine and monitor and change anything on the Web site from that computer.

Application Construction

Application construction uses Web editors and extensions to produce Web pages—either static or dynamic. Some Web development systems provide simple tools to create Web pages, while others include an extensive and rich developmental engine in which you can create dynamic features without the need to know CGI or use API coding. Capable Web page creation software can detect any HTML code that varies from the current standard or that is Web browser–specific. Some Web development packages can create Web pages that *do* recognize which browser is requesting a Web page and then respond with a dynamically generated page that takes advantage of the requesting browser's unique capabilities.

Dynamic Content

Dynamic content is nonstatic information constructed in response to a Web client's request. For example, if a Web client inquires about the status of an existing order by entering a unique customer number or order number into a form, the Web server will search the customer information and generate a dynamic page based on the customer information it found, thus fulfilling the client's request. Successful Web sites employ dynamic Web content, are attractive enough to keep customers coming back, and hold users' interest for as long as possible. (The longer someone stays at a particular site, the "stickier" the site is said to be. Web sites strive to increase their stickiness.) Assembled from backend databases and internal data on the Web site, the successful dynamic page is tailor-made to the requester's query. Any server that can handle dynamic content can handle information from a variety of databases. Using **Open DataBase Connectivity** (ODBC), the Web server can assemble information from disparate database systems, such as Oracle, SQL Server, and Informix. ODBC, developed by Microsoft, makes it possible for a program to access any data from an application, regardless of which database management system is dispensing the data.

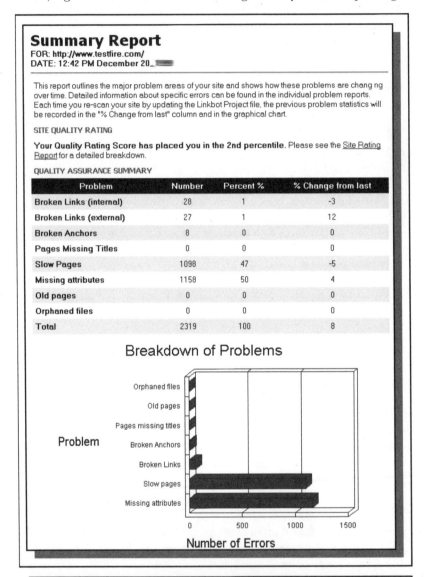

Summary Report
FOR: http://www.testfire.com/
DATE: 12:42 PM December 20_

This report outlines the major problem areas of your site and shows how these problems are changing over time. Detailed information about specific errors can be found in the individual problem reports. Each time you re-scan your site by updating the Linkbot Project file, the previous problem statistics will be recorded in the "% Change from last" column and in the graphical chart.

SITE QUALITY RATING

Your Quality Rating Score has placed you in the 2nd percentile. Please see the Site Rating Report for a detailed breakdown.

QUALITY ASSURANCE SUMMARY

Problem	Number	Percent %	% Change from last
Broken Links (internal)	28	1	-3
Broken Links (external)	27	1	12
Broken Anchors	8	0	0
Pages Missing Titles	0	0	0
Slow Pages	1098	47	-5
Missing attributes	1158	50	4
Old pages	0	0	0
Orphaned files	0	0	0
Total	2319	100	8

Breakdown of Problems

Figure 3-7   *Typical link report*

Active Server Pages (ASP), developed by Microsoft, is a server-side scripting mechanism to build dynamic sites and Web applications. With ASP you can use a choice of programming languages, such as VBScript, Jscript, and Perl, to produce dynamic pages within your HTML documents. Java, a programming language introduced by Sun, can produce dynamic pages by running server-side programs or by running on client computers.

Site Development

Site development tools comprise features such as an HTML/visual Web page editor, software development kits, and Web page upload support. Exactly which tools are bundled with Web server software varies from one server to another. The best known of these tools are the HTML editors and visual Web page editors. Early in the development of the Web, HTML editors provided very little capability. Their Notepad-style text editors provided a few toolbar buttons to assist in inserting tags and creating tables. Today, visual Web development tools are sophisticated and complete. Many capable Web page and Web site development tools abound. Examples include FrontPage, Dreamweaver, Cold Fusion, PageMill, HoTMetaL Pro, and Netscape Composer. Today's Web development tools support the latest HTML specification. Software development kits usually take the Web master beyond mundane Web page creation. Kits typically contain sample code and instructions for developing server- and client-side programs in one or more of the Java, Visual Basic, WinCGI, or Perl languages. The kits include sample code and code development instructions for the selected language.

Wizards are a quick and easy way to create content and handle otherwise difficult or time-consuming Web page construction and oversight tasks. Wizards can aid you in creating a "What's New at the Site" page, an "About the Company" page, a site index that users can search, search forms for users to fill in, and portal construction wizards.

One-button publishing, the ability to move a modified page from a local PC to its remote home, is very handy. Most Web server software suites include this feature. Dreamweaver, for example, lets you develop and change a site's page on a local PC and then easily transfer all changed pages onto the remote Web server host machine. A complete copy of the Web site can reside on the PC, and that provides built-in redundancy if the server should crash. Collaboration built into packages such as Dreamweaver allows several people to modify Web pages simultaneously without the danger of one person modifying a page that another is currently working on. This is done through a file checkout system that prevents another developer from inadvertently modifying an in-process page.

Electronic Commerce

While electronic commerce servers and Web servers are distinct (a Web server handles Web pages, and a commerce server deals with buying and selling goods and services), a Web server should support electronic commerce software. WebSite Professional, the Web server software from O'Reilly, comes bundled with electronic commerce templates and other tools to add electronic commerce capabilities to a Web site. In the best of circumstances, turnkey software provides electronic commerce templates that simplify the creation of graphics, product and company information, shopping carts, and even credit card processing. Ideally, the number of products that the electronic commerce software supports is unlimited, and new products present no challenges to the software. If the electronic commerce software allows you to use a simple browser-based interface to manage or modify the Web store, then you don't need to learn an entirely new software system. Like Web server hardware and software, electronic commerce software that can grow as the site's product offerings and customer hit rates increase is better than electronic commerce software that is not scalable.

The best electronic commerce software will generate sales reports on demand, allowing store managers to see the latest figures on what is selling, what is popular, and other important sales information. Also, Web advertisements, a fact of Web life, should be rotated and replaced automatically by your software. Electronic commerce software may allow you to "weight" certain advertisements, so they show up more or less frequently than other advertisements.

In the next section, we will take a look at examples of available Web server software. Now that you have learned about several Web server features, you can examine particular Web server software offerings and see how they stack up. The next section describes a handful of the most popular Web server packages. Bear in mind that there is not a best package for all cases. Several factors affect your choices, not the least of which are the hardware and operating system you choose and the Web server features that are the most important to you.

WEB SERVER SOFTWARE AND TOOLS

The Web server market divides into two distinct areas: intranet servers and public Web servers. There are several Web server software packages from which to choose. Some servers run on only one computer operating system while others run on several operating systems. This section describes three popular Web server programs, where "popularity" is the estimated number of Web sites with the particular software installed. These estimates were accumulated through surveys done by **Netcraft**, a networking consulting company in Bath, England, known throughout the world for its Web Server Survey. Netcraft frequently conducts surveys to determine the number of Web sites and to measure the relative popularity of Internet Web server software. Naturally, the percentages change regularly as new software becomes available and other software falls out of favor. A recent Netcraft report indicates that three of the most popular Web server programs are Apache HTTP Server, Microsoft Internet Information Server, and Netscape Enterprise Server. Figure 3-8 shows the market share of several Web servers.

Server software from Apache and Microsoft accounts for the majority of the installed sites. While these numbers are volatile and will be different as you read this, the three packages highlighted in this section are likely to be popular for some time. (Click the **Netcraft surveys** link on the Online Companion to check out the latest survey results.) According to a *PC Magazine* survey (see the Alwang reference listed in the "For Further Study and Research" section at the end of the chapter), the percentages for intranet Web servers are quite different than for public Web servers. While the Web server software packages briefly described in this chapter are all top selections among intranet servers, Microsoft IIS and Netscape Enterprise Servers (all types) together account for 75 percent of the installed intranet server programs. Recently Netscape and Sun formed an alliance called iPlanet. The iPlanet software and hardware combination will enable enterprises and service providers to create and operate what the alliance calls "the next generation of digital marketplaces—Open Digital Marketplaces." The iPlanet backers hope to convince the vast number of businesses—both large and small—that their platform is highly scalable, follows existing open software and hardware standards, and is highly reliable.

Though it is too early to judge whether iPlanet will be a huge success in the Web server and electronic commerce marketplace, you should keep a sharp eye on iPlanet in the next few years.

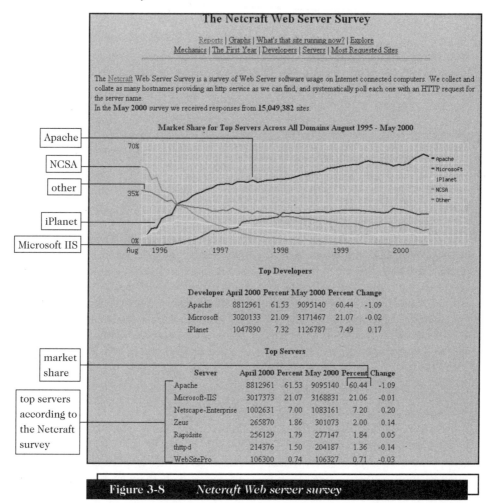

The Netcraft Web Server Survey

Reports | Graphs | What's that site running now? | Explore
Mechanics | The First Year | Developers | Servers | Most Requested Sites

The Netcraft Web Server Survey is a survey of Web Server software usage on Internet connected computers. We collect and collate as many hostnames providing an http service as we can find, and systematically poll each one with an HTTP request for the server name.
In the May 2000 survey we received responses from 15,049,382 sites.

Market Share for Top Servers Across All Domains August 1995 - May 2000

Apache
NCSA
other
iPlanet
Microsoft IIS

Top Developers

Developer	April 2000	Percent	May 2000	Percent	Change
Apache	8812961	61.53	9095140	60.44	-1.09
Microsoft	3020133	21.09	3171467	21.07	-0.02
iPlanet	1047890	7.32	1126787	7.49	0.17

Top Servers

market share

top servers according to the Netcraft survey

Server	April 2000	Percent	May 2000	Percent	Change
Apache	8812961	61.53	9095140	60.44	-1.09
Microsoft-IIS	3017373	21.07	3168831	21.06	-0.01
Netscape-Enterprise	1002631	7.00	1083161	7.20	0.20
Zeus	265870	1.86	301073	2.00	0.14
Rapidsite	256129	1.79	277147	1.84	0.05
thttpd	214376	1.50	204187	1.36	-0.14
WebSitePro	106300	0.74	106327	0.71	-0.03

Figure 3-8   Netcraft Web server survey

Recall from the previous section that the performance of one Web server differs from that of another based on workload, operating system, and the size and type of Web pages being loaded. *PC Magazine* evaluates computer products regularly. Reports of tests the magazine performed using **WebBench** on various combinations of Web server software and operating systems under stress conditions revealed some surprises. Some Web software fared well when delivering static HTML pages, but other Web server software performed better when delivering dynamic Web page content. The differences between servers are dramatic, suggesting that you should carefully consider the mix of dynamic and static pages your site may produce. Whatever hardware and software combination you choose, one thing is clear: Picking the right server for each different business need is critical (see the Machrone reference listed in the "For Further Study and Research" section).

The sections that follow contain descriptions of Apache HTTP Server, Microsoft Internet Information Server, and Netscape Enterprise Server. Each description examines the program's general characteristics, its configuration and management tools, log files and reporting tools, security and directory support, applications development, and database connectivity.

Apache HTTP Server

Apache is an ongoing group software development effort. Rob McCool developed Apache while he was working at the University of Illinois at the **National Center for Supercomputing Applications (NCSA)** in 1994. Several Web masters from around the world created their own extensions to the server and formed an e-mail group so that they could coordinate their changes (known as "patches") to the system. The system became known as Apache because it consisted of the original core system with a lot of patches—thus, it became known as "a patchy" system, or simply, "Apache."

Apache HTTP Server dominates the Web in numbers, in part because it is free (even from CNET Shopper.com, which normally charges for software) and performs very efficiently—it is powerful enough that IBM has licensed it for its own WebSphere application server package. In the period from 1996 through 2000, Apache has enjoyed the highest increase in Internet Web sites of all Web servers, according to a Netcraft survey. Currently, Apache is more widely used than all the other Web servers combined. Apache runs on many operating systems (AIX, BSD/OS, FreeBSD, HP-UX, Irix, Linux, Microsoft NT, QNS, SCO, and Solaris) and the hardware that supports them. Apache has a built-in search engine and HTML authoring tools and supports FTP.

Apache can be managed either from a server console or a Web browser. A **server console** is a terminal that is in the same room as the server and that is directly attached to it. Wizards are available to create new sites and directories, and the server provides for multiple logs that can be automatically cycled or archived. (**Cycling** a log means replacing the oldest log with the newest, thus recycling the space it occupies. **Archiving** a log means saving it, perhaps on a large backup storage device.) The log entries conform to the established, standard NCSA common log format to which many servers adhere.

Apache's security is well thought out, with support for password authentication and digital certificate authentication. (Chapter 6 discusses digital certificates and authentication when using them.) Access can be restricted by domain name, by IP address, or by user and group. Apache can prohibit access by directory or file, and supports SSL.

Apache's application development tools support CGI and several proprietary APIs. Once the API blocks are built, programmers can invoke the code blocks to perform their duties by using the common API interface. Apache supports **Server Side Includes (SSI)**, a type of HTML comment that directs the Web server to dynamically generate data for the Web page when it is requested. An example of an SSI is the following comment that directs a program, which is identified by its filename and extension (.cgi), to execute. (Click **Server Side Includes (SSI) tutorial** for more information.)

```
<!--#exec cgi="filename.cgi"-->
```

SSIs can be used to execute programs and insert the results into returned Web pages. (Web pages that contain SSIs often end with the extension .shtml.) There is no official standard for SSIs—each Web server can support different SSIs in different ways.

Apache also supports Active Server Pages and Java servlets. Similar to CGI, ASPs generate dynamic content using either Jscript code or the Visual Basic programming language. When a browser requests an ASP page (which will have the extension .asp), the Web server generates the HTML page and sends it back to the browser. The browser does all the heavy lifting, in other words. (See the **ASP** link in the Online Companion.) **Java servlets** are applications that run on a Web server and generate dynamic content. When the Java application runs on the browser (client) side of the client/server pair, it is called a **Java applet**. (Java **servlets** are a type of API that is very popular as an alternative to CGI, because they are more efficient than CGI.) Apache supports the ODBC standard and can access Oracle, Sybase, Microsoft SQL Server, and IBM's DB2 databases. Figure 3-9 shows Apache's home page.

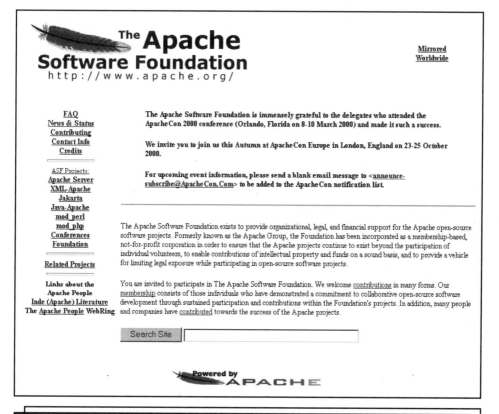

Figure 3-9 *Apache's home page*

Microsoft Internet Information Server

Microsoft Internet Information Server (IIS) comes bundled (free) with Microsoft's Windows NT Server and 2000 Server operating systems. IIS serves equally well as an intranet Web server or a public Web server program. IIS is very popular Web server software for both public Web sites and corporate intranet Web sites. Of course, popularity can be fleeting in the fast-paced Internet world. Recent tests performed by *PC Magazine* revealed that Microsoft's IIS obtained top scores compared with several other servers in delivering static HTML pages. It also performed very well on similar tests involving dynamic Web page tests. In fact, it scored second (just behind Apache) in the dynamic tests while running on a Solaris operating system. A robust and capable Web server program, IIS is suitable for small sites right up to enterprise-class sites doing high transaction volumes.

Currently, IIS runs only on the Windows NT and 2000 operating systems. IIS includes an integrated search engine that allows users to create custom search forms with a variety of tools, including ASP, ActiveX Data Objects, and SQL database queries. The IIS Web server software also includes Microsoft's FrontPage HTML development tool and reporting tools from Crystal Reports. (**Crystal Reports** is a visual reporting tool that lets you create presentation-quality reports and integrate them into database applications.) IIS supports FTP, allowing users to download files and data from the IIS server site with the FTP protocol.

IIS creates CERN/NCSA common log file formats and allows writing to multiple logs. Like most other Web server products, IIS supports automatically cycling or archiving log files. The Microsoft Management Console (MMC), which is included in IIS, provides central server management from any server on the network. IIS also permits administration from a remote browser. Because Windows NT lets you associate additional IP addresses with a single network interface card (NIC), IIS supports multiple virtual hosts. That is, IIS permits each virtual server to have its own IP address.

Security in IIS is tightly integrated with Windows NT's operating system security. Thus, NT basic access control mechanisms (usernames and passwords) and SSL software encryption are also provided in IIS. IIS includes a built-in certificate server that allows organizations to issue and manage digital certificates verifying identities. Access control can limit use by groups or by individuals and can be applied to directories and files. Parts of documents can be hidden from users who do not have clearance to access them.

IIS's inclusion of ASPs provides an application environment where you can combine HTML pages, ActiveX components, and scripts to produce dynamic pages. Microsoft also includes its own Internet Services API (ISAPI), which is an application programming interface for creating programs that run as processes. Database support includes ODBC and Microsoft SQL. Figure 3-10 shows a Microsoft Internet Information Server information page complete with links to technical documents.

Netscape Enterprise Server

Another very popular Web server program, and the descendant of the original Web server program, is **Netscape Enterprise Server** (NES). There are several versions of the software, so the family of programs is described as a whole in this text. Over the coming months, Netscape will migrate Netscape Enterprise Server to iPlanet, partnering with Sun Microsystems, to create one of the few 64-bit combinations of hardware and server software available. (See the Dragan and Derfler article reference in the "For Further Study and Research" section for a review of iPlanet and the forecast for its future.)

Anyone developing a sophisticated, enterprise-strength Web site will appreciate Netscape's extensive server features. Though Netscape is not free, its $1300 to $2000 licensing fee is very reasonable, and it allows a free 60-day trial. The Netscape server software runs on a representative collection of operating systems: AIX, Digital UNIX, HP-UX, Irix, Solaris, and Windows NT. Some of the busiest and best-known sites on the Internet, including BMW, Dilbert, E*Trade, Excite, Lycos, and Schwab, are running some version of NES (or were at the time this book was written).

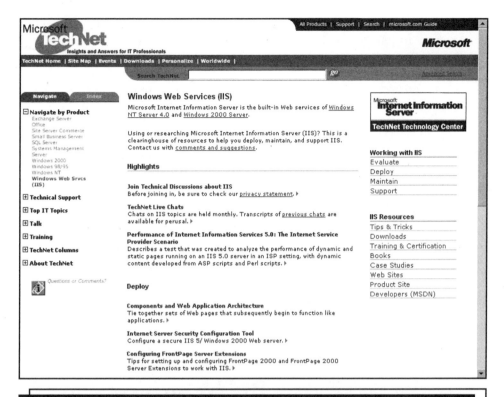

Figure 3-10 *Microsoft Internet Information Server information page*

NES provides a powerful development environment that supports development of Web-based applications that can be run on the Internet, an intranet, or an extranet. NES's content management allows users to create their own Netshares—personal home directories—using an interesting method that provides services including link management, Web publishing, agent services, and access and version control. The NES Web Publisher enables even novice users to upload their finished pages from their PCs to the Web host, which means users don't have to learn FTP to perform upload and download operations. NES is unique in providing this local publishing mechanism.

NES comes with document conversion and indexing utility programs, and these programs bundle a Verity search engine. The Verity search engine is versatile because it can index documents in various formats, including Adobe PDF (Portable Document Format), Microsoft Word, and Microsoft PowerPoint. NES also provides a utility program to convert common document file formats to HTML. (Of course, Microsoft PowerPoint and Word provide that feature also.)

NES's management tools allow administrators to manage users and monitor server activity interactively. Using centralized control, a manager can use the integrated Netscape Directory Server to add, delete, or change user information. Netscape provides cluster management, which is a way for an administrator to manage multiple remote servers as a single group. This allows the administrator to update configuration files remotely or to start and stop a group of servers.

Both password/challenge user authentication (see Chapter 6) and digital certificate authentication are found in Netscape Enterprise Server. Netscape Directory Server (NDS), bundled into NES, provides basic security through username/password-based authentication mechanisms for discretionary access control. **Discretionary access control** allows you to specify which users have access to which computer files and other computer resources. NDS is robust, providing support for over 50 million user entries and supporting 5000 queries per second on one system. Network Directory Server's Certificate Management System (CMS) provides digital certificate authentication management and is integrated into the enterprise server. Netscape also works with SSL performance enhancement devices, which increase the efficiency of the server while it is performing SSL functions.

Like most other server programs, NES supports dynamic application development, including CGI and Netscape's own version of an application program interface: Netscape Server API (NSAPI). NES supports the Java Servlet API for server-side applications. A Netscape product called the LiveWire runtime environment is included in NES and allows you to write server-side scripts that, among other things, provide connectivity to a rich variety of databases, including Oracle, Sybase, and Informix. Its ODBC conformance means that NES provides connectivity to other database sources as well. Figure 3-11 shows a Netscape page listing several Netscape browser and server products.

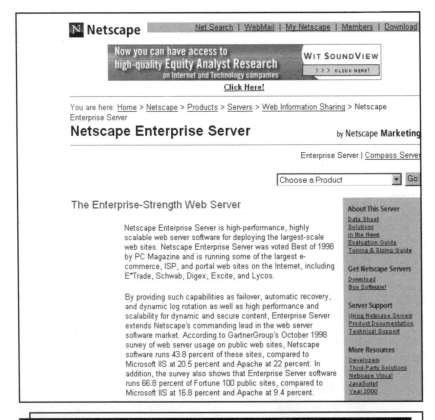

Figure 3-11   *Netscape Corporation browser and server products*

Determining Web Server Hardware and Software Information

You can determine the type of hardware and software most Web sites are running by visiting **Netcraft**. On Netcraft's home page is a link named "What's that site running?" (Or, you can click **What Web software is running on a site?** in the Online Companion.) The "What's that site running?" page opens and displays a prominent Hostname textbox. Type any Web address (for example, www.ibm.com) in the text box and click the Examine button. Netcraft software examines the designated Web site and returns both Web server hardware and software information (see Figure 3-12). Occasionally, the information is not readily available, but Netcraft is almost always successful. If you were to continue doing the hardware/software search repeatedly for 15 sites or so, then you would generate the same type of information that Netcraft collects and analyzes as shown in Figure 3-8.

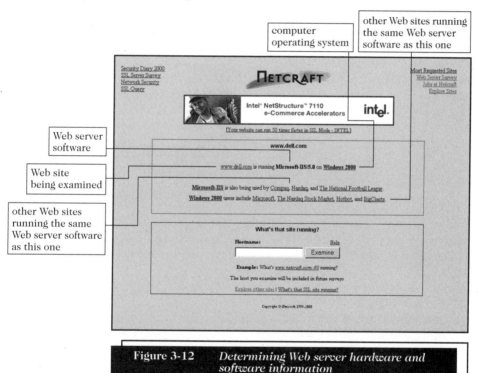

Figure 3-12 *Determining Web server hardware and software information*

OTHER WEB SERVER TOOLS

Besides Web server hardware and Web server software, there are other tools that are part of Web servers that you should know about. Three particularly important tools are portals, search engines, and intelligent agents.

Web Portals

Any organization intent on attracting people to its Web site will be aware of the growing attention and importance that portals are receiving. A **Web portal** (or simply a **portal**) is a "cyber door" on the Web; it serves as a customizable home base from which users do their searching, navigating, and other Web-based activity. Your selected portal loads automatically when you launch your Web browser, giving you a familiar starting point with elements that always appear in the same place on the page. Portals usually include general interest information and can help you find just about anything on the Web. Portals are not aimed at a particular target audience, which is part of their attraction, and you can customize a portal to display exactly what is important to you (such as the latest sports headlines, updated Dow figures, or current lowest plane fares).

Consumer Portals

Web managers are discovering that increased sales and advertising income result from attracting more people and *retaining* them longer. In fact, Web ratings companies (**Media Metrix**, for example) have sprung up—akin to television's Nielson ratings—that measure the number of people who visit a site and how long they linger. Knowing which pages visitors view and how long they linger there helps companies design more effective and attractive Web sites. Web sites with successful portals attracting large numbers of browsers who linger (the previously mentioned "stickiness" factor) can charge more for Web advertising. Examples of successful portals include the **About.com**, **Amazon.com**, **Excite**, **Netscape Netcenter**, and **Yahoo!** sites. Click **Top 10 online portals** for information on portals that were judged by critics as the best on the Internet. The site also describes how the company derived the ratings (see Figure 3-13).

You should study and emulate successful portals if you want to turn your Web site into a portal. Most portals include free e-mail; links to search engines and to categories of information; membership services; news, sports, and business headlines and articles; personalized space with a user's selections; links to chat rooms; links to virtual shopping malls; and Web directories.

Figure 3-14 (on page 106) shows a Yahoo! page that Barb Goldberg personalized just the way she wants it. Because she spent a lot of time getting all the page elements just right, she will likely return to her own Yahoo! page frequently. A few portals have been trying another ploy to retain people. They offer credits for shoppers who purchase from their site. These "frequent shopper" credits accumulate until the consumer can purchase something with them. Newer portals include online address books and calendar programs that allow portal visitors to store one global copy of their important contact information or appointments so that they are Web-accessible from anywhere in the world. You will see more of this personal information manager software all over the Web in the future.

Business Portals

In terms of total value exchanged, the most important portals in the last few years are the business-to-business portals. These portals, most of which can be accessed only by member enterprises, specialize in business commodities and materials such as steel, gasoline, or chemicals. Until recently, the largest players were portals that housed many different portals from different industries. Lately, many new B2B portals arranged around single industries have emerged. Dubbed "vortals," these groups of closely related industries exchange goods and services in a particular market, such as steel or lumber. The term **vortals** is an abbreviation of "vertical portals" and represents vertical industries that are part of a value chain beginning with raw materials and ending with finished products. Examples of vortals include the multicompany portal called **Petrocosm Marketplace** introduced by Chevron. Its purpose is to bring together on the Web large numbers of gas and oil suppliers that are willing to offer competitive prices on their goods and services for the "prize"—a huge audience of petroleum industry players—all on one Web site.

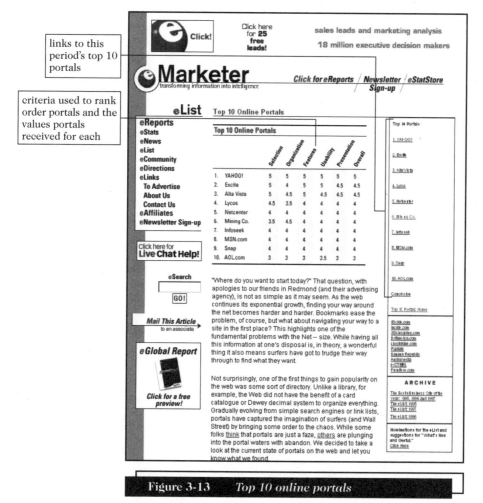

links to this period's top 10 portals

criteria used to rank order portals and the values portals received for each

Figure 3-13   *Top 10 online portals*

Excite@Home launched the **Work.com** B2B portal, which is designed to meet the diverse needs of business professionals. Work.com provides a plethora of links and services to help business managers run their businesses, including voice mail, calendar services, and storefront hosting. The site also contains links to comprehensive information about electronic commerce, financing, and sales.

In the field of steel, the best-known B2B portal is **e-STEEL**. The e-STEEL portal's proclaimed purpose is to bring together steel buyers and sellers in order to

"...streamline operations and reduce costs, connect existing customers in new ways, extend market reach, increase convenience and choice, generate new opportunities for business, and create new market efficiencies." Figure 3-15 (on the next page) shows e-STEEL's home page.

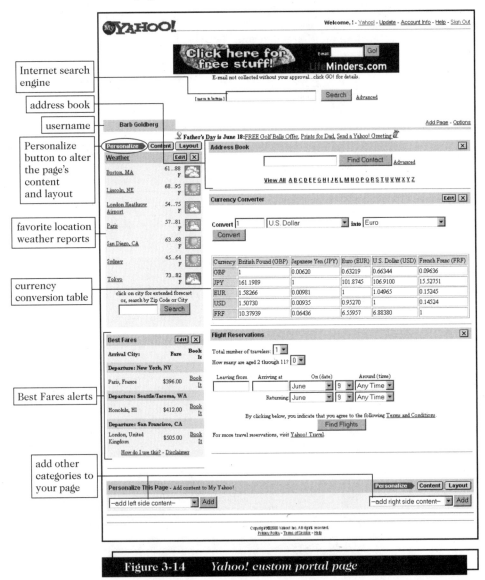

Figure 3-14   Yahoo! custom portal page

Each month, you can find new B2B portals addressing the needs of a variety of businesses. For example, **FoodUSA** is a leading online exchange for the meat industry. An international timber exchange is found on the portal **Timber.org**. Similarly, **BigTray** claims to be the world's leading Internet supplier of food service equipment and supplies at competitive prices. **TurboStaff.com** is a global B2B portal offering staffing and related services. TurboStaff.com's purpose is to attract a critical mass of buyers and sellers in the staffing services business, thus providing an economical and timely supply of staffing services.

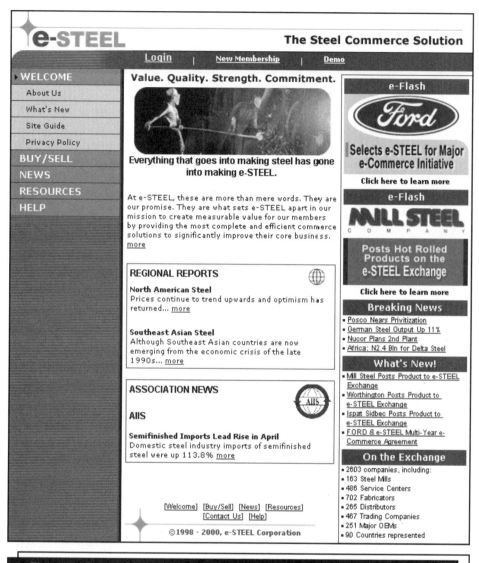

Figure 3-15   e-STEEL's home page

Search Engines

Many Web sites traditionally include a Web search engine that lets you seek and find topically related Web sites. A **search engine** is a special kind of Web page software that finds other Web pages that match a word or phrase you entered. The word or phrase you enter, called a **search expression** or **query**, might include instructions that tell the search engine how to search. For example, you can enter a phrase and search for all or any of the words in the phrase. A Web site's search engine does not examine every Web page to find a match; it only searches its *own* database of Web pages and Web page information. That explains, in part, why you can get two different results using two different search engines. A **hit** is a Web page that is indexed in the search engine's database and that contains text that matches your search expression.

Some Web sites also have directories. A **Web directory** is a listing of hyperlinks to Web pages that is organized into hierarchical categories. The difference between a search engine and a directory is that *people* select the Web pages to include in a Web directory—a very labor-intensive process. Sites must submit their links to a reviewer who assigns them to the appropriate categories. Yahoo! is the oldest and most famous example of a directory-based site. It built its fame on being an excellent directory, and its directory contains links to more than 1 million human-categorized Web pages.

Search engines contain three major parts. The first part, called a **spider**, a **crawler**, or a **bot**, is a program that automatically and frequently searches the Web to find Web pages and updates the information about old Web sites that already are in its database. One of the more important duties of the spider is to delete old information about Web sites that no longer exist. Everything that the spider finds goes into the second part of the search engine—the index. An **index** is like a gargantuan book containing every Web page the spider finds. When a Web page changes, the spider notices the change and updates the "book," or index. Updates to the index with page changes found by the spider do not occur quickly, and that is why you often find search engine results that have links that are incorrect or stale. The search engine utility itself is the third part of the search engine. When you request a search, a search engine combs through millions of pages to locate matches and then ranks them in order by the engine's judgment of each link's relevance.

All search engines have these three components, but the differences among them lie in how each engine performs its tasks. For one thing, each search engine includes different Web pages in its database. And the search results you receive depend on whether the search engine looks for *all* query words in the Web page. Detailed discussion of the differences among search engines is beyond the scope of this text. (The Online Companion contains links to popular search engines.)

There are organizations that keep track of search engines and provide information about how they work on their own sites. One of those organizations is **Search Engine Watch**, which compiles mountains of information about search engines, including ratings of features, an "EKG" (which shows how frequently the search engines refresh their indexes), and reports on the popularity of various search engines. Figure 3-16 shows a search engine popularity report compiled by **Media Metrix**.

Intelligent Agents

Software agents have been around for a few years. It wasn't until scholars began building them and studying them—and the Internet grew rapidly—that Web agents in particular started to become a topic of public interest. An **agent**, **intelligent agent**, or **software agent** is a program that performs functions such as information gathering, information filtering, or mediation running in the background on behalf of a person or entity.

Dr. Pattie Maes, the founder of the MIT Software Agents Group, has produced fundamental research in the area of software agents (see her paper listed in the section "For Further Study and Research"). She and her team built the first truly successful agents for personalized information filtering. Her **Software Agents Group** investigates how to put computer systems to work on behalf of people, by delegating tasks to software. The group describes agents as different from conventional software because "...they are long-lived, semi-autonomous, proactive, and adaptive." In this section, you are introduced to this large and very important field that will have an increasing impact on electronic commerce.

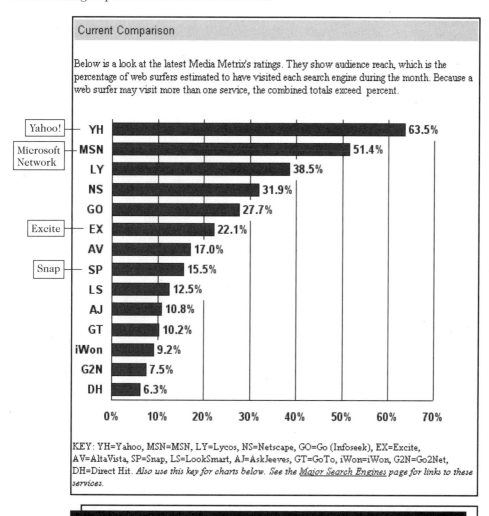

Current Comparison

Below is a look at the latest Media Metrix's ratings. They show audience reach, which is the percentage of web surfers estimated to have visited each search engine during the month. Because a web surfer may visit more than one service, the combined totals exceed percent.

KEY: YH=Yahoo, MSN=MSN, LY=Lycos, NS=Netscape, GO=Go (Infoseek), EX=Excite, AV=AltaVista, SP=Snap, LS=LookSmart, AJ=AskJeeves, GT=GoTo, iWon=iWon, G2N=Go2Net, DH=Direct Hit. *Also use this key for charts below. See the Major Search Engines page for links to these services.*

Figure 3-16   A Media Metrix report

Research reveals that software agents will become extremely important in the electronic commerce field sooner rather than later. Imagine sending your personal Web agent out on the Internet to search for the best price and availability of 500 personal computers for your company. Such an agent, armed with all the critical specifications you supply, would first locate electronic commerce sites that sell what you want. Then, the agent would collect information about the price and characteristics of the equipment that the site sells from comparable selling agents that work on behalf of the commerce site. Armed with similar information from other Web sites selling computer equipment, the agent then performs its next critical task—determining from which of several computer-selling

Web-Based Tools for Electronic Commerce

agents you should purchase your equipment. Once your software agent identifies the best vendor, the agent can then proceed to negotiate any remaining terms of the transaction (procurement officers would love to automate this part to save them time). Finally, the buying and selling software agents agree upon purchasing and delivery details. According to Maes, the preceding activities are part of six stages of the buying process.

Because software agents are always running in the background, ready to help you when you need them, they can help reduce the workload that people normally take on in locating, thinking about, negotiating, and purchasing goods and services on the Internet. (Of course, the same is true for selling behaviors also.) Currently, several agent systems are available on the Internet. Several of them are helpful in some, but not all, of the processes involved in buying on the Internet. Examples of more widely known agent systems include **AuctionBot**, **BargainFinder**, Firefly (Maes' own company), and **Kasbah**. These and other agent systems are listed on the Online Companion. Figure 3-17 shows an Excite Jango agent searching for a particular coffee.

A simple example of where an intelligent agent can save a lot of drudgery and time is in procurement, which involves the systematic process of deciding what, when, and how much to purchase. Broader than just purchasing, procurement also includes ensuring that what you receive is the correct quantity and that you receive it on time. Simplifying the example just a bit, imagine that one of your jobs as a procurement officer is to ensure that your fabricating department has a continuous supply of rolled steel—enough so that the department can produce its expected weekly output of 55-gallon galvanized steel drums. If the department runs out of rolled steel midweek, 75 people and dozens of machines sit idle while they wait for raw materials. Rather than always being on the lookout for the best prices and delivery schedules for the raw materials, would it not be much better to employ an Internet agent that monitors your company's inventory and then automatically orders new stock in just the right amount when needed? And, better yet, this agent could win you the employee-of-the-month award if you could have the new rolled steel delivered and unloaded as the last 50 yards of existing stock are being used up—just in time! That is what agents do best. They notice the need for more materials, go out on the Internet to known suppliers, locate stock, negotiate the best price, and then arrange for delivery. Meanwhile you are attending to less mundane tasks and handling procurement matters that cannot be given to agents.

Another example of agents at work is a stock alert. Customers of E*Trade, for example, can establish criteria for purchasing or selling stocks when selected conditions occur. An intelligent agent monitors the stock and sends an alert when the specified conditions occur. For instance, you might want to purchase 100 shares of Microsoft if the stock price drops below $61 per share, or you might want to sell MP3.com stock when the price-to-earnings ratio exceeds 3. Agents, those tireless software helpers, can perform either task. As researchers are able to make more headway in artificial intelligence and learning, you will see software agents enhanced with advances such as self-modifying behaviors and altered strategies that adapt to changing conditions and results.

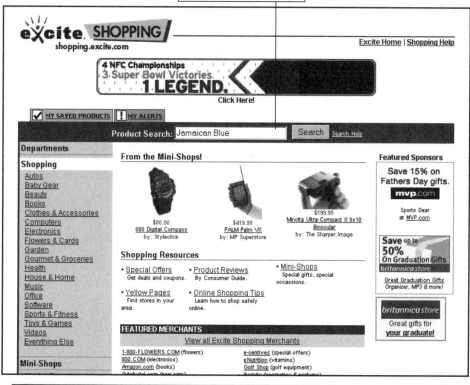

type the product or name
in the search text box

Figure 3-17   *Using an Excite Jango agent*

Summary

In this chapter, you learned about specifications of Web server hardware and software and the duties of both. To create a Web site, a company must consider whether to purchase Web server hardware and software or to use a Web hosting company. The most important consideration after you decide to host a site yourself is to select a hardware and software combination that will allow the Web site to grow—to be scalable—as business grows. Be sure to choose a Web server hardware platform and a software package that have good track records for performance. Before making a Web server purchase decision, you should run benchmark tests simulating both heavy and light traffic to observe the Web server's performance. Will it hold up under the worst-case scenarios? Each Web server software package offers slightly different capabilities. Some are very easy to set up and run, but may have poor documentation. Others, such as Apache, may be a bit more challenging to set up but are wonderfully efficient and run smoothly. Web development tools are also very important, and site management tools can make a Web administrator's job easier if the administration tools are full-featured and straightforward to use.

Web portals can be a good choice for setting up a Web site. A portal, if successful, will attract and retain customers and provide them with a positive Web browsing and shopping experience. Business-to-business portals are attractive to industries such as food service, steel, or staffing services. While intelligent agents are not yet the norm, their importance is enormous and will grow over time. Agents provide buyers with an automatic and persistent way to filter and collect information, locate goods and services, and negotiate price and delivery agreements. Though not yet in widespread use, companies must consider sellers' agents in their future commercial Web site plans. Links in the Online Companion will help you locate more details about all the topics covered in this chapter.

Key Terms

Active Server Pages (ASP)

Agent

Application Construction

Application Program Interface (API)

Application Server

Archiving

Backbone

Benchmarking

Bot

Cache

Common Gateway Interface (CGI)

Crawler

Cycling

Dead Link

Discretionary Access Control

Dynamic Content

Dynamic Page

Gopher

Hit

Index

Intelligent Agent

Java Applet

Java Servlet

Log File

Managed Hosting

Managed Server Hosting

National Center for Supercomputing Applications (NCSA)

Open DataBase Connectivity (ODBC)

Operating System

Portal

Query

Response Time

Scalable

Search Engine

Search Expression

Secure Sockets Layer (SSL)

Server Console

Server Side Include (SSI)

Software Agent

Spider

Static Page

Throughput

Unmanaged Hosting

Unmanaged Server Hosting

Virtual Host

Virtual Server

Vortal

Web Directory

Web Portal

Review Questions

1. What advantages and disadvantages are there to having an independent ISP host a Web site?

2. Discuss the two most important measures of a Web site's performance.

3. Beginning with the links provided in the Online Companion, locate more information about two of the three Web servers mentioned. Select between Apache, Internet Information Server, and Netscape Enterprise Server. Write approximately 250 words about six features of each Web server and indicate the computer platforms and operating systems on which each runs. Also indicate three other companies that use both of the Web servers you have selected by using the Online Companion link **What Web software is running on a site**.

4. What is it about Web portals that makes them attractive for advertisers? Name six consumer portals and list at least five elements that they all have in common. Similarly, list three business portals that are not mentioned in the text and provide a two- or three-sentence explanation of the purpose and target industries of each of the three portals.

5. Describe in 200 words or fewer which search indexes are considered medium-sized and which search engines are considered large. One of the Online Companion links will lead you to that information. Compare four of the search engines. To compare them, indicate how they collect their indexes and from what part of the document they do so. Also indicate whether the search engines allow the use of AND and OR to restrict or enlarge the query.

Exercises

1. Your friend Nancy Moore wants to set up a small Web site devoted to gardening. She believes her many years of experience in gardening give her an understanding of the kinds of gardening tools, fertilizers, soil amendment products, herbicides, pesticides, and plants that will appeal to the serious gardener. Right now Nancy doesn't want to sell anything. She merely wants to display lots of pages of plant photography, write and store short how-to papers for novice gardeners, and provide links to other gardening tips and traps on the Web. She wants your advice on whether to build the Web site from the ground up or use an ISP to start her endeavor. Use the links supplied on the Online Companion and then do some research on the Web to locate information on the cost of using an Internet Service Provider to host a Web site. Then, estimate what a small Web site might cost in terms of the minimal configuration of hardware and software. Estimate the design and development costs and the annual maintenance costs. Select, for example, a simple midrange PC running Windows NT. Then, select one of the Web server programs. Estimate the cost of a Web connection. Write one or two brief paragraphs about everything you think Nancy will need using either of the two options (she builds it or she uses an ISP). Also, list the URLs of at least three other sites specializing in gardening. What features do these sites have?

2. Your boss, Felicity Freedman, has sent you on a scavenger hunt. You must visit three prominent electronic commerce sites and three university sites and report back to Felicity what hardware platform each of these six sites is running and what Web server software they are using. Also note, by visiting the sites, whether they are portal sites (you can customize their home pages) and whether they have a search engine. Using the links on the Online Companion, investigate these corporate sites: Yahoo!, Apple Computer, and Oracle Corporation. Visit these three universities and

answer the same question: Purdue University, the University of Queensland (Australia), and Mount San Antonio College (it is *not* in Texas).

3. Create your own custom portal on any one of the following sites: Yahoo!, Excite, or NetCenter. Print out a copy of the Web portal once you have customized it to your liking. At a minimum, have the page display your login name (you will have to create an account, but they are free), weather in three cities around the world (only one can be in North America), a stock panel that displays stock information, current news, an address book (if possible), and a currency converter. If you cannot locate each of these exact panel options, then select any other custom choices.

4. You have created a Web site for International Paper Products and Pulp complete with links to other pages on your site and to pages on the Internet. Bob Pardee, your supervisor, wants you to check periodically that the links on the corporate site are still valid. Instead of purchasing and installing a link-checking program, you decide to investigate online link checkers—Web sites that allow you to enter a Web site's root or home address and that then check all the links that emanate from that site. Do light research on link checkers and locate a few that provide the link-checking service online. One such service is **Site Check** (see "Exercises" in the Online Companion to go to that site). Use Site Check's services to check the links on any site of your choice. Alternatively, you can check the links on the Online Companion that accompanies this site. Print a few pages of the report and be prepared to turn them in to your instructor. Be sure to click the Link Checker option button before clicking the Submit button in order to check a site's links. Be patient. The program takes a little time to complete its work—especially on a large and complex Web site.

For Further Study and Research

Alwang, G. 1998. "Internet Web Servers," *PC Magazine*, 17(9), May 5, 184–208.

Cohen, A. 1999. "Shopping Bots," *PC Magazine*, 18(13), July, 35.

Dragan, R. and F. Derfler. 2000. "Sun Microsystems Solaris 8.0 with iPlanet Web Server Enterprise Edition 4.1," *PC Magazine*, 19(10), May 23, 42.

Lee, Y. 2000. "Low-Cost Dedicated Servers," *Web Techniques*, 5(7), July, 88–89.

Lipschutz, R. 1999. "Internet in a Box," *PC Magazine*, 18(12), June 22, 189–202.

Maes, Pattie, R. H. Guttman, and A. G. Moukas. 1999. "Agents that buy and sell," *Communications of the ACM*, 42(3), March, 81.

Machrone, W. 2000. "Picking the Right Server Is Key," *PC Magazine*, 19(10), May 23, 52.

Morgan, C. 2000. "Web Content Management," *Computerworld*, 34(17), April 24, 72.

Murphy, K. 1999. "'Stickiness' Is the New Gotta-Have," *Internet World*, 5(12), March 29.

Schroeder, E. 1999. "The Internet rises and sets with Sun," *PC Week*, March 8.

ELECTRONIC COMMERCE SOFTWARE

INTRODUCTION

WebMethods produces software that helps facilitate commerce between customers and their suppliers. It hopes to reap sizeable profits in what nearly every electronic commerce researcher and practitioner predicts to be the most fertile field in electronic commerce—the business-to-business market. Wall Street pundits believe now is a prime time for initial public offerings in the business-to-business software market. Headquartered in Fairfax, Virginia, **webMethods** was founded by Australian Phillip Merrick and has attracted an impressive list of customers, including FMC Corporation and Dun & Bradstreet Corporation.

WebMethods' software solves a difficult problem that businesses on the Internet face when trying to exchange information. The webMethods software allows companies to exchange business information with each other—information such as invoices and inventory information—using a special Web page coding protocol called XML (see Chapter 2). With XML and webMethods' software, a machine tool company's request for a **Giddings and Lewis boring mill** is translated into a Web page that the supplier's software and Web server can interpret. One of webMethods' largest customers, Dun & Bradstreet, compiles financial and credit information and uses the webMethod software to translate data from proprietary systems into a common format that any Dun & Bradstreet customer's computer can understand. WebMethods' software has saved Dun & Bradstreet and its customers money because each customer no longer needs a customized program to interpret Dun & Bradstreet's data. Moreover, Dun & Bradstreet no longer has to worry about supporting its regional data center's many different financial data and credit information formats—webMethods takes care of translating the different formats into a single form.

When your company goes from a mere Web presence to electronic merchant status— actually selling goods and services on the Web—you will begin to see real dollars coming in to offset the expenses of building a Web presence. To become an electronic merchant, you must answer some important questions: What sort of electronic store should you have? Should you build a storefront yourself, or is it better to employ an Internet Service Provider (ISP) to host your electronic commerce site? In either case, what electronic commerce software is best for you or your ISP? Should you build a big site right away and grow into it, or is it possible to start small and scale up when business grows? What electronic commerce single-package solutions are available?

All of these questions and more are addressed in this chapter, which groups the options for electronic commerce software into three basic size categories: small, midrange through large, and enterprise. You will learn about the fundamental attributes of any electronic commerce site (or cyberstore) and the models that exist for such sites. The chapter also briefly describes several software packages and points out their strengths and shortcomings. Armed with the information in this chapter, you will be able to make an informed decision about which electronic commerce software is right for you.

LEARNING OBJECTIVES

In this chapter, you will learn about:
- Basic functions that an electronic commerce package should provide
- Characteristics to look for in an ISP-hosted electronic commerce solution
- Types of traditional and electronic store models
- Software packages available for small electronic commerce sites
- Software packages suitable for medium-sized to large electronic commerce sites
- Electronic commerce solutions for large organizations with an existing infrastructure and legacy software in place
- Several electronic commerce sites and their characteristics

WHAT KIND OF SOFTWARE SOLUTION DO YOU NEED?

Whether you are a huge business that looks at the Web as the main outlet for nearly all that you have to sell or a startup business that views the Web as merely an ancillary sales outlet, the options for software and hardware solutions to set up your electronic business vary greatly. At the inexpensive end of the spectrum of electronic commerce solutions are choices like Yahoo! Store, which costs next to nothing for a small store. At the other end are high-end choices, such as Netscape's CommerceXpert, which can cost more than $100,000, handle high traffic volumes, and provide a rich assortment of facilities and tools.

The type of electronic commerce software you need depends on several factors. One of the most important factors is the expected size of your enterprise and its projected traffic and sales. A high-traffic electronic commerce site with thousands of catalog inquiries each minute requires more muscle than a small shop selling a dozen items. Another determining factor is budget. Small budgets require low-cost solutions—at least until significant Web revenue rolls in—and larger budgets can accommodate more elaborate software and hardware. Whichever solution you pick, the good news about electronic commerce is that it is much less expensive than the traditional "brick-and-mortar" stores that surround us. The start-up costs of electronic commerce sites typically are a fraction of the costs of even the smallest conventional stores. A traditional store requires a physical location with leases, employees, utility payments, and maintenance. The cost of entering the electronic commerce market can be as low as a few dollars.

Deciding on your target commerce audience will help you select the best electronic commerce software choices for your site, or storefront. If you are setting up a business-to-consumer (B2C) commerce site, your software choices will be different from those of a business-to-business (B2B) site. For example, business-to-consumer software accounts for and looks up sales tax rates for different states. On the other hand, business-to-business software is wholesale and can use existing extranet connections, does not include sales tax, and has a different set of rules than traditional business-to-consumer software. Business-to-business electronic commerce can incorporate electronic data transfers between trading partners involving invoices, purchase orders, and other accounting transactions.

Another early decision you must make is whether your company should use a full-service ISP or host the electronic commerce site in-house. The presence or absence of company talent—programmers, Web-savvy personnel, and so on—can be a critical factor in your decision. If you do not have or cannot easily hire the skills required to set up and maintain an electronic commerce site, you may want to let an experienced ISP do the heavy lifting for you. Do you want the software to run "out of the box," or can you afford to do some customization and tweaking? If your company has an array of Internet resources—including in-place hardware, servers, knowledgeable staff, and database systems—you may find that in-house commerce hosting is neither difficult nor cost-prohibitive.

Electronic Commerce Requirements

Electronic commerce software must be hosted on a Web server. Chapter 3 describes Web server software alternatives. Some of the more reliable packages include

Microsoft Internet Information Server, Apache, and Netscape Enterprise Server. Once you have located or built a host server, you can investigate and install electronic commerce software.

The exact duties you can expect electronic commerce software to perform range from a few fundamental operations to a rich and complete solution—spanning catalog display to fulfillment notification. Simply stated, though, all electronic commerce solutions must provide at least:

- A catalog display
- Shopping cart capabilities
- Transaction processing
- Tools to populate the store catalog and to facilitate storefront display choices

Catalog Display

A small commerce site with only a few dozen items to sell can have a very simple **catalog**, which is a static listing of goods and services. Larger catalogs can feature photos of items, descriptions, and a search feature that allows you to search for an item and determine its availability. Large catalogs almost always store their data in databases that are on separate machines accessible to the commerce server machine (the machine containing the electronic commerce software). Figure 4-1 shows the Web page of a large commerce site, **MP3.com**, selling music that can be downloaded as digital files. Note the features present on the page and its professional look. Figure 4-2 shows the Web page of a small commerce site, **Women in Music**, selling apparel. This site requires less sophisticated, less expensive electronic commerce software and has a more "down home" appeal.

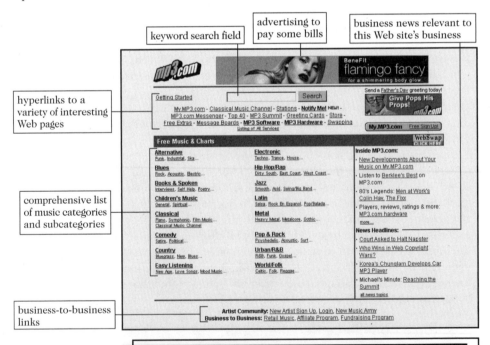

Figure 4-1 *Large electronic commerce site with many features*

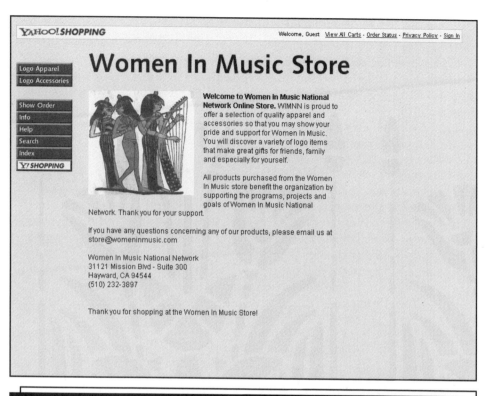

Figure 4-2   *Relatively small electronic commerce site*

Small storefronts (Figure 4-2) selling fewer than 35 items can get away with a simple list of products or categories not organized in any particular way. For small inventories of merchandise, the site can supply a photo of each item, which is a link to the product and more information about it. Larger sites, as shown in Figure 4-1, cannot get by with a simple listing of products—there are just too many of them and they are difficult to locate. Larger stores require more sophisticated navigation aids and better product organization. That is where a catalog, perhaps delivered from a powerful backend database system, becomes a must.

A catalog organizes the goods and services being sold. Frequently, departments are a convenient way to organize your offerings even more. As in a physical store, merchandise in an online store can be grouped within logical departments to make locating an item, such as a camping stove, simpler. Web stores often use the same department names as their physical counterparts. **L.L. Bean**'s Web site product guide, for instance, categorizes outdoor gear and apparel into groups such as Men's Clothing, Sporting Gear, and Home & Camp.

It is important to give buyers alternative ways to find products. Besides offering a well-organized catalog, large sites with many products should provide a search engine that allows customers to enter a description, such as "men's shirts," so they can quickly find the Web page containing what they want to purchase. This follows the implicit and important rule of all commerce: Never stand in the way of a customer who wants to purchase something.

Shopping Cart

In the earlier days of electronic commerce, shoppers selected items they wanted to purchase by filling out online forms. Using Windows text boxes and list boxes to indicate their choices, users entered the quantity of an item in the quantity text box, the SKU or product number in another text box, and the unit price in yet another text box. This system quickly proved to be awkward. One problem with forms-based shopping was that the shopper usually had to write down product codes, unit prices, and other information about the product before going to the order form, which was inevitably on another page. Another problem was that customers sometimes forgot whether they had clicked the submit button to send in their order. As a result, they either sent the same order twice (pressing the submit button when they had already done so) or thought they had submitted the order when they really hadn't. The forms-based method of shopping was confusing and error-prone. Figure 4-3 illustrates this problem. For starters, it is difficult to remember the often-challenging spelling of the coffee names. In addition, coffee prices, which are located on a different page, must be entered in the text boxes, requiring the customer to write them down, unless he or she has memorized them.

Shipping Information

     ⊙ Ship the order to me.
     ○ Ship to the address below.

Name: Otis Toadvine

Address: 1212 Pond Place

City, State, Zip: Lincoln, Ne 47906

Order Information (Please complete entire row)

Coffee Name	Type	Grind	Qty.	Price
Zimbabwe	Caffeinated	Whole Bean	2 lbs	
Yemen Mocha	Caffeinated	Whole Bean	4 lbs	
Kopi Luwak	Caffeinated	Perc	1/2 lb	

text boxes that must be filled out

S&H: 4.95

Total:

Comments

NOTE: *Please do not enter your credit card number.*
We will call back to verify the order and receive the credit card number.

Credit Card Information

Name on credit card   **Card Type**   **Expires**

Figure 4-3   *Using a form to enter an order*

The forms-based method of ordering has given way to electronic shopping carts. Today, they are a standard of electronic commerce. A **shopping cart** keeps track of the items you have selected to purchase and allows you to view what is in your cart, add new items to it, or remove items from it. To order an item, you simply click that item. All the details about it—including its price, product number, and other identifying information—are stored automatically. If you later decide to remove one or more items from the cart, you can do so by viewing the cart's contents and removing the unwanted items. Then, when you are ready to pay for your purchases, you click a button (usually labeled "Proceed to checkout") and commit your purchase transaction. Once committed, you cannot reverse the transaction; even though your browser's back button may take you to the screen displayed prior to committing your transaction, the actual data already have been sent to the fulfillment center and your purchase has been made. Figure 4-4 shows an example of a virtual shopping cart full of computer equipment. Because the consumer has not gone to the checkout counter, no shipping costs or applicable taxes appear on the form.

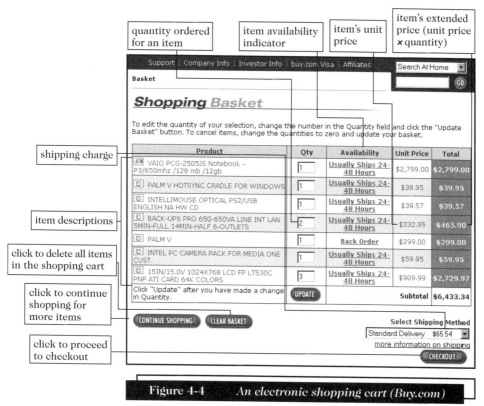

Figure 4-4   *An electronic shopping cart (Buy.com)*

Clicking the "Checkout" button displays another screen that usually asks you to fill out billing, shipping, and payment (credit card, check, money order, and so on) information, and to confirm your order. As you can see from the figure, the software keeps a running total of each type of item. A total appears on the display, and includes sales tax and shipping costs that are looked up in tables or database records. Shopping cart software at some Web commerce sites allows you to fill your

Electronic Commerce Software

shopping cart with purchases, put the cart in virtual storage, and come back days later to confirm and pay for the purchases. **QuickBuy** is one company that makes this type of shopping cart software.

Because the Web is a *stateless* system—unable to remember anything from one transmission or session to another—shopping cart information must be explicitly stored for the shopper to retrieve later. Furthermore, it must distinguish one shopper from another so that the purchases are not mixed up. One way to uniquely identify users and to store information about their choices is to create and store **cookies**, which are bits of information stored on a client computer. When a customer returns to a site that issued a particular cookie, the shopping software reads either the cookie from the customer's computer or the database record from the merchant's server. In the likely event that a shopper's browser does not allow storage of cookies, there must be another way to preserve shopping cart information from one browser session to another. OpenMarket's **ShopSite** does this is by automatically assigning a shopper a temporary number. The number is added to the end of the shopper's URL and persists as he or she navigates from one Web site to another. When the customer returns, the URL still contains the bits of information about his or her shopping cart. When the customer closes the browser, the temporary number is discarded and thus cannot be reused—even if the customer later reopens the browser and returns to the same Web site.

Transaction Processing

Transaction processing occurs when the shopper proceeds to the virtual checkout counter by clicking a checkout button. Then the electronic commerce software performs any last-minute calculations, such as volume discounts, sales tax, and shipping costs. At checkout, the browser normally switches into a secure state of communication. Unless you have disabled it, a dialog box appears indicating that your browser is entering or leaving a secure state. Today, all browsers support SSL (secure sockets layer) and other protection protocols. (Security is described in detail in Chapters 5 and 6 of this book.) Transaction processing is the trickiest part of the electronic sale. Computing taxes and shipping costs is an important part of this process, and commerce administrators must continually check the tax and shipping tables to make sure they are current. Some programs simplify shipping calculations by connecting directly to shipping companies such as **UPS**, **FedEx**, and **Airborne** to retrieve shipping costs. One site, **SmartShip**, determines which shipper offers the lowest price to deliver your package.

Transaction processing software must also handle additional details, such as tax-free sales. Many business-to-business transactions involve tax-free sales when items are purchased for resale or under other conditions. Some business-to-consumer sales are also tax-free. However, California is one state that requires many of the larger retailers to charge sales tax—even when the sale originates outside the state. Other calculation complications include provisions for coupons, special promotions, and time-sensitive offers (for example, "purchase a round-trip ticket before May 25th and receive a 50 percent discount"). (The U.S. government may cause a lot of changes in software to occur if it passes an Internet sales tax law.) Most Web electronic commerce software provides internal connections to popular accounting and legacy systems so that Web sales can be tallied simultaneously in the company's internal accounting system.

Electronic Commerce Tools

Solutions for setting up shop online vary from inexpensive and simple to very expensive and complicated. You can choose one of the several very inexpensive storefronts that are offered by established portals, such as Yahoo!. Targeted at smaller stores, these systems offer hosting services and software that let merchants quickly create storefronts. Large business-to-consumer and business-to-business systems have vastly different software and hardware requirements. Huge online stores, such as Amazon.com, require robust electronic commerce software suites that run on large, dedicated computers and interact with database systems to display catalogs and process orders. These systems are expensive to create and operate and require a dedicated staff to oversee and maintain them.

Inexpensive, hosted electronic commerce stores provide the tools to create a storefront in less than an hour. The package includes Web store space as well as catalog creation, shopping carts, and transaction processing. Fees vary, with some sites letting you do online business for a few hundred dollars a month (or less). Some sites charge a one-time setup fee of up to $200. Such simple, inexpensive Web hosting services all charge a monthly fee that is based on the number of items for sale. **Yahoo! Store** and **ShopBuilder** are two low-cost hosting services offering complete services for small to medium-sized commerce sites. See the section "Cost Analysis of Using a Shared Web Host" later in this chapter for details on their pricing structures.

Commerce systems in the middle tier serve a large number of companies—both small and large—whose commerce needs are well beyond that of mom-and-pop storefronts, but well below that of an MP3.com or Amazon.com. These commerce software suites range in price from $1000 to over $5000. Commerce software packages appropriate for firms doing this level of business are described later in this chapter, in the "Midrange Packages" section. INTERSHOP enfinity is a typical member of this group.

Setting up commerce software suites is not always simple. Some midrange packages can take only an hour or two to install and tune, whereas others are more complicated and can take several days and several calls to technical support people to get them up and running. Commerce servers frequently tie into database servers (such as SQL Server and Oracle), but the commerce server price does not include the cost of the database software.

On the high end of the scale are large businesses—having large online transaction volumes or several business-to-business trading partners engaging in commerce—that require larger and more expensive commerce systems. Amazon.com and L.L. Bean are two such commerce sites with huge business-to-consumer transaction volumes.

Business-to-business electronic commerce conducted over the Internet, extranets, and intranets requires tools and capabilities different from those required to implement most business-to-consumer Web sites. Business-to-business systems, for example, usually require tools not standard in business-to-consumer systems, such as encryption, authentication, digital signatures, and signed receipt notices. Many business-to-business commerce systems must be able to connect to existing legacy systems, including **enterprise resource planning** (**ERP**) software packages. ERP systems are business software packages, such as SAP-R3 or Baan, that integrate all facets of a business, including planning, manufacturing, sales, and marketing.

123

(Increasingly, ERP firms are losing ground to the Web as online businesses discover online B2B relationships, Web-based procurement and supply chain management, and electronic marketplaces—vortals.) Netscape's ECXpert, for example, is an Internet-based electronic commerce software package designed for use by medium-sized to large companies for implementing business-to-business commerce. Figure 4-5 shows an overview of a business-to-business commerce server that connects existing legacy systems, including EDI and ERP systems, of trading partners.

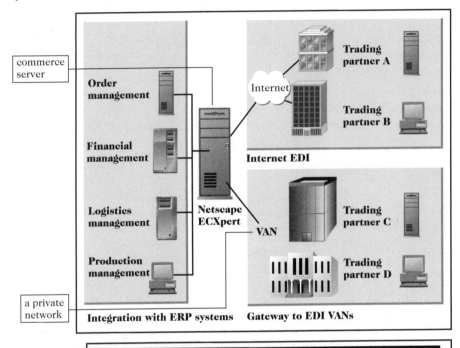

Figure 4-5 *Business-to-business commerce server topology*

With the exception of the very smallest stores with small, off-the-shelf, Web-hosted sites, companies must carefully consider the features of commerce software packages before choosing one to test and use to run an online business. An important feature that most firms will need to consider is database support. Most Web storefronts operate with a database that stores product information, including size, color, type, and price details for the many products that midsized and larger sites sell. Usually, the database that serves an online store is the same one that is used by the existing corporate clients. It is better to have one database serving two disparate entities, because it eliminates parallel but distinct databases—something you should avoid if possible. If you do not have database expertise, then you must either hire the talent to handle that part of your Web business or stick with Web products that do not use database backends. If you do have existing inventory and product databases, then you should not consider any Web commerce software that does not support your system. For example, if you are looking at a commerce package that connects only to an IBM DB2 database but you currently keep inventory in an Oracle database, the IBM-supporting system would be a bad choice—you should consider packages that support Oracle and eliminate those that do not.

Another important factor is the type of transaction-processing system that you are using or considering. That software should be compatible with the commerce suite you are reviewing. **Commerce suites** offer credit card processing, and large systems allow you to "bolt on" a variety of independent transaction-processing systems offered by other vendors. However, you must determine whether the commerce system allows other forms of payment, such as electronic cash (see Chapter 7), or is limited to only one payment type.

While most commerce software comes with wizards and other automated helpers that create template-driven pages—including the home page, about pages, and contact pages—most merchants will probably want to modify those pages. If you don't feel comfortable customizing such pages, then find someone who does. You will undoubtedly want to customize Web pages with company and product pictures and text. Because you will have to perform Web site maintenance—for example, adding new products to your catalog—test the software before you commit to it to make sure that adding new categories of products and new items to an existing product does not take an entire afternoon. For instance, creating sale item specials and displaying end-of-quarter sale items should not be difficult or time-consuming tasks.

Many commerce software packages have online demonstrations that walk you through adding items to your inventory and creating pages. Several of these are listed in the Online Companion and are mentioned explicitly in this chapter.

MARKETING SMARTS

Plopping your well-designed and savvy electronic commerce site smack in the middle of the Internet is not enough. You cannot expect customers to come charging into your storefront if they do not know it exists. You have to get the electronic word out—visibility is the key. A special subdivision of Web hosting, called **Web malls**, will list your Web site in a portal-style directory, though Web mall hosts don't supply all the services of a full Web host. If your site is in a Web mall, however, you should take advantage of the mall host's business listing services. Yahoo! hosts a Web mall that lists its member stores in categories. Locating your online store in a Web mall can increase your store's visibility.

The smallest and least expensive step you can take for visibility is obtaining and registering your own domain name. Some services described here do not provide a domain name in their electronic shopping malls, but you can maximize a store's vitality if you go ahead and obtain one anyway. It helps your business if the domain name you choose says something about your business. For instance, www.StuffToSell.com might be a catchall name that many people will select, but it hardly attracts the target audience you want. If you are selling sailboats, then try to snag a meaningful domain name such as www.sailboats.com or www.sailmakers.com.

Another clever tactic is to make your site search-engine-friendly by including a META tag in your store's home page. A **META tag** is a special HTML tag, invisible to the displayed Web page, that contains keywords that represent your page's content and that search engines use to build indexes pointing to your store's pages. A more aggressive approach to marketing your store to search engines is to actively seek out search engines and submit your Web pages to them for inclusion in the search engine's index. If you are unsure about which search engines to use, you can use search engine submission services to present your pages for you. Services such as **NetCreations** and **SubmitIt!** provide free submission services. Any actions that make Web browsers aware of your store and that drive shoppers breathlessly to your site are time and money well spent. Check with other online merchants about what they have found works best for them. Often, fellow electronic commerce storeowners are happy to share their knowledge with others.

HOSTING SERVICES

If you are not General Motors with a committed investment in existing intranets, extranets, and in-house Web sites, you may not want to spend a lot of money to start an online business. The total costs of setting up your own in-house Web commerce site can exceed $50,000; you have to purchase hardware, Web server software (see Chapter 3), electronic commerce software, and a T1 Internet connection—just for openers. Is this in your budget? If not, an attractive alternative is to turn to a Web hosting service. **Web hosting** services allow businesses to jump into electronic commerce without spending a fortune. ISPs provide a connection to the Internet and some handy services such as FTP programs. Web hosting services provide all the services that an ISP does, as described in Chapter 3. In addition, they provide electronic commerce software, store space, and electronic commerce expertise that ISPs do not supply.

Web hosting services have many advantages, including spreading the cost of a large Web site over several "renters" hosted by the service. Web hosts provide expertise, do all the work of registering and supplying you with your own domain name, maintain the Web service 24 hours a day, process payments, authorize credit cards, and take care of shipping and taxes. And that's just for starters. Of course, the biggest single advantage—low cost—occurs because the host provider has already purchased the server and configured it. The host provider has to worry about keeping it working through lightning storms and power outages. In all cases, look for Web hosts that offer reliability, good security, system simplicity, and widespread visibility.

There are several Web hosting choices whose prices reflect the number of services supplied and the size of the Web or commerce site you require. Hosting choices loosely fall into one of the types mentioned in Chapter 3: self-host, shared host, or dedicated host. A slight variation of dedicated hosting is colocated hosting. Each has its pros and cons.

Self-hosting, where the online business owns and maintains the server and all its software, implies full control, instant hardware access, and complete flexibility—all advantages. At the same time, self-hosting means that the business must have additional staff, Web expertise, expensive equipment, and a high-speed direct Internet connection.

Shared hosting, where your Web or commerce site resides on the same server as several other sites, is inexpensive, requires very little of an online store's time to maintain, and may have a very high-speed connection to the Internet. The downsides of shared hosting are loss of direct control, sometimes slow updates due to increased traffic on the Web site by other online businesses, and security concerns arising from unrelated online businesses sharing the same server.

With **dedicated hosting**, a Web host provides a server for your Web site alone. Advantages include more Web and commerce software options, a good high-speed connection, and more decision-making and site design control for the Web site owner. Disadvantages include higher software costs, higher maintenance costs, and very little control over the hardware containing your store.

Colocated hosting is self-hosting with a twist—the server is owned by the online store but is located at the Web host's site, and the Web host provides maintenance based on the level of service the online business requires. Advantages match those of self-hosting, while disadvantages include maintenance costs that are higher than self-hosting, somewhat more difficult access to the server when the online business wants to implement site changes, and more expensive software.

ValueWeb, operating since 1996, is an example of a Web hosting service. ValueWeb offers businesses comprehensive Web site hosting solutions including shared hosting, dedicated hosting, and colocation services. Hosting more than 65,000 Web sites for customers in over 136 countries, ValueWeb has become one of the largest Web hosting companies in the world (see Figure 4-6).

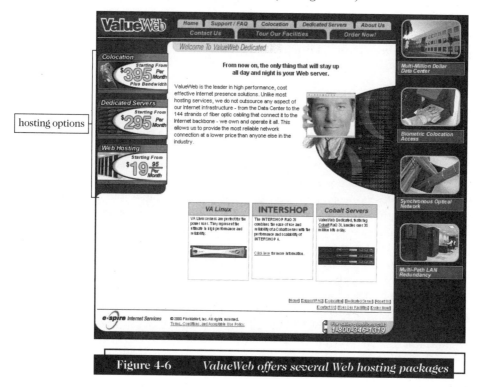

hosting options

Figure 4-6 *ValueWeb offers several Web hosting packages*

One of these general hosting arrangement types is a good fit for most types of online organizations. Cost, easy access, and levels of service determine the type of hosting a business will choose. Some of these are described further in this section. Click **HostCompare.com** for a comprehensive presentation about Web hosting. HostCompare.com contains hundreds of links and lots of good information, and walks you through finding an appropriate Web host in five easy steps.

What Is Web Hosting?

When you sign up for a Web hosting service for your budding electronic commerce site, the hosting service already has the robust computer and all the software that you require. You get a variety of services for a low monthly fee, because you and other customers, or Web host "tenants," share the infrastructure, computer, network, routers, switches, and high-speed Internet connections. In addition, all tenants benefit from the cadre of Web computing professionals running the Web server—most of whom are experts in their fields. There is one important difference between these professionals and the Web staff at the customer's company: someone else is paying them.

Along with access to hardware, software, and personnel, Web hosting customers have their own domain name and IP (Internet Protocol) address. For a fee of approximately $25 to $200 per month, customers get disk storage where they can put their store's Web pages. A small store will have from 25 to 60 MB of allocated space; more space is available for larger stores paying higher fees.

With Web hosting, you can create Web pages with any page designer you like and then upload them to your site, or you can use stock store templates that many Web hosts supply. Full Web hosting services provide e-mail service, the ability to use FTP to upload and download data and Web pages, basic shopping cart software, Web page multimedia extensions (sound, animation, movies), trustworthy credit card processing, and a staff of technically savvy personnel.

Advantages of a Shared Web Host Over Dedicated Hosting or Self-Hosting

What do you gain by letting someone else host your Web commerce site? Well-established Web hosts (not your cousin Vern's fish, bait, and Web-hosting service) provide you with a reliable server, reasonable cost, great functionality, and full-time technical support. When you sign up for the monthly service, there usually is no fixed-length contract, so you can terminate the contract at any time. Your commitment is month to month. Setup fees range from zero to a few hundred dollars and include your own officially registered domain name. Your customers can find you at an address that is indistinguishable from names of the truly big sites, such as http://www.bigshot.com. Without your own domain name, you would have to use Web mall hosts, such as Yahoo! Store, that supply you with a domain name that consists of the host's domain name followed by a short name you supply, as in http://stores.yahoo.com/bigshot.

Because you don't have to hire a staff to run your site or invest any capital in hardware, costs are very low. Web hosts charge from $25 to few hundred dollars per month, whereas your own hardware, software, staff, and infrastructure costs could be easily several thousand dollars per month if you decide to self-host. And don't forget about maintenance; it can be costly if you do it yourself. But if you choose a Web host for your commerce site, it is covered by the monthly service fee.

Technical support supplied by Web hosting services is invaluable. No matter how savvy you or your company Web master are, you can count on some odd, arcane error occurring at the worst time. Having good support available 24 hours a day can save you from excessive site downtime. In addition, most Web hosting technical support personnel can help you integrate database access to your Web commerce application. That can be difficult for a novice.

Reputable Web hosts guarantee minimum availability times—the least amount of time that the site will be online. You should locate a Web server whose host site is available 99.5 percent of the time or better. Even 95 percent reliability, which *sounds* good, means your host site is down more than eight hours per week. That is not acceptable and Web hosting services know that. In fact, most Web hosts strive for "five nines" availability: 99.999 percent. Happily, there are many good, reliable, and reasonably priced Web hosting services and Web malls out there. Click **Top Hosts** and **Ultimate Web Host List** on the Online Companion for lists of Web hosting services. We will discuss several of these later in this chapter.

Cost Analysis of Using a Shared Web Host

Web hosting services provide a large cost advantage over self-hosting. For example, GeoShops charges no setup fee, and the basic service is $24.95 per month. There are no per-transaction costs for the basic service. Its more robust service bundles in transaction processing and required security and costs $99.95 per month, plus 55 cents per transaction. Comparable Web hosting services such as **Yahoo! Store** or **ShopBuilder** have similar arrangements. Yahoo! Store charges $100 per month for up to 50 items and $300 per month for a 1000-item store. Yahoo! Store charges no setup or per-transaction fees. A ShopBuilder store containing 1000 items costs $250 per month and has no setup or per-transaction fees.

Sites that provide you with your own domain name charge a separate fee—usually between $40 and $150—to register your domain name with InterNIC[1]. (**InterNIC** is one of a handful of World Wide Web official domain-name registration services, or **registrars**. Everyone seeking his or her own domain name must apply with InterNIC or one of the few other registrars. The registration fee covers two years and is renewable annually after that.) At least one site, **B-City**, charges no fees whatsoever for a very small store with fewer than 10 items. Some sites will pay you to set up a storefront if you allow their banner advertising to appear on your store's Web pages. That deal is hard to beat!

The next sections of this chapter contain details about services that are offered by full-service, small, and midrange Web hosts. First, we will look at the basic packages, which provide fundamental commerce services. Then, midrange packages are discussed. Finally, enterprise-class commerce solutions are presented.

BASIC PACKAGES

Basic packages are free or low-cost electronic commerce software supplied by the Web host for building electronic commerce sites that will be kept on the host's server. Services in this category usually cost less than a few hundred dollars per month, and the software is available on the host site so that you can immediately begin building and storing your storefront on the host's server. Many good packages fall into this basic packages group. Several are listed on the Online Companion. This section will first examine the low-cost host sites offering very humble facilities. Then, banner advertising arrangements are described. Next, you will move up the capability ladder and look at full-service mall-style host facilities. Finally, operating expenses will be reviewed.

---

[1]InterNIC® is a registered service mark of the U.S. Department of Commerce.

Fundamental Host Services

Several Web hosting services offer free or low-cost commerce services designed for small online businesses selling only a small number of items (fewer than 50) and having relatively low transaction rates (orders placed per unit of time). Hosting services in this class offer adequate space for your Web store and forms-based shopping (no shopping carts here). Often, they do not include transaction processing. The free sites in this group can offer a great value for Web store first-timers unsure about taking the electronic plunge. The host makes money from advertising banners that are placed on the storefront's Web pages. The merchant does have some control over which banners are displayed, because you wouldn't want a banner advertisement about another store's beef products if your site were a vegetarian store.

A typical example of these types of hosts is **B-City**, which offers free space in exchange for the merchant's agreeing to follow its rules. Besides allowing advertising on your site, you must agree to create a business site (not a personal one), and agree not to present adult content, distribute illegally licensed software, or send bulk or spam e-mail. That is a pretty good deal.

Other sites in this category are **BizLand.com** and **HyperMart**. HyperMart's site contains a six-part tutorial, "Setting Up Shop," that guides users through the steps to setting up their store and running it once it is online. HyperMart is home to over 190,000 active businesses. It allows you to use any HTML editor you like and offers a turnkey, browser-based Web creation package to make creating store pages straightforward. BizLand.com offers an impressive list of tools for the small Web entrepreneur, including support for Microsoft FrontPage, a unique domain name, analysis reports providing statistics about visits to your store, and registration with the major search engines. Figure 4-7 shows a list of features that BizLand.com offers its hosted storefronts.

Signing up for a free B-City Web store is a simple process. From the B-City home page, you proceed to the registration page (click the "Sign Up Here!" link, and then click the "Register" link on the Sign Up page). Then, fill in the text boxes with your business name, enter a password, and click the submit button. If the business name (and thus the account name) is already taken, B-City will inform you. Once B-City accepts the proposed business name, it displays another form with several text boxes for you to complete. You supply standard information such as your first and last names, your e-mail address, a password reminder phrase and answer, and country citizenship information. Other list boxes that you must complete include your birth date, highest education level completed, occupation, number of children, and approximate household income. (All of this information is kept in confidence and provides B-City a great deal of marketing information.) Once you complete the form and submit it, B-City sends an e-mail message to the e-mail address that you entered. The message is important, because it confirms your username (you must remember your password) and contains a cryptic activation code.

Logging on to your account at the B-City home page activates your site and allows you to begin building Web pages. When you click the "Member Login" button, you enter your username, password, and—one time only—your account activation key. Subsequent trips to your store to maintain it do not require you to use the activation key.

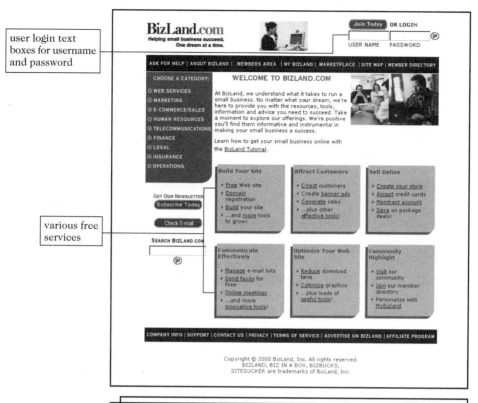

user login text boxes for username and password

various free services

Figure 4-7   BizLand.com's hosting service

If the business name you choose is "Toadvine," then your permanently assigned URL on the B-City site would be http://www.bcity.com/Toadvine. That is not a personal domain name, because B-City is the host and your business is a folder (directory) on the B-City computer. If you want your own domain name, then you will have to register and pay for it yourself or move to a full-service host that charges monthly fees.

The host sites in this category all provide templates for constructing a commerce site. Everything is template-driven, and most users can cobble together a site in 30 minutes or less. One major drawback of host sites in this group is that customer purchase transactions are handled by e-mail. When a customer orders items from a store, the hosting service groups and e-mails the orders to the merchant, who must then fill the orders and process the payment, including dealing with credit cards, checks, or money orders. Another drawback (more of a distraction really) is the banner advertising that appears on your site. Because the banner advertisements pay for your site, you cannot remove them. However, for a monthly fee, you can make your storefront banner-free. The final drawback is the set of tools provided by Web hosting sites in this group. They are more difficult to use and lack the variety of styles and looks that you can get with more expensive Web hosting services.

Banner Advertising Exchange Sites

Banner exchange sites (BESs) are Web sites that help electronic merchants promote their stores online. Banner exchange agreements between group members allow one member to place a banner ad on another member's commerce site at no charge as long as the second store can place its banner ad on yet another member's commerce site. This mutual advertisement sharing means that no member pays for advertising and that all members get commerce-site promotion.

The BES organizes the banner exchange among members, enforces the banner exchange rules, collects statistics about customers who click the advertisements on each merchant's site, and changes ("rotates") ads on members' sites. The underlying principle of this system is that if a storefront advertises its services with a banner advertisement on a high-traffic Web site, then more Web customers will visit the site being advertised.

A banner advertising exchange site provides ad monitoring software that helps members determine their advertisement click-through count and thus helps them determine the optimal Web sites for a particular banner ad. A **click-through count** is the number of visitors who click a Web advertisement link to go to the advertiser's Web site. Click-through counts are a useful measure of the effectiveness of placing an advertisement on a particular Web site. Examples of BESs are **BannerExchange.com**, **Eurobanner**, **Exchange-it**, **LinkExchange** (part of bCentral), **SmartClicks**, and **Web Publisher's Advertising Guide**. BESs have strict rules about membership and the content that is appropriate for selected members' sites. Often, the BESs have helpful articles on a variety of advertising topics, including how to market your Web site and how to increase traffic to your site. Figure 4-8 shows the home page of bCentral's LinkExchange. If you decide to join a banner exchange group, check with your hosting provider before you join to make sure placing banner ads does not violate the host's rules.

Full-Service, Shared Mall-Style Hosting

Full-service shared hosting sites provide online stores with good service, good Web creation tools, and little or no banner advertising clutter. Web hosts in this group charge a monthly fee (which is often higher than that of lower-end providers), may charge one-time setup fees, and may also charge a percentage or fixed amount for each customer transaction. These Web hosts also provide high-quality tools, storefront templates, an easy-to-use interface, and quick Web page generation capabilities and page maintenance.

There are several significant differences between free Web host sites, described in the previous section, and these full-service host sites. Full-service hosts provide shopping cart software or the ability to bolt on another vendor's shopping cart software. Secondly, they furnish comprehensive customer transaction processing through one of a few merchant services. Such a service allows your customers to choose to purchase their goods and services with a credit card, electronic cash (sometimes), or other forms of payment. The Web host processes the acceptance and authorization of credit cards for you. Another benefit is that because you are paying a monthly fee to the Web hosting service, your site does not have to display any Web banners, which can sometimes be unattractive and distracting. The fourth benefit of the full-service Web hosts is that they provide much higher quality Web store building and maintenance tools than do the free sites.

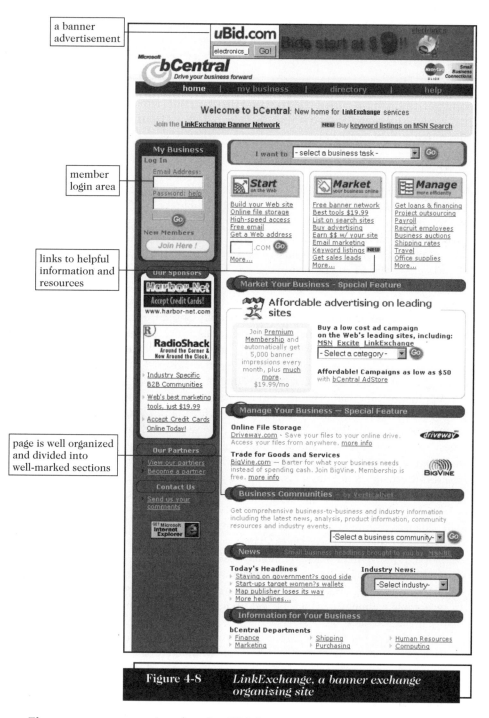

Figure 4-8   *LinkExchange, a banner exchange organizing site*

There are many examples of quality Web hosting services in this category, including **Yahoo! Store**, **GeoShops**, **ShopBuilder**, and **Virtual Spin Internet Store**. These four are representative of most of the full-service Web hosts and are quite popular. Some list your site, if you desire, in a directory listing on the host Web's opening page. The Yahoo! directory, for example, categorizes all the storefronts it hosts into fewer than a dozen major groups (Gifts and Collectibles, for example) and then subdivides each group into three to five subgroups, such as Arts and Crafts.

You can learn how capable these Web hosting services are by trying them out. Yahoo!, for instance, allows you to test its hosting service and store-builder software for a free 10-day trial period. The next section examines the merits of Yahoo! Store and contrasts it to the free Web host sites described previously. Then Geoshops and ShopBuilder will be discussed.

Yahoo! Store

Not long ago, the prominent Web site Builder.com (a Web site of CNET, which owns a popular collection of Web sites and is highly visible) called Yahoo! Store the clear winner among affordable electronic commerce solutions. Yahoo! Store is clearly a good value. It charges a monthly fee that varies from $100 to more than $700, depending on the number of items you are selling. Selling up to 50 items, for example, will cost a flat $100 per month.

Yahoo! Store is the most expensive of this group of Web hosts. It serves as the business Web host not only for hundreds of small storefronts you've never heard of, but also for big name stores, such as the Kennedy Space Center Space Shop, The Sharper Image, PalmPilotGear, *Rolling Stone* magazine, and that wonderfully different circus, Cirque du Soleil. Click the link **Yahoo! Store stores listing** in the Online Companion (in the subsection called "Full Service Hosting" of the "Basic Packages" section) to view Yahoo! Store's top-level store directory.

Merchants can create, change, and maintain their Yahoo! storefronts through a Web browser. On its own site (which is a shared hosting site), Yahoo! holds all the stores' pages in a proprietary format. Yahoo! processes merchant transactions on a secure server to minimize fraud and snooping. When you sign up for a Yahoo!-hosted storefront, your URL will be a subdomain of the Yahoo! URL. For example, if your business/username is BoldImpressions, then your URL—the one that your customers go to in order to do business—is http://store.yahoo.com/BoldImpressions (a subdomain of Yahoo!). A different URL takes you to the "employee's entrance" when you want to make changes to your store. That URL is http://store.yahoo.com. Once you log on with your secret store-owner username and password, you can edit or manage your store.

There are two main ways to manage a storefront. You can edit every aspect of your store's contents, behavior, and appearance, or you can use Manager (a Yahoo!-supplied software product) to examine a large amount of statistical information, set up acceptable payment methods, select preferred customer shipping methods, and change global site settings.

Yahoo!'s Web editing and page creation interface is clean and intuitive. You can create major product categories and then add items to the categories. Whenever you create a new category (or a department in a large store), the Yahoo! software places a button on your home page. In addition to entering information manually, you can upload graphics and data in a text-delimited file.

Figure 4-9 shows the editing session for modifying one Web page of the Coffee and Tea Merchant's store, a Yahoo! Store site. Near the top of the page are the editing links that allow the merchant to create new items, modify links, move items around, change the page layout, and alter the entire color scheme and font of all Web pages within the storefront. Each item listed in the main section is a link to a separate page containing details about the item, including its description and its price in various quantities.

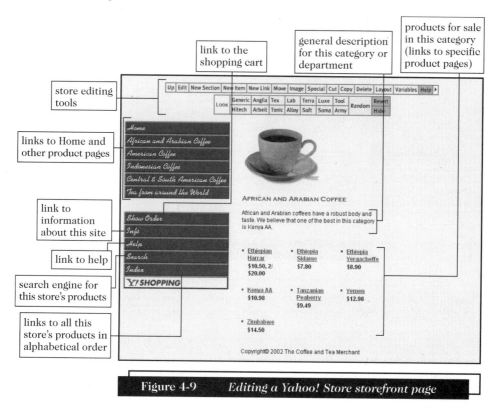

Figure 4-9   *Editing a Yahoo! Store storefront page*

Figure 4-10 (on the next page) shows the Yemen coffee bean page, which contains an order button that places the item in your shopping cart. A copyright notice such as the one appearing on this page appears on all storefront Web pages. Also notice the editing bar near the top of the page. The editing bar is available on all Yahoo! Store pages. For example, the "Special" button is a convenient way to highlight selected products that are on sale.

Merchants should display a sufficiently detailed description of each item they are selling. In a recent study conducted by Yahoo! Store and reported in *Internet World*, research suggests that longer product descriptions yield higher sales. In fact, Yahoo! Store reveals that for every order placed in a Yahoo! Store, there are approximately 1.5 orders for items whose description is longer than 1000 characters (approximately 175 words). At the same time, products with no descriptions—typically those with photographs instead—also sell more frequently than the average item.

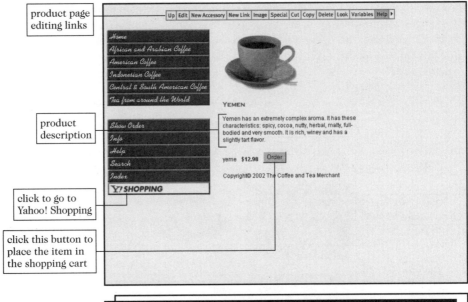

product page
editing links

product
description

click to go to
Yahoo! Shopping

click this button to
place the item in
the shopping cart

Figure 4-10   *An item's order page*

Yahoo's management page, shown in Figure 4-11, contains many great management, reporting, and global site settings tools. Yahoo's Statistics reports are especially well designed and full of useful information for merchants' analyses of their stores. The reports display the number of items you have sold, the number of page hits (only 3067 page hits for this demonstration storefront—all from people reading this book), total revenue (sadly, no sales for the demonstration store shown here), and the total number of orders. Reports can group sales by customer so that you can determine who is buying what from your store. Interestingly, selecting "Click Trails" under the Statistics heading (see Figure 4-11) provides detailed information about the paths that your customers take through your site. These paths list the trail of buttons that customers click. Click trails can help you identify particularly difficult or enigmatic parts of your store and locate places where customers sometimes get lost.

One of the few shortcomings of Yahoo! Store is that Yahoo! produces storefront Web pages in a proprietary format. This means that you cannot back up your store onto your own server. If you attempt to download your pages by executing a File Save operation from Internet Explorer or Netscape Navigator, the resulting saved page is gibberish—not HTML-coded pages. Yahoo! does this to prevent you from building a store on its site and then someday moving it to your own computer. If you decide to leave Yahoo! Store and host your own site, you will have to build your store from scratch using someone else's Web editing software. Yahoo! Store is both representative of its class of hosting solutions and a very good choice for a full-service Web host.

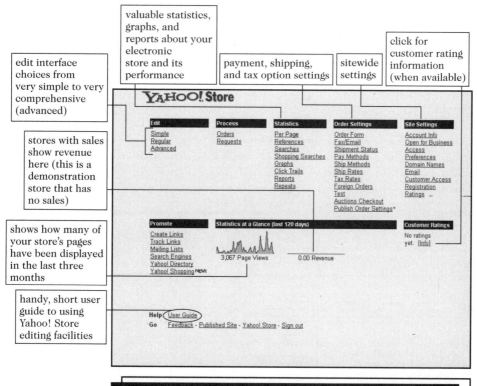

The figure contains the following annotations and interface text:

- edit interface choices from very simple to very comprehensive (advanced)
- valuable statistics, graphs, and reports about your electronic store and its performance
- payment, shipping, and tax option settings
- sitewide settings
- click for customer rating information (when available)
- stores with sales show revenue here (this is a demonstration store that has no sales)
- shows how many of your store's pages have been displayed in the last three months
- handy, short user guide to using Yahoo! Store editing facilities

YAHOO! Store

Edit
Simple
Regular
Advanced

Process
Orders
Requests

Statistics
Per Page
References
Searches
Shopping Searches
Graphs
Click Trails
Reports
Repeats

Order Settings
Order Form
Fax/Email
Shipment Status
Pay Methods
Ship Methods
Ship Rates
Tax Rates
Foreign Orders
Test
Auctions Checkout
Publish Order Settings*

Site Settings
Account Info
Open for Business
Access
Preferences
Domain Names
Email
Customer Access
Registration
Ratings

Promote
Create Links
Track Links
Mailing Lists
Search Engines
Yahoo Directory
Yahoo Shopping NEW!

Statistics at a Glance (last 120 days)
3,067 Page Views   0.00 Revenue

Customer Ratings
No ratings yet. [Info]

Help User Guide
Go Feedback - Published Site - Yahoo Store - Sign out

Figure 4-11   *Management, reporting, and site settings page*

Bigstep.com

Bigstep.com has received a number of industry awards, including *PC Magazine's* prestigious Editor's Choice award in 1999. Bigstep.com provides a good commerce Web site and a well-designed storefront package—without charging hosting fees. You can become a member, build a store with unlimited number of Web pages and as much as 12 MB of graphics all for no monthly charge. Bigstep.com imposes no limit on the number of items you can sell or list on your store.

It sounds too good to be true, right? If a Bigstep store is free, how does Bigstep.com make money? The Bigstep.com hosting company makes money from as-yet-undefined optional add-on services. To process transactions, you must obtain a merchant account through Bigstep.com that is supplied by Cardservice International, even if you already have a merchant account for your brick-and-mortar store(s). The Cardservice International merchant account provides real-time credit card transaction processing, and automatic calculation of sales taxes and shipping costs. The card processing service charges are $14.95 per month plus $0.20 per transaction. In addition, there is a per-transaction fee of 2.35 percent. There are no merchant account setup or application fees.

Similar to Yahoo! Store, Bigstep enables merchants to create, change, and maintain a storefront through a Web browser. Bigstep provides a comprehensive set of page building tools and a series of wizards that provide assistance all along the way.

To create a store, you must register with Bigstep. The registration process identifies you with an e-mail address and password. If you want to build more than one storefront, you must use a different e-mail address. Once you become a member of the Bigstep merchant family, you can select any store name that is not already taken. Your store will be a subdomain of Bigstep.com. For example, if you choose BoldImpressions as your store name, your store's URL would be http://www.BoldImpressions.Bigstep.com. Once you log on with your storeowner e-mail address and password, you can create and manage your electronic store.

Like Yahoo!'s store-building software, Bigstep.com's Web editing and page creation interface is comprehensive and easy to use. During catalog population and store setup, Bigstep.com maintains a convenient list of tasks and their current status. Figure 4-12 shows the Site Building page of the **ExoticFruit demonstration store** during construction. Of course, a customer never sees this view of a store.

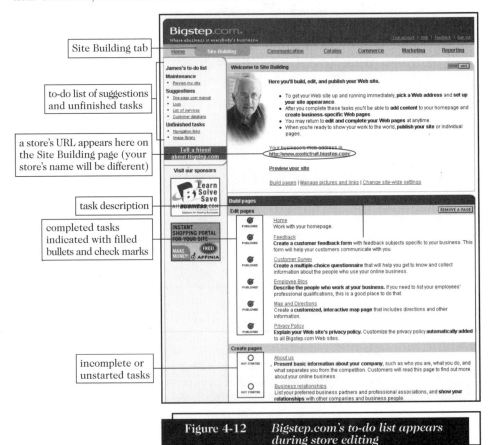

Figure 4-12   *Bigstep.com's to-do list appears during store editing*

While you are editing your store, Bigstep.com displays seven tabs corresponding to the sections of your store that define various aspects of the store's catalog, reporting and utility functions. They are the Home, Site Building, Communication, Catalog, Commerce, Marketing, and Reporting pages. The Site Building section is where you build, edit, and publish the core pages of the electronic commerce store. In the

Communication section, you use your customer database to send out newsletters and build relationships with visitors to your site. The Catalog section contains all of your goods and services. You connect your catalog to the store's commerce system to begin selling online. The Commerce section is where you establish a merchant account and create a checkout process for your Web site. In the Marketing section, you submit your Web site to various search engines so a wide audience can get to know your site. Other promotion activities begin in this section, including announcing Web site changes to visitors and customers. The Reporting section provides a variety of reports about the pages your visitors view, the path they take through your site, and how long they stay. Bigstep automatically gathers statistics that help you identify trends and help you make informed decisions about the future of your electronic business.

Figure 4-13 (on next page) shows the editing session for modifying one Web page of the ExoticFruit store. The page lists the items currently available. Clicking the "Your sections" tab displays different store "departments." Editing links allow you to add, modify, or delete items. You also can add, modify, or delete sections. (There is no limit to the number of sections or items you can create.) Items you create are not automatically added to any section—you do that with a separate action. Graphics, such as pictures of your merchandise, can be linked to items listed on your store. (Bigstep limits graphics to 12 MB.)

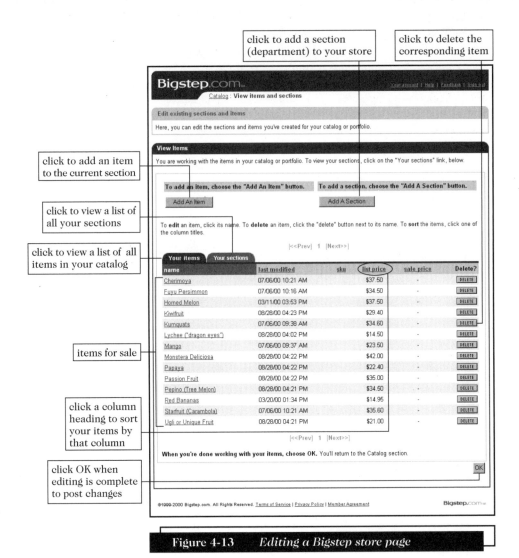

Figure 4-13   *Editing a Bigstep store page*

Bigstep's reports provide data mining capabilities that plow through site data collected in log files. **Data mining**—looking for hidden patterns in data—can help businesses find customers with common interests and discover previously unknown relationships among the data. Reports can indicate problematic pages in your store's design where, for example, a large number of customers get stuck and then leave your Web site. Other data that Bigstep.com reports can reveal are the number of pages an average customer must load and display before locating the merchandise he or she wants. If customers have to load too many pages (three to seven pages is usually considered excessive), customers often become impatient and leave without making a purchase. In other words, available reports can answer the following questions:

- How many visitors are coming to your site?
- What is the average length of stay for each visitor on each page?
- Which pages lead to actual purchases?

- What advertisements or links have brought qualified visitors to your page or site?
- What is the average number of pages that each visitor views?
- Are repeat customers attracted to the site?

If your electronic commerce site complements a brick-and-mortar store, you can use Bigstep.com's built-in map locator to display the location of your store. The map locator uses Mapquest technology to map the location based on the address you supply in the Map and Directions page of the Site Builder section. Additional Bigstep.com store features include automatic calculation of taxes and shipping, collection of customer data, merchant e-mail notification of sales, and customer e-mail confirmation when products ship. Figure 4-14 shows a typical product page for the ExoticFruit store as a customer would see it on Bigstep.com.

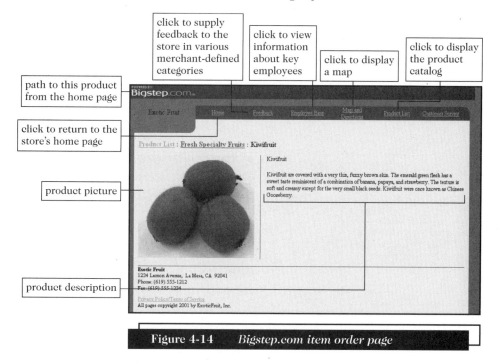

Figure 4-14   *Bigstep.com item order page*

ShopBuilder

Like the other two Web store hosts described in this section, **ShopBuilder** provides the electronic commerce tools you need to build a shop on its site. You create a store using your browser, and you do not need programming, HTML, or site design experience. ShopBuilder processes credit card transactions, tracks sales trends, and computes and graphs statistics on both the items you sell and the customers who visit and purchase items on your site. ShopBuilder automatically generates and sends receipts by e-mail to customers after they have completed their transactions. It also creates titles, navigation buttons, and thumbnail pictures for you. ShopBuilder lets you try out its software—take a test drive—before you make a commitment. You can even visit an actual working store's "back office," where you can explore the tools to track customers and sales.

Figure 4-15 displays an example of the information that ShopBuilder tracks. You can sort the data various ways and display daily, weekly, monthly, or yearly information totals. Merchants commit to one month at a time and pay a low $20 monthly fee to sell as many as five items. For a 250-item store, the monthly fee is $250. There are no setup charges. There is a one-time merchant account setup fee, and per-transaction fees accrue based on the volume of business—higher volumes have lower unit transaction charges.

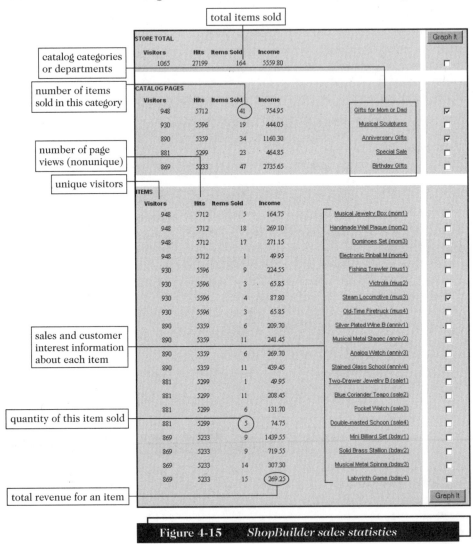

Figure 4-15   ShopBuilder sales statistics

Setting up ShopBuilder is straightforward. If you simply follow a wizard-like procedure, you get a set of standard store pages, including a home page, catalog, contact information page, and store information. You can add as many pages as you want. Unlike Yahoo! Store, ShopBuilder currently has no catalog search capability. Like other products in this class of electronic commerce storefront packages, you have to enter product information manually. The store cannot be updated from a database, worksheet, or ASCII file, for example. This can be time-consuming the first time you set up your store's catalog. Once products are in the catalog, adding new ones or removing a few is simple and takes very little time. Using a typical shopping cart system, the store calculates taxes and shipping charges at checkout time. A customer's order is confirmed by e-mail and you, the merchant, also receive an e-mail confirmation of each transaction. ShopBuilder supports real-time transaction processing, which includes the usual steps of credit card authorization and verification over secure communications facilities.

Estimated Operating Expenses

What does it cost for a small entrepreneur to put up a store on the electronic frontier and offer, for example, 50 items for sale? That is a difficult question to answer accurately, but there are some estimates you can use to assist in your planning. If you advertise your store using banner ads, costs can go even higher. On the other hand, if you don't mind someone else's advertisement on your site, you can drive marketing costs down to zero by using any of several banner exchange services. The following table breaks down electronic commerce costs that merchants can expect to incur during the first year of business. The total omits variable transaction costs, which might average 55 cents per transaction and 2.5% of each sale's total.

Operating Costs	Cost Estimate
Initial site setup fee	$200
Annual maintenance fee (12 × $100)	1200
InterNIC domain name registration	70
Scanner for photo conversion	250
Photo touchup software	100
Occasional HTML and design help	300
Merchant credit card setup fee	200
Total first-year cost	**$2320**

The preceding costs are typical, but they can vary because different Web hosting sites charge varying fees for various services. Use these numbers as a guideline. Additional transaction processing fees can run into hundreds and thousands of dollars, but those fees occur only when you make sales. A good guideline for processing fees is to multiply your expected annual gross sales by 3%. So, if your annual gross sales are $50,000, then the transaction fees should be approximately $1500. That value will adequately cover both the per-transaction fixed costs and the percentage of total sales costs charged by most merchant credit card processing agencies.

Contrast the preceding costs with those estimated costs for self-hosting your Web site. Setup and Web site maintenance costs include equipment, communications, physical location and staff. Equipment—a server and communications gear—has a one-time cost ranging from $5000 to $15,000. A T1 connection or fraction thereof (see Chapter 2) costs from $8000 to $12,000 per year. A server cannot be placed in your backyard. It must be housed in a room that is both secure and convenient to communications access. The cost to secure a room, properly air-condition it, and install a chemical fire-extinguishing system can be near $5000 a year—or more. A self-hosted system requires staff—experts in a variety of Web-related languages, electronic commerce packages, and database management systems. Other technicians will likely be required. Staff costs are minimally between $50,000 and $100,000 annually. In total, annual operating costs for self-hosting approach $60,000 to $100,000 or more the first year. Costs for subsequent years will be about the same. Carefully compare self-host cost estimates with the costs charged by various hosting services.

Next, we will examine midrange electronic commerce packages. Midrange packages are suitable for running larger businesses and are more robust (and more expensive) than the Web hosting, template-driven packages described previously.

MIDRANGE PACKAGES

The line between basic electronic commerce software and midrange electronic commerce software can be blurry. However, midrange electronic commerce packages distinguish themselves in several clear ways. Midrange packages allow—indeed, expect—the merchant to have explicit control over merchandising choices, site layout, internal architecture, and remote and local management options. In addition, the midrange and basic electronic commerce packages differ on price, capability, database connectivity, software portability, software customization tools, and computer expertise required of the merchant.

Buying and using midrange electronic commerce software is significantly more expensive than using a hosted site, with prices ranging from $2000 to $9000. While not cheap, midrange software is far from expensive when you consider that the high-end, enterprise-class systems can cost as much as $100,000. The midrange commerce packages provide more options and operate much more efficiently and capably than their low-end cousins. Midrange software traditionally has connectivity with sophisticated database systems and store catalog information. Having the catalog stored in a database makes product maintenance much simpler. Several of the midrange systems provide connections, sometimes called "hooks," into existing inventory and ERP systems. This can yield savings because there is no need to run duplicate inventory systems and the cost of the existing systems (often called legacy systems) is spread across several software systems.

All the midrange systems, including those described in this section, are hosted on the merchant's own computer. As with Web server software, not all electronic commerce software runs on every computer and operating system. Commerce systems, Web servers, and computer hardware must be matched carefully. Interestingly, some of these midrange packages can be turned into Web hosting packages that can serve many small storefronts, allowing you to recapture some of your investment costs. Midrange systems are all highly customizable, allowing an almost dizzying number of facets to be altered, from the number of templates available to the type of credit cards and shipping combinations you want to allow. Perhaps the biggest difference between a typical midrange package and a basic electronic commerce system is that a midrange package usually requires part-time or full-time programming talent. In addition, expert advice may be needed to extend the package beyond its standard settings and capabilities.

Three midrange electronic commerce systems are described in this section. They are representative of the whole group, yet are different from one another in important ways. The systems are INTERSHOP enfinity by INTERSHOP Communications Inc., WebSphere Commerce Suite by IBM, and Commerce Server 2000 by Microsoft. You can find links on the Online Companion to other equally capable commerce software packages. These include **AbleCommerce Developer**, **Cat@log, iCat Electronic Commerce Suite Professional Edition**, and **Maestro Commerce Suite Professional Edition**.

INTERSHOP enfinity

INTERSHOP enfinity is one of four electronic commerce programs, produced by INTERSHOP Communications Inc., that provide search and catalog capabilities, electronic shopping carts, online credit card transaction processing, and the ability to connect to existing (backend) business systems and databases. Sometimes called **commerce service providers** (**CSPs**), Web-hosting services can provide INTERSHOP's business hosting, storefront building, and management tools in much the same way as Yahoo! Store provides tools for its electronic commerce mall customers.

The INTERSHOP enfinity has setup wizards and good catalog and data management tools. It provides many storefront templates. Management and editing of a storefront are done through a Web browser—either locally at the server or remotely through any Internet connection. The functionality included in INTERSHOP's inventory manager is noteworthy: It tracks inventory levels and allows merchants to view the quantity available, create a list of inventory transactions, and enter new products into the inventory. Discount rules are also easy to enter. You define the business rules for a discount (for example, a quantity over 50 is discounted 10 percent) and dates during which special discounts apply. Shipping costs can be entered through a built-in method that is part of the shipping package. Bundled with the software is a database management system (currently, it is Sybase). Alternatively, you can opt to use INTERSHOP's ODBC access to DB2 (IBM's relational database) or Oracle. If you own several online stores, then each store requires a separate database, however. Happily, moving product information from the existing corporate databases into INTERSHOP database files is not difficult. You simply use INTERSHOP's Data Import wizard to load the product database from existing files. INTERSHOP contains all the standard features of commerce software, including a shopping cart and the automatic calculation of shipping charges and local and state sales taxes. Customers receive e-mail to verify their orders. INTERSHOP supports secure transactions. A wide variety of site and customer reports are available. These track Web page visits and customer activities. Figure 4-16 shows an INTERSHOP enfinity diagram illustrating how the product connects to other parts of an enterprise's systems and serves customers.

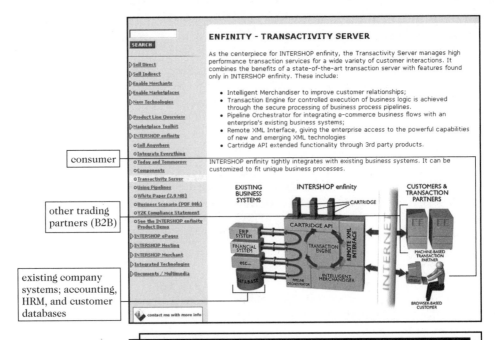

consumer

other trading
partners (B2B)

existing company
systems; accounting,
HRM, and customer
databases

Figure 4-16   INTERSHOP enfinity diagram

WebSphere Commerce Suite

IBM produces the **WebSphere Commerce Suite**. Starting at $9500, the WebSphere Commerce Suite is an excellent family of electronic commerce packages that scales well. IBM WebSphere Commerce Suite is a set of software components that provides software suitable for medium-sized to large businesses to sell goods and services on the Internet. It comes complete with catalog templates, setup wizards, and advanced catalog tools to help Web designers create attractive and efficient electronic commerce sites. WebSphere Commerce Suite can be used both for business-to-business and business-to-commerce applications and provides a smooth connection to existing corporate systems such as inventory databases and procurement systems.

WebSphere Commerce Suite, Start Edition is available for Microsoft Windows NT and 2000 operating systems. You can begin with a small store and then move up to a bigger, more capable store as necessary. Building a store is straightforward. WebSphere Commerce Suite, Pro Edition is available for the IBM S/390 platform, replacing IBM Net.Commerce and running on a variety of operating systems for larger systems, including IBM AIX, IBM AS/400, Windows NT, and Sun Solaris. A wizard leads you through the process of creating a starter store. Once that is up and working, you can add more functionality by executing commands and writing code (you'll want to get that consultant ready). The store creation wizard can create three types of stores: a simple catalog service called One Stop Shop, a more capable and powerful Personal Delivery store, and a Business-to-Business store, whose target market is corporations who conduct business with one another. The wizard asks various questions—such as store name, tax information, preferred and allowed shippers, and contact information—that help it build the merchant's storefront. With the basic pages built, you then can populate the catalog with your products, prices, and product pictures. The WebSphere Commerce Suite also accommodates electronic download products, such as audio tracks or software.

If you are interested in providing a more complete storefront, WebSphere offers a very large collection of functions, utility programs, and commands that allow you to further customize your site. However, you will want to have someone nearby who is handy with JavaScript, Java, or C++. As you would expect from a commerce program in this class, you can connect WebSphere to existing databases and other legacy systems through DB2 or Oracle databases. You can administer a single store or several distinct stores from the same browser-based interface.

The system has all the features you would expect. These include a shopping cart, e-mail notifications upon sale completion, secure transaction support, promotions and discounting, shipment tracking, links to legacy accounting systems, and browser-based local and remote administration. The system runs on AIX, Solaris, and Windows NT operating systems. For large organizations needing more muscle, IBM makes a scaled-up version for enterprise systems (WebSphere Commerce Suite, Pro Edition). Figure 4-17 shows the **WebSphere Commerce Suite** home page.

| e-commerce tab

| WebSphere Commerce Suite editions

Figure 4-17   *IBM WebSphere Commerce Suite home page*

Commerce Server 2000

Microsoft's **Commerce Server 2000** allows you to sell products or services on the Web using tools such as user profiling and management, transaction processing, product and service management, and target audience marketing. Commerce Server 2000 is not an out-of-the-box solution. Though the wizards help you build a site in several steps, you will want to customize the code to fit your needs.

The wizards set up the administrative back office parts of the system, including the site name, a connection to the merchant's database of choice, and a specified administrator or group access to the site's management tools. Commerce Server 2000's Store Builder Wizard helps you construct the store. Merchants can access the tool any time to change properties of their site. Because Commerce Server 2000 tools are Web-based, they can be run remotely. A wizard helps you create or copy an existing site, enter merchant information (contact information and site name, for example), select currency formats (Euro, Portuguese Escudos, U.S. dollars, and so on), and establish the basic style of the site (location of the navigation bar, type-faces, and color scheme). It helps you select products for promotion, sales, and specials. With the Store Builder Wizard's help, merchants can choose to have customers register when they enter the site or when they purchase something. Product types and site structure are important aspects of the store, and the wizard provides basic management choices here too. Finally, you can set up shipping and handling charge options, set up tax information, and select the methods of payment your customers can use.

You can customize your site with the Visual InterDev tools bundled with Commerce Server 2000. Microsoft uses pipelines as a visual model of how the commerce server operates. Reminiscent of pipeline processing computers, the Microsoft pipelines model a series of business processes. Each process in the pipeline operates on the data in some way. Two pipelines exist: the Commerce Interchange Pipeline for business-to-business transactions, and the order-processing pipeline for business-to-consumer processing. With a connector in the pipeline for each business process, a transaction can pass through several pipeline connections and be manipulated. There are connections to attach business rules to each step. There are a small number of predefined stages for the order processing pipeline, beginning with product information and extending to shipping, handling, tax, and order total information. Site Server stores product details—prices, descriptions, sizes, and so on, in a database.

Like other electronic commerce systems, Commerce Server 2000 has tools for the commerce cycles of engaging the customer (through marketing and advertising), transacting an order, and analyzing the sales information after the sale. While some systems focus their efforts on transactions, Commerce Server 2000 has mature tools for engaging and analyzing activities as well. Commerce Server 2000 also has powerful tools for advertising, promotions, cross-selling, and customer targeting and personalization.

To aid in sales and site analysis, Commerce Server 2000 provides many predefined reports for analyzing site activities and product sales data. Reporting tools include one that produces graphs of site and product activities. Commerce Server 2000 can grow with increasing business demands. The system provides the usual list of electronic commerce capabilities you would expect from this product, including several storefront templates, wizards for setting up and initializing a store, and database connections (Oracle, SQL Server, and Sybase). In addition, Commerce Server 2000 provides a shopping cart, verifies completed sales transactions by e-mail, and supports secure transactions. Besides accepting credit cards, Commerce Server 2000 accepts CyberCash (see Chapter 7). It can connect to existing accounting systems, and the administrator can oversee the site through a Web browser. Commerce Server 2000's home page is shown in Figure 4-18.

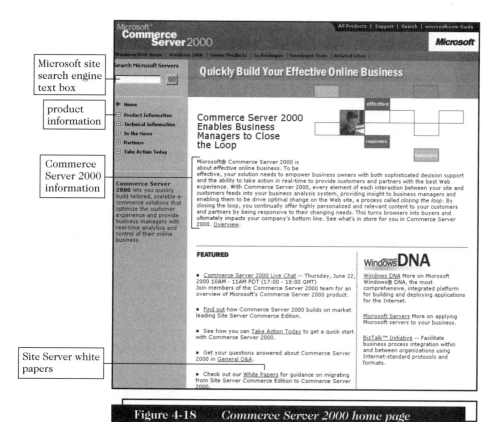

Microsoft site search engine text box

product information

Commerce Server 2000 information

Site Server white papers

Figure 4-18   *Commerce Server 2000 home page*

ENTERPRISE SOLUTIONS FOR LARGE FIRMS

The line separating midrange and enterprise-scale electronic commerce software is much clearer than the one between basic systems and midrange systems. The tell-tale sign is price. Other factors, such as extensive support for business-to-business-commerce, also indicate that the software is enterprise-level commerce software. Commerce software in this class is frequently called e-business (for electronic business) software. E-business software provides tools for both business-to-business and business-to-consumer commerce. In addition, e-business software interacts with a wide variety of existing back office systems, including several database, accounting, and ERP systems.

E-business software running large online organizations requires one or more dedicated computers—in addition to the customary Web server front-end system and any necessary firewalls. An enterprise-scale solution also requires a Domain Name Server (DNS), an SMTP system to handle electronic mail, an HTTP server (Web front-end), an FTP server for upload and download capability, and a database server (Oracle or SQL Server, for example). The last section of this chapter discusses a few typical enterprise-class systems. Examples of powerful e-business systems capable of running a large online company with high transaction rates include IBM's **WebSphere**

Commerce Suite, **Pro Edition**, Netscape's **Netscape CommerceXpert**, Oracle's **iStore**, and **Pandesic Web Business Solution** (a joint venture between Intel and SAP). Other examples include **Commerce Exchange**, **One-To-One Enterprise**, and **Transact**.

E-business software typically provides good tools for linking to and supporting supply and purchasing activities. A large part of business-to-business commerce is ordering supplies from trading or business partners and issuing the appropriate documents, such as purchase orders. For a selling business, e-business software provides standard electronic commerce activities, such as secure transaction processing and fulfillment, but it can also do more. For instance, it can interact with the firm's inventory system and make the proper adjustments to stock, issue purchase orders for needed supplies when they reach a critically low point, and generate other accounting entries in ERP or legacy accounting or file systems. In contrast, both basic and midrange electronic commerce packages usually require an administrator to manually check inventory and explicitly place an order for items that need to be replenished.

A standout example of the use of e-business software is the mutually beneficial relationship forged between Wal-Mart and its suppliers. Wal-Mart's success can be traced, in part, to the economies of scale and cost savings in inventory. The company persuaded several of its suppliers to make their own decisions about resupplying Wal-Mart when stocks ran low. By allowing suppliers to perform the monitoring and automatic restocking functions, Wal-Mart and its suppliers save money. Efficient business-to-business (e-business) systems that are connected and that interact with each other generate the savings.

In business-to-consumer situations, customers use their Web browsers to locate and browse a company's catalog. For electronic goods (software, research papers, music tracks, and so on), customers can download the items directly from the site or a mirror server that is in close proximity to the customer, or they can choose to complete order forms and have the hard copy versions of the products shipped to them. The Web server is linked to backend systems, including (almost always) a relational database management system (RDBMS), a merchant server, and an application server. An RDBMS usually contains millions of rows of information about products, prices, inventory, user profiles, and user purchasing history. The history provides a way to recommend to a user on a return visit related items that he or she might wish to purchase. A merchant server houses the e-business system and key back office components. It processes payments, computes shipping and taxes, and sends a message to the fulfillment department when it must ship hard goods to a purchaser. Figure 4-19 (on the next page) shows an e-business system architecture.

What are the ranges of system software costs, exclusive of any other administrative costs, for large e-business systems? Netscape's CommerceXpert starts at $75,000 and can exceed $250,000 for a typical system. Pandesic ranges from $25,000 to $100,000. Other equivalent systems (see the Online Companion) start at $50,000 for more basic systems and top $1 million for encompassing solutions. Netscape's CommerceXpert commerce system, for example, provides a catalog management system that connects to a backend order management and payment-processing system consisting of a Merchant Server and a Transaction Server. The Merchant Server is a catalog server that can keep track of customers' preferences and track orders. CommerceXpert supports encryption, authentication, digital signatures, and signed receipt notices. Like smaller systems, CommerceXpert can be administered locally or remotely through a browser interface. IBM's WebSphere Commerce Suite, Pro Edition the larger and more capable cousin of WebSphere Commerce Suite, provides advanced catalog tools that include search methods and shopping advisers. On the back office side of the system, WebSphere Commerce Suite, Pro Edition connects to legacy systems handling inventory and logistics and to EDI software.

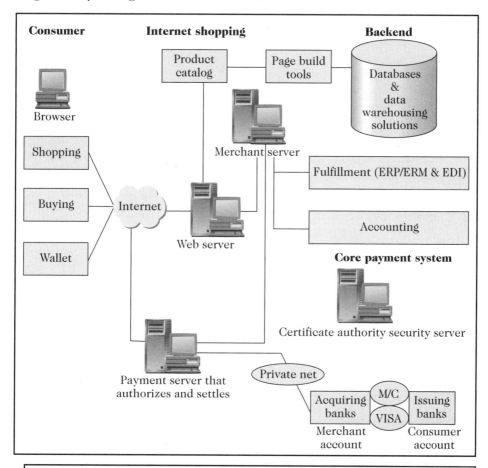

Figure 4-19   *Typical, large e-business system architecture*

Summary

In this chapter, you learned about electronic commerce software for small, medium-sized and large businesses and the functions provided by each software type. Choosing electronic commerce software requires making some major decisions; it is not a trivial process. You must first choose between paying a commerce service provider (CSP) to host your site or supplying the computer and running the requisite software on your own machine. Small enterprises that are not sure whether electronic commerce is a good idea should consider using a CSP. Banner exchange programs provide free marketing for sites willing to display banner advertisements from other stores in exchange for other storefronts displaying their advertisements.

CSPs are great if you don't want to invest in machinery and people, but if you already have computing equipment and staff in place, purchasing a midrange package will give you more control over your site. Large enterprises that have high transaction rates, B2B partnerships, and a large investment in ERP and other backend systems will want to invest in robust and capable enterprise-class systems. Whichever class of Web commerce site

software you choose, your system should provide straightforward storefront construction and catalog creation facilities. In addition, commerce software should contain shopping cart software or provide the system "hooks" to interface with a third-party shopping cart software package. Finally, and most importantly, commerce software should have transaction processing functionality that provides full service, from credit card authorization to calculation of shipping and taxes. The Online Companion has links to a large number of Web sites that supply commerce software ranging from free hosting to $300,000 turnkey systems.

Key Terms

Banner Exchange Site (BES)
Catalog
Click-Through Count
Colocated Hosting
Commerce Service Provider (CSP)
Commerce Suite
Cookies
Data Mining
Dedicated Hosting
Enterprise Resource Planning (ERP)

InterNIC
META Tag
Registrar
Self-Hosting
Shared Hosting
Shopping Cart
Transaction Processing
Web Hosting
Web Mall

Review Questions

1. List and describe the three fundamental requirements of an electronic commerce site.

2. How can the Internet, a so-called "stateless" machine, keep track of a person's preferences between browsing sessions? Are there any security restrictions that a browser could impose that would force a change in the way information about users is recorded? Explain one way to record information about a user that doesn't require accessing the user's computer.

3. List two disadvantages of hosting your commerce site on a host that is free. What is missing from the host's services that would make your job as an online entrepreneur more difficult?

4. Describe in a paragraph the differences between basic electronic commerce software and midrange electronic commerce software. Discuss at least four differences and give examples of each type of software.

5. What are the characteristics of large firms conducting both business-to-business and business-to-consumer transactions that require more robust and capable electronic commerce systems? Consider the volume and types of transactions and store maintenance activities that differ between a small storefront operation and, for example, an Amazon.com-caliber store.

Exercises

1. Yahoo! Store allows you to create and save a storefront free for 10 days. You are to create a small store, called The Coffee and Tea Merchant, using Yahoo! Store as your Web host. Because your site will disappear on the 11th day after you create it, plan to complete the assignment within 10 days. When you have completed building your store to the specifications here, visit the store as a customer and print several of your store's Web pages. Hand in to your instructor your printed storefront Web pages. Here are the details of your storefront's contents: Your coffee store should have at least two departments, called "sections" by Yahoo! Store, that correspond to the coffee-growing regions "African and Arabian Coffee" and "Indonesian Coffee." You may add more growing regions if you want. Figure 4-9 shows the African and Arabian coffees and their prices in your store. Figure 4-20 shows the Indonesian coffees and their prices. All prices are per pound.

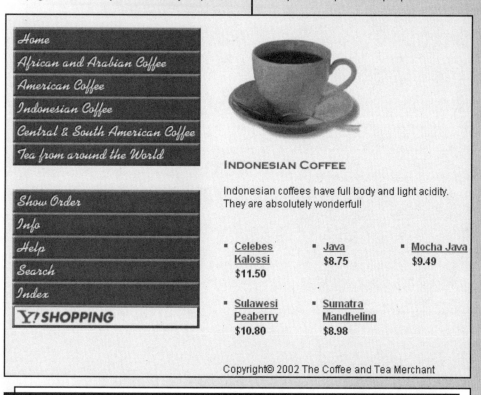

Figure 4-20   *Indonesian coffees*

Figure 4-21 shows the "Info" page—the one that appears if you click the Info button. Your store should have an information page with your name, address, and e-mail address. See if you can figure out how to include a copyright notice on every page of your Web store similar to the one in the figures. (*Hint*: You only need to enter it in one place.) To help you in the design of your site, you can download coffee cup graphics of five types that we created for your use. Click the Online Companion links **Cup1** through **Cup5** to view each of these graphics.

(We have provided these graphics for you to use in this exercise.) When you find a graphic that you like, right-click the link and select Save As (Netscape) or Save Target As (Internet Explorer) to download the graphic to your PC. If you want to visit The Coffee and Tea Merchant demonstration store we built, click the Online Companion link **The Coffee and Tea Merchant** and wander around the store. Place some coffee and tea in your shopping cart and then view the contents of your cart. (Just don't actually check out and enter payment information!)

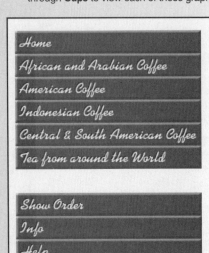

We can ship any product the same day for all orders received before 4:00 PM, U.S. Pacific time zone.

All orders are shipped UPS ground unless you specify otherwise. Overnight delivery is available.

The Coffee and Tea Merchant
2234 Roasting Lane
San Diego, California 90000-1234
(619) 555-8500
OTToadvine@yahoo.com

The first Coffee and Tea Merchant was established in 1978 in Setubal, Portugal. When the Toadvine family moved to the United States, they sold their Setubal store and opened one in San Diego, California.

Privacy Policy: At The Coffee and Tea Merchant, we are committed to protecting your privacy. We use the information we collect about you to provide a more personalized shopping experience. The information is never sold nor disclosed to anyone else. Our secure server software (SSL) encrypts all information you input before it is sent to us. Furthermore, all customer data we collect is protected against unauthorized access.

Figure 4-21 *Sample information page*

Here are a few pointers on how to get started: To register your Yahoo! Store temporarily and start building your free demonstration storefront, click the **Build your own Yahoo! Store** link in the Online Companion. Next, click the "Test Drive" link on the Yahoo! Store page and then click the "Build your own store" link. A long form opens with several text boxes. Fill out the text boxes and click the "Create" button at the bottom of the page. Yahoo! will respond in a few seconds, indicating that your page has been created. If not, then you may have to pick a different business name (username); in this case, be creative about your business name. Click the button labeled "I Accept" on the next page to proceed. You can take a guided tour or go directly to your site. Your site's name is the URL http://store.yahoo.com/ followed by your username. To edit your site, click the "Yahoo! Store" link and then click the "Edit" button. When you have finished with an editing session, be sure to go back to your home page and click the "Publish" button. Otherwise, none of your changes are saved on your Yahoo! Store site. You can pause during the construction of your store and come back later and finish it. Keep the 10-day limit in mind, though.

2. Annette Leonard owns a small crafts store in central Missouri. She wants to expand her store's reach outside the region to increase her profits and simultaneously reduce her inventory. Annette has been watching her teenage daughter, Kelly, using the Internet to order music CDs and books. After learning from Kelly how simple it is to order from online stores, Annette decided that she needs to create an online store. She has asked you to do a little research on how much it might cost in the first year to create a simple store selling 100 items.

Annette wants you to investigate three CSPs and report back to her what you have found. Because her store is small, limit your research to fundamental host services and full-service hosting, and ignore standalone packages such as midrange packages and large B2B packages. You may want to begin your research with sites such as **B-City**, **Freemerchant.com**, or **Yahoo! Store**. Here's what Annette wants to know about the three host systems you examine:

- What are the costs: initial setup fee, monthly fee, and transaction fees?
- How much disk space does her 100-item store receive?
- Does the CSP provide a search engine?
- What promotion and marketing opportunities does it provide?
- How are customers notified following a transaction?
- Does the CSP provide a shopping cart? If not, how do customers enter orders?
- Are storefront-building wizards available to help create a new store?
- Are the transactions conducted in a secure environment?
- Does a store get its own domain name? Is there an extra charge for a domain name and its registration?
- Can you upload product names, descriptions, images, and costs from files or databases, or must you enter each item manually?
- Does the CSP provide some sort of online user manual for the merchant?

Produce a written report of no more than 1000 words describing what you have learned about the CSPs. Clearly label each section, grouping your information by service provider.

3. Investigate and write a 400-word report summarizing the costs and features of any enterprise-class commerce package for large businesses. Pick 10 or fewer characteristics about the commerce package and describe them in some detail. Begin your

research using the links provided in the Online Companion. You may want to use a good search engine such as **AltaVista**, **Google**, **Hotbot**, or others listed under Exercise 3 in the Online Companion.

For Further Study and Research

Buss, D. 2000. "The Big 'Vortal' Payoff," *Internet World*, April 15, 35.

Callaghan, D. 2000. "Keeping e-carts moving/Startup eBSure prepares to roll out site usage measurement suite," *eWeek*, 17(20), May 15, 43.

Callaghan, D. 2000. "Web tools track site users' actions," *eWeek*, 17(22), May 29, 43.

Derfler, F. 2000. "Finding a Hassle-Free Host," *PC Magazine*, 19(19), November 7, 135–154.

Du Bois, G. 2000. "Ask Jeeves bids banner ads adieu, welcomes new method," *eWeek*, 17(24), June 12, 33.

Graven, M. 2000. "e-Store Solutions," *PC Magazine*, 19(18), October 17, 146–167.

Herel, H. 2000. "E-Tailing: More than a Catalog," *PC Magazine*, 19(13), July, 147.

Internet World. 1999. "Longer Product Descriptions Pay Off," 5(16), April 26, 15.

King, J. 2000. "Companies Aren't Rushing to Conduct Business Online," *eWeek*, 17(13), March 27, 20.

King, J. 2000. "Filling Orders a Hot e-Business," *Computerworld*, 34(24), June 12, 3.

King, N. 1998. "Sell it on the Web," *PC Magazine*, 17(20), November 17, 163–200.

King, N. 1999. "To Host or Be Hosted?," *Internet World*, 10(10), October 15, 64–66.

Luh, J. 1999. "Carnival Time for Ads," *Internet World*, November 15, 51. Available online at: (http://www.internetworldnews.com/idx_article.asp?inc=111599/ebusiness/advertising/11.15Advertising1&issue=11.15).

Marx, A. 1999. "Dissatisfaction With Banner Ads Is On the Rise," *Internet World*, 5(24), June 28, 13–14.

Munro, J. and D. Lidsky. 1999. "Web Storefront Software," *PC Magazine*, 18(1), January 5, 155–194.

Rush, L. 2000. "Data Mining For Dummies," *Internet World*, June 19. Available online at: (http://ecommerce.internet.com/solutions/ectips/article/0,1467,6311_397341,00.html).

Schwartz, M. 2000. "Constructive Web Critics," *Computerworld*, 34(21), May 22, 48–52.

Seminerio, M. 2000. "Click-and-Mortar Brigade Born," *eWeek*, 17(24), June 12, 58.

Stanek, W. 1999. "Meta Tags Target Your Pages," *PC Magazine*, 18(13), July, 253–255.

Wang, N. 1999. "Net Bigger Than Billboards as '98 Ad Total Hits $1.9B," *Internet World*, 5(18), 15.

SECURITY THREATS TO ELECTRONIC COMMERCE

INTRODUCTION

On November 3, 1988, thousands of computer systems operators and systems administrators across the United States came to work and found their computer systems inoperable. Seemingly catatonic, the computers would not respond no matter what the administrators tried.

This disastrous event was eventually traced to a 23-year-old Cornell graduate student named Robert Morris Jr., who had unleashed the **Internet Worm** on the Internet community—arguably the most infamous Internet attack ever. The worm caused thousands of computers around the country to slow down or cease production altogether. The program traveled across the Internet in a matter of minutes by exploiting a flaw in a popular UNIX e-mail program called sendmail. It managed to invade and infect over 6200 computers (10% of the computers on the Internet at the time) and cause a widespread slowdown. As news of the incident spread, several sites that were not yet infected disconnected themselves from the Internet. The exact cost of the shutdown and associated damage could not be accurately calculated. Several sources have estimated that the lost computing time (called a denial attack) was valued

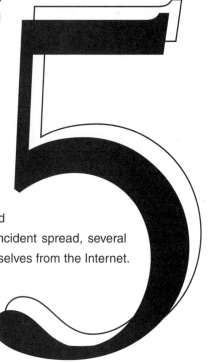

at $24 million and direct costs to eradicate the virus and bring machines back onto the Internet were estimated at roughly $40 million, although other estimates put the cost closer to $100 million.

Industry experts who studied the worm discovered that it had no destructive code. The damage was caused by its uncontrolled, cancer-like growth inside each machine. It eventually absorbed all computing resources until each infected machine failed. The Internet Worm was a wake-up call to the computer community. Since the Internet Worm attack, computer security has become an important and integral consideration in all computer hardware and software acquisitions and deployment plans.

LEARNING OBJECTIVES

In this chapter, you will learn about:
- Important computer and electronic commerce security terms
- The reasons that secrecy, integrity, and necessity are three parts of any security program
- The roles of copyright and intellectual property and their importance in any study of electronic commerce
- Threats and countermeasures to eliminate or reduce threats
- Specific threats to client machines, Web servers, and commerce servers
- Methods that you can use to enhance security in back office products, such as database servers
- The ways in which security protocols help plug security holes
- The roles that encryption and certificates play in assurance and secrecy

SECURITY OVERVIEW

When the Internet was very young, electronic mail was one of its most popular uses. Since e-mail's inception, people have often worried that a business rival might intercept e-mail messages and turn them into a weapon. Another fear has been that an employee's nonbusiness correspondence—such as a Monday morning analysis of the weekend social scene—might be read by his or her supervisor, with negative repercussions. These were fairly significant—and realistic—concerns.

Today, the stakes are much higher. The Internet has matured, and the ways we use it have changed too. The consequences of a competitor having unauthorized access to messages and digital intelligence are now far more serious than in the past. Electronic commerce, in particular, brings to the forefront long-held information security concerns. A typical concern of Web shoppers contemplating first-time purchases on the Internet is that their credit card numbers will be exposed to a few million people as they travel across the network to "who knows where." Of course, this echoes the same concern shoppers have expressed for the last 30 years about credit card purchases over the phone: "How can I trust the person on the other end of the phone who is writing down my credit card number?" These days, consumers are more comfortable giving their credit card numbers and other information over the phone to strangers, but many of those same people fear providing the same information through a computer. This chapter examines the broad topic of computer security in the context of electronic commerce. Though many aspects of computer security are complex and the subject of ongoing research, this chapter presents an overview of the important security issues and their current solutions.

Computer security is the protection of assets from unauthorized access, use, alteration, or destruction. There are two general types of security: physical and logical. **Physical security** includes tangible protection devices, such as alarms, guards, fireproof doors, security fences, safes or vaults, and bombproof buildings. Protection of assets using nonphysical means is called **logical security**. Any act or object that poses a danger to computer assets is known as a **threat**.

Countermeasure is the general name for a procedure, either physical or logical, that recognizes, reduces, or eliminates a threat. The extent and expense of the selected countermeasures can vary, depending on the importance of the asset at risk. Threats that are deemed low risk and unlikely to occur can be ignored when the cost to protect against the threat exceeds the value of the protected asset. For example, it would make sense to protect from tornadoes a computer network in Oklahoma City, where there is a lot of tornado activity, but not to protect one in Los Angeles, where tornadoes are rare. The risk management model shown in Figure 5-1 illustrates four general actions that you could take, depending on the impact (cost) and the probability of the physical threat. In this model, a tornado in Kansas or Oklahoma would be in quadrant II, whereas a tornado in Southern California would be in quadrant III or IV.

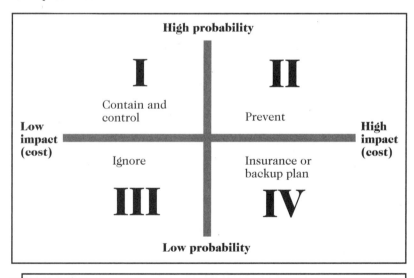

Figure 5-1   *Risk management model*

The same sort of risk management model applies to protecting Internet and electronic commerce assets from both physical and electronic threats. Examples of the latter include impostors, eavesdroppers, and thieves. An **eavesdropper**, in this context, is a person or device that is able to listen in on and copy Internet transmissions. To implement a good security scheme, you must identify the risk, determine how you will protect the affected asset, and calculate how much you can spend to protect the asset. In this chapter, the primary focus in risk management protection is not on the protection costs or value of assets. Instead, the focus is on the central issues of identifying the threats and determining the ways to protect assets from those threats.

Computer Security Classification

Experts in this area generally agree that you can classify computer security into three categories: secrecy, integrity, and necessity (also known as denial of service). **Secrecy** refers to protecting against unauthorized data disclosure and ensuring the authenticity of the data's source. **Integrity** refers to preventing unauthorized data modification. **Necessity** refers to preventing data delays or denials (removal). Secrecy is the best known of the computer security categories. Every month, newspapers report on illegal break-ins to governmental computers or unauthorized uses of stolen credit card numbers, which are used to order goods and services. Integrity threats tend to be reported less frequently and thus may be less familiar to the public. For example, an integrity violation occurs when an Internet e-mail message's contents are changed—possibly negating the message's original meaning. There are several instances of necessity violations, and they occur relatively frequently. For example, delaying a message or completely destroying it can have huge consequences. Suppose that you send an e-mail message at 10:00 a.m. to E*Trade, an online stock trading company, with an order to purchase 1000 shares of IBM at market. Then, imagine the stockbroker does not receive the message (because an enemy delays it) until 2:30 p.m.—after the stock's price has increased 15 % in the interim. The delay has cost you 15 % of the value of the trade.

Copyright and Intellectual Property

Copyright and intellectual property rights also are security issues, although they are protected with different countermeasures. **Copyright** is the protection of expression—some entity's intellectual property—and it typically covers items such as literary and musical works; pantomimes and choreographic works; pictorial, graphic, and sculptural works; motion pictures and other audiovisual works; sound recordings; and architectural works. **Intellectual property** is the ownership of ideas and control over the tangible or virtual representation of those ideas. Similar to computer security violation, copyright infringement causes damage. However, unlike computer security breaches, the damages as a result of copyright violation are narrow and have a smaller impact on an organization or individual.

In the United States, the **U.S. Copyright Act of 1976** protects items, such as those in any of the preceding categories, for a fixed time period. For items published before 1978, the copyright expires 75 years from the item's publication date. For items published after January 1, 1978, the copyright expires 50 years beyond the life of the author for an individual holder, or 75 years after the date of the publication for employers of the author. Every work is protected when it is created. The work does not require a copyright notice, such as *Copyright©2002 So and So Company*, to be protected, and this fact is the most misunderstood part of the U.S. copyright law. Unless you have received permission to reproduce a graphic that you found on an electronic commerce site that is subject to copyright protection, you may be violating the Copyright Act if you reproduce the item on your Web site. Using any search engine, you can type the term *copyright* and find hundreds of Web sites that discuss copyright issues. One example is the very informative **Copyright Clearance Center**, which is chock full of U.S. copyright information. Figure 5-2 shows its home page.

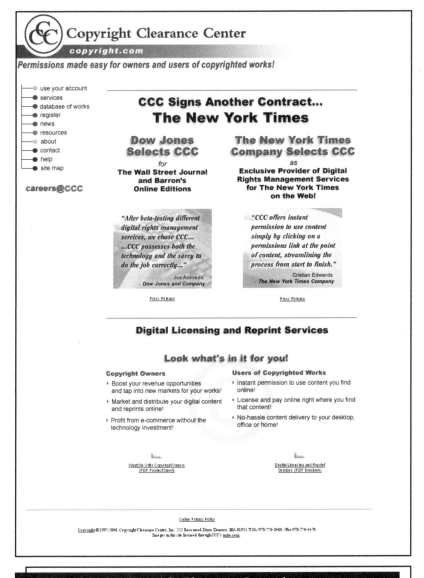

Figure 5-2 *Copyright Clearance Center home page*

Security Policy and Integrated Security

Any organization concerned about protecting its electronic commerce assets should have a security policy in place. A **security policy** is a written statement describing which assets to protect and why they are being protected, who is responsible for that protection, and which behaviors are acceptable and which are not. Mainly, the policy addresses physical security, network security, access authorizations, virus protection, and disaster recovery. The policy develops over time and is a living document that the company and security officer must review and update at regular intervals.

To create a security policy, you start by determining what to protect from whom (for example, you want to protect credit cards from eavesdroppers). Then, you determine who should have access to various parts of the system and who should not. Next, you determine what resources are available to protect the assets you have identified. Given the information the security team has acquired, the team develops a written security policy. Finally, you commit resources to building or buying software, hardware, and physical barriers that implement your security policy. For example, if your security policy disallows any unauthorized access to customer information, including credit card numbers and credit history, then you must either write the software that guarantees end-to-end secrecy for electronic commerce customers or purchase that software (programs or protocols) to enforce that security policy.

The **Center for Security Policy** (**CSP**) is a nonprofit organization that stimulates national and international debate about security policy—particularly security policies that address defense and technology interests of the United States. The CSP can provide guidance about what a security policy should contain and what it should not contain. Figure 5-3 (on the next page) displays the CSP home page.

While absolute security is difficult—even impossible—to achieve, you can create enough barriers to deter most intentional violators. If the cost to an electronic thief of attempting an unauthorized activity exceeds the value of accomplishing the illegal action, then you significantly lower the probability of the attempt. With good planning, you can also greatly reduce the impact of natural disasters.

Integrated security means having all security measures working together to prevent unauthorized disclosure, destruction, or modification of assets. A security policy covers many security concerns that must be addressed by a comprehensive and integrated security "blanket." Specific elements of a security policy address the following points:

- Authentication: Who is trying to access the electronic commerce site?
- Access control: Who is allowed to log on to the electronic commerce site and access it?
- Secrecy: Who is permitted to view selected information?
- Data integrity: Who is allowed to change data and who is not?
- Audit: What and who causes selected events to occur, and when?

In this chapter, you will explore these issues with a special focus on how these security policy issues apply to electronic commerce in particular. Next you will learn about threats to digital information, beginning with intellectual property threats.

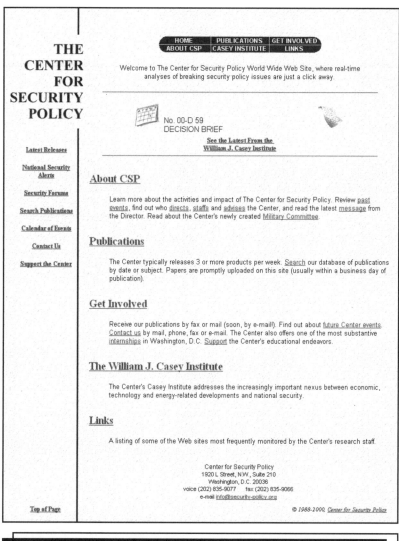

Figure 5-3   *Center for Security Policy home page*

INTELLECTUAL PROPERTY THREATS

Intellectual property threats are a larger problem than they were prior to the wide-spread use of the Internet. It is relatively easy to use existing material found on the Internet without the owner's permission. Actual monetary damage resulting from a copyright violation is more difficult to measure than damage from secrecy, integrity, or necessity computer security violations. However, the damage can be just as significant.

The Internet presents a particularly tempting target for two reasons. First, it is very easy to reproduce an exact copy of anything you find on the Internet, regardless of whether it is subject to copyright restrictions. Second, many people are simply naïve or unaware of the copyright restrictions that protect intellectual property. Instances of both unwitting and willful copyright infringements occur daily on the Internet. For example, fans of *Dilbert* cartoons often put up Web storefronts and fan club sites that feature Scott Adams's cartoons. Though the gesture may be flattering, it is nevertheless a serious copyright violation. Most experts agree that copyright infringements on the Web occur because users are ignorant of what they can and cannot copy. Most people do not maliciously copy a protected work and post it on the Web.

Although copyright laws were enacted before the creation of the Internet, the Internet itself has complicated publishers' enforcement of copyrights. While recognizing unauthorized reprinting of written text is relatively easy, perceiving when a photograph has been borrowed, cropped, or illegally used on a Web page is a more difficult task. The **Berkman Center for Internet and Society** at the Harvard Law School recently introduced a course comprising a series of week-long modules collectively titled "Intellectual Property in Cyberspace 2000." **The Copyright Website** tackles the issues of copyright and newsgroup postings and fair use. **Fair use** allows limited use of copyright material when certain conditions are met. Figure 5-4 shows The Copyright Website home page.

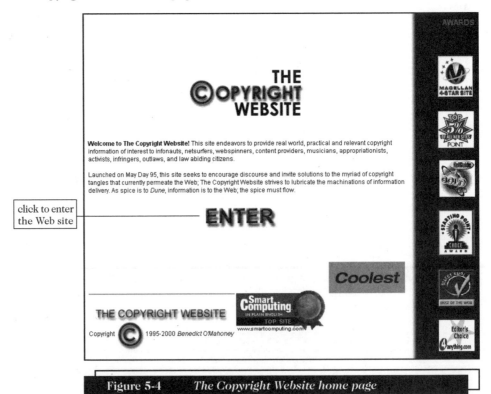

click to enter
the Web site

Figure 5-4   *The Copyright Website home page*

Music Online

No single industry better illustrates the copyright and intellectual property issues that can arise than the music industry. Music fans and the music industry have been waging a battle over the rights of recording artists to control the distribution of their music and receive royalties for their work. A community of music fans organized around a service called **Napster** has dramatically changed the way music is delivered and has attracted the attention of record producers and artists. Bearing the nickname of its founder, Napster is both a special interest group and a service that combines Internet chat, file transfer, and file searching services. With Napster software installed on individual computers, music fans can collect and store music stored in the MP3 format and make it available for other Web surfers to download— all for no charge.

Billing itself as a digital revolution, Napster lets relatively unknown musicians make their music available online. In theory, new artists gain acceptance and publicity by letting their music be downloaded and passed around—all without the help of the music recording industry. This approach has worked well both for well-known microcomputer operating systems and other software known collectively as shareware—sort of a "try it now and buy it later if you like it" system. While a good idea for new artists, the system works for well-known artists too. That's where the conflict arises. The act of **ripping** a song—extracting tracks from CDs using software commonly known as a **ripper** and storing the music in digital format on a computer—without proper permission is a copyright violation. Of course, artists whose music is reproduced illegally do not receive any royalties for its use—and that is the real issue for musicians. Famous artists including Metallica have taken action to prevent their music from being downloaded from Napster servers. Napster is facing litigation brought by several other music industry organizations, as well. The last chapter of the Napster story has yet to be written. Figure 5-5 shows Napster's home page.

Domain Names

Quite a bit of controversy has sprung up in the last couple of years about intellectual property rights and Internet domain names. A growing number of cases are swirling around the controversy of cybersquatting. **Cybersquatting** is the practice of registering a domain name that is the trademark of another person or company in the hopes that the owner will pay huge amounts of money to acquire the URL. In addition, successful cybersquatters can attract many Web browsers and consequently charge high advertising rates. A related problem, called **name changing**, occurs when someone registers purposely misspelled variations of well-known domain names. These variants sometimes lure consumers who make typographical errors in entering a URL. **Name stealing** occurs when someone, posing as a site's administrator, changes the ownership of the domain name assigned to the site to another site and owner.

Cybersquatting

On November 29, 1999, the **U.S. Anticybersquatting Consumer Protection Act** (**ACPA**) was signed into law. Also known as the Trademark Cyberpiracy Prevention Act, it protects the trademarked names owned by corporations from being registered as domain names by parties that do not own the trademarked names. Under U.S. law, parties found guilty of cybersquatting can be found liable for damages of up to $100,000 per hijacked trademark. If the registration of the domain name is found to be "willful," damages can be as much as $300,000 (see the Null reference listed in the "For Further Study and Research" section at the end of the chapter). Recent U.S. cases that were finally settled out of court illustrate the problem. Three cybersquatters made headlines a few years ago when they tried to sell the URL barrydiller.com for $10 million. Barry Diller, the CEO of USA Networks, sued the trio and won.

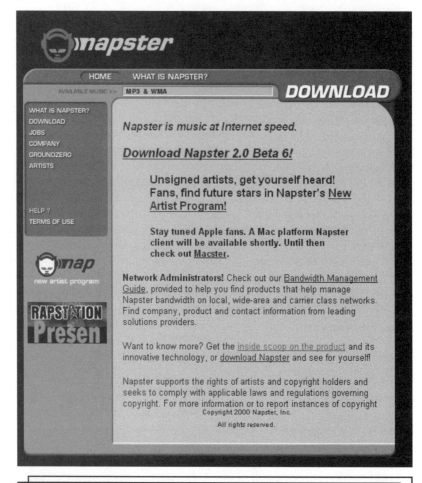

Figure 5-5   *Napster's home page*

Registering a generic name such as Wine.com is very different from registering a trademarked name in bad faith—cybersquatting. The former act is merely prospecting and is legal, whereas the latter is clearly extortion. If someone has already registered a domain name that would be an appropriate URL for another company, then that other business desiring the name may have to pay a high price to obtain it. That is both legal in the United States and quite common around the world. Anyone selling port wine online would value the domain name PortWine.com, because the name alone attracts online customers seeking port wine. Similarly, a name such as Shoes.com is a valuable name for an online shoe store. See the "Creating and Maintaining Brands on the Web" for further discussion of domain names and their role in a firm's success.

Name Changing

After obtaining the domain name, or several related variations, companies still face the possibility that a high-tech hustler may steal unsuspecting customers by registering a domain name that is a slight variation or even a misspelling of a company's well-known domain name. A simple typo in a Web address by caffeinated Web surfers can lead them to LLBaen.com instead of LLBean.com. The Anticybersquatting Consumer Protection Act will help define which cases are true cybersquatting and which are simply cases of permissible competition. Most businesses agree that the practice of name changing is annoying to affected online businesses and confusing to their customers. A company's best defense is to register as many variations in product and company spellings as possible. Unfortunately, there is no permanent or 100% fix to this problem, because when additional proposed high-level domains such as .shop become available, the name changing problem will recur.

Name Stealing

Perhaps the most flagrant case of domain name malevolence is name stealing. Name stealing occurs when someone other than a domain name's owner changes the ownership of the domain name. A **domain name ownership change** occurs when owner information maintained by a public domain registrar is changed in the registrar's database to reflect the new owner's name and business address. This usually happens only when safeguards are not in place. Once domain name ownership is changed, the name stealer can manipulate the site, post graffiti on it, or redirect online customers to other sites selling substandard goods. The purpose of name stealing is to harass more than anything else. The temporary loss of its domain name can cut off a business from its Web site for several days.

The ownership change can occur without notice because it is automated and because some domain name registrars' security procedures are sometimes lax. Not long ago, six domain names disappeared quietly from the registry of the Virginia registrar **Network Solutions Inc. (NSI)**. In one case, an electronic commerce site was unavailable to the domain's owner for five days. Of course, that can add up to tremendous losses in sales.

Name stealing is fairly uncomplicated. A savvy perpetrator can type in a registered administrator's address and change Web site ownership. Then, he or she can register the stolen domain name at another registrar's site. Of course, the name stealing does not go unnoticed for long.

ELECTRONIC COMMERCE THREATS

You can study electronic commerce security requirements by examining the overall process, beginning with the consumer and ending with the commerce server. When you consider each logical link in the "commerce chain," the assets that must be protected to ensure secure electronic commerce include client computers, the messages traveling on the communication channel, and the Web and commerce servers—including any hardware attached to the servers. Many movies portray the business of espionage as comprising activities that focus on wiretapping and listening in on various communications devices, such as telephone lines and satellite communication links. While telecommunications are certainly one of the major assets to be protected, the telecommunications links are not the only concern in computer and electronic commerce security. For instance, if the telecommunications links were made secure but no security measures were implemented for either client computers or commerce and Web servers, then no communications security would exist at all. Or if the client computer contained a virus, contaminated information could be securely delivered to a Web or commerce server. In this case, the commerce transactions would be only as secure as the least secure element—the client machine.

This chapter's remaining sections are organized around a three-element theme: protecting client computers, protecting the transmission of information on the Internet, and protecting the electronic commerce server. First we will examine threats that exist for clients.

Client Threats

Until the debut of executable Web content, Web pages were mainly static. Coded in HTML, the Web's standard page description language, static pages could do little more than display content and provide links to related pages with additional information. The widespread use of active content has changed that.

Active Content

Active content refers to programs that are embedded transparently in Web pages and that cause action to occur. Active content can display moving graphics, download and play audio, or implement Web-based spreadsheet programs. Active content is used in electronic commerce to place items you wish to purchase into a shopping cart and compute your total invoice amount, including sales tax, handling, and shipping costs. Developers embrace active content because it extends the functionality of HTML and adds spark and excitement to Web pages. It also offloads some data processing chores from the busy server machine to the user's mostly idle client computer.

Active content is provided in several forms. The best-known active content forms are Java applets, ActiveX controls, JavaScript, and VBScript. (See the Online Companion links **Java applet central** and **Java security** for Web information about Java applets and Java security questions.) JavaScript and VBScript are known as scripting languages; they provide scripts, or commands that are executed. VBScript, a subset of Microsoft's Visual Basic programming language, is a fast, portable, lightweight interpreter for use in World Wide Web browsers and other applications that use Microsoft ActiveX controls or Java applets. An **applet** is a program that executes within another program and cannot execute directly on a computer. Applets typically run within the Web browser. See **Cool applets** for many examples of eye-catching Web applets. Figure 5-6 shows an example of one of the applets you can launch from the Cool applets Web page.

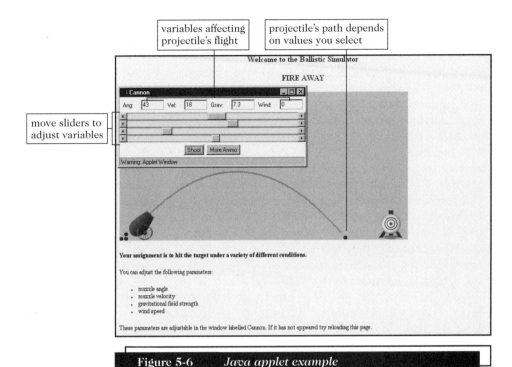

Figure 5-6   *Java applet example*

Other ways to provide Web active content may not be as well-known or as obvious to many people. These include graphics and Web browser plug-ins. Graphics files can contain embedded and invisible instructions that execute on the client computer when they are downloaded. Programs or interpreters that execute the instructions found in graphics programs and a variety of other formats can cause malicious instructions—hidden within legitimate graphics instructions—to be executed. Recall that plug-ins are programs that interpret or execute instructions embedded in downloaded graphics, sounds, and other objects. Active content, including all forms, enables Web pages to take action. Buttons on a form, for instance, can activate embedded programs to calculate and display information or to send data from the client computer to a Web server. Active content gives life to static Web pages.

By using the divide and conquer approach, a computational activity that could take 10 years to accomplish on one computer might be reduced to a few months when several computers simultaneously work on parts of the calculation. Active content, delivered by Web pages, provides an ideal way to divide up a computationally intense activity among many computers. Searching the heavens for recognizable signals is an example using this technique. Volunteers agree to allow their computers to download and work on signals that have been received from space and collected by the federal government. Idle cycles of volunteers' computers—time during which the computers are simply sitting and not executing any program—are used to filter radio waves from space to separate noise from potential signals from intelligent life. Once the pattern recognition program is downloaded, volunteers can regularly download and process data in an attempt to ascertain whether anyone out in space is attempting to communicate.

How is active content launched? You simply use your Web browser and view a Web page containing active content. The applet automatically downloads along with the page you are viewing and begins running on your computer. Therein lies the problem. Because active content modules are embedded in Web pages, they can be completely transparent to anyone browsing a page containing them. Anyone intent on doing mischief to client computers can embed malicious active content in seemingly innocuous Web pages. This delivery technique, called a Trojan horse, immediately begins executing and taking actions that cause harm. A **Trojan horse** is a program hidden inside another program or Web page that masks its true purpose. The Trojan horse could snoop around your computer and send back, to a cooperating Web server, information that is private—a secrecy violation. Worse yet, the program could alter or erase information on a client computer—an integrity violation. Zombies are equally threatening. A **zombie** is a program that secretly takes over another computer for the purpose of launching attacks on other computers. Zombie attacks cannot be traced to their creators.

Adding active content to Web pages involved in electronic commerce introduces several security risks. Malicious programs delivered quietly via Web pages could reveal credit card numbers, usernames, and passwords that are frequently stored in special files called cookies, which you learned about in Chapter 4. Because the Internet is a stateless machine and cannot remember a response from one Web page view to another, cookies help solve the problem of remembering customer order information or usernames and passwords. (**Cookie Central** is a Web site devoted to Internet cookies and everything about them.) Malicious active content delivered by means of cookies can reveal the contents of client-side files or even destroy files stored on client computers. In one recent example, a computer virus successfully examined users' e-mail address books and mailed malicious content to other people on the Internet. In this case, the evil program gained entry through electronic mail accessed through a Web browser. While cookies are not inherently bad, some people dislike the idea of storing cookies on their computers. You can accumulate large numbers of cookies as you browse the Internet, and some cookies could contain sensitive, personal information. Thankfully, there are a variety of free and shareware programs that help you identify, manage, display, and eliminate cookies. Examples are **Cookie Crusher**, which controls cookies *before* they are stored on your hard drive, and **Cookie Pal**. Figure 5-7 shows Cookie Pal's display of cookies stored on a computer. Notice that each cookie's origin (server), expiration date, and name are displayed. You can remove cookies from your computer by selecting the undesired cookie(s) and then clicking the Delete button. You can download Cookie Pal for a free 30-day trial by clicking the **Download Cookie Pal** link in the Online Companion and then clicking the appropriate "Download from…" button.

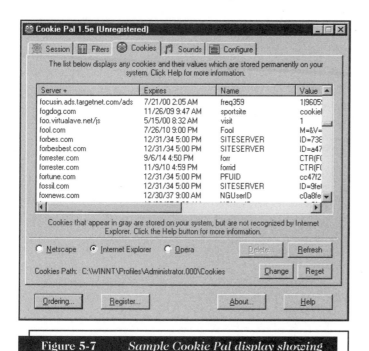

Figure 5-7   *Sample Cookie Pal display showing some cookies' contents*

Java, Java Applets, and JavaScript

Java is a high-level programming language developed by **Sun Microsystems**. Originally called OAK, Java was invented for embedded systems. Many Java supporters believe that Java code eventually will be embedded in the computer chips of household appliances to run them more intelligently. (Perhaps the toaster can wake up the coffee maker at half past seven!) Java's most popular use to date, however, is in Web pages where thousands of applets implement a wide variety of client-side applications. The applets are downloaded along with the requested Web pages and run on the client computer as long as the Web browser is Java-compatible. Both Netscape Navigator and Microsoft Internet Explorer are Java-compatible. Figure 5-8 shows Sun's Java applet page, which contains links to several applets you can purchase or receive for free.

Java is a true object-oriented language. Its object-oriented nature is desirable because it encourages code reuse (a sort of software recycling). Java can run on an operating system outside the confines of a Web application too. Another reason Java is so wildly popular is that it is platform-independent—it will run on anyone's computer. This "develop once, deploy everywhere" feature reduces development costs because only one source copy needs to be maintained for all machines.

Java adds functionality to business applications and can handle transactions and a wide variety of actions on the client computer. That relieves an otherwise very busy server-side program from handling thousands of transactions simultaneously. Once downloaded, embedded Java code can run on a client's computer. That means

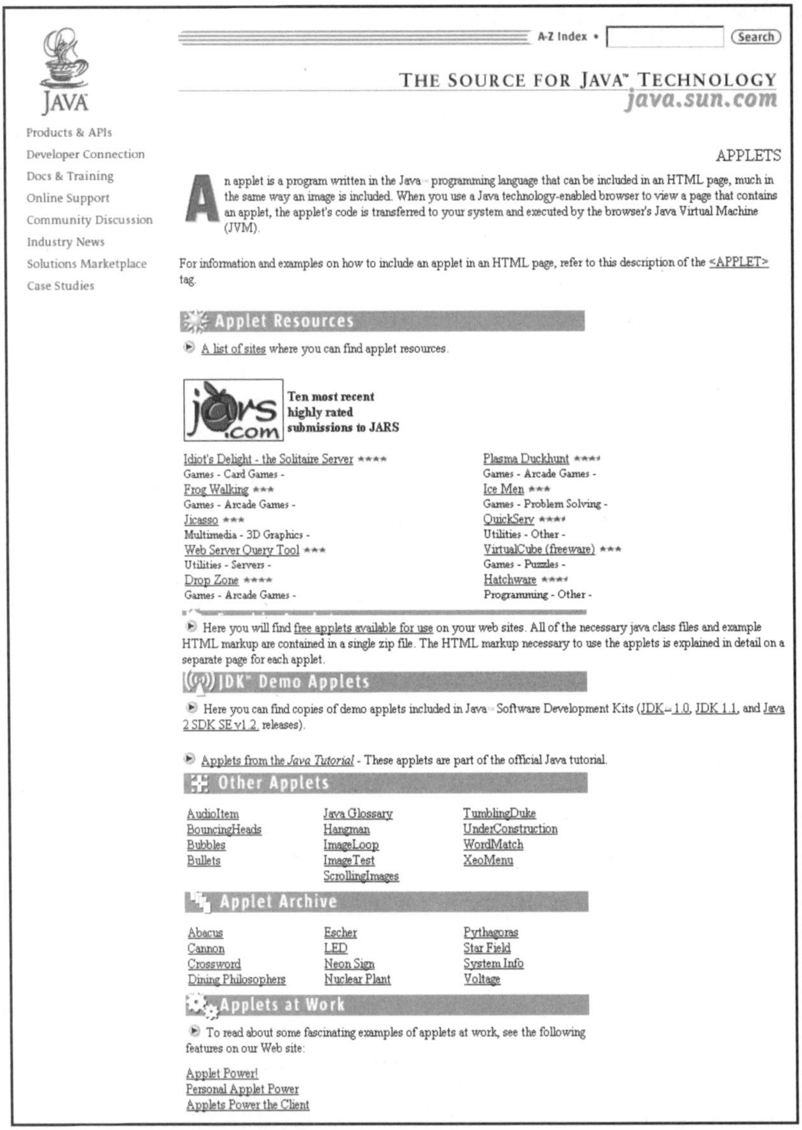

A-Z Index • [   ] (Search)

THE SOURCE FOR JAVA™ TECHNOLOGY
java.sun.com

JAVA™

Products & APIs
Developer Connection
Docs & Training
Online Support
Community Discussion
Industry News
Solutions Marketplace
Case Studies

APPLETS

An applet is a program written in the Java™ programming language that can be included in an HTML page, much in the same way an image is included. When you use a Java technology-enabled browser to view a page that contains an applet, the applet's code is transferred to your system and executed by the browser's Java Virtual Machine (JVM).

For information and examples on how to include an applet in an HTML page, refer to this description of the <APPLET> tag.

Applet Resources

▶ A list of sites where you can find applet resources.

Ten most recent highly rated submissions to JARS

Idiot's Delight - the Solitaire Server ★★★★
Games - Card Games -
Frog Walking ★★★
Games - Arcade Games -
Jicasso ★★★
Multimedia - 3D Graphics -
Web Server Query Tool ★★★
Utilities - Servers -
Drop Zone ★★★★
Games - Arcade Games -

Plasma Duckhunt ★★★★
Games - Arcade Games -
Ice Men ★★★
Games - Problem Solving -
QuickServ ★★★★
Utilities - Other -
VirtualCube (freeware) ★★★
Games - Puzzles -
Hatchware ★★★★
Programming - Other -

▶ Here you will find free applets available for use on your web sites. All of the necessary java class files and example HTML markup are contained in a single zip file. The HTML markup necessary to use the applets is explained in detail on a separate page for each applet.

JDK™ Demo Applets

▶ Here you can find copies of demo applets included in Java™ Software Development Kits (JDK™ 1.0, JDK 1.1, and Java 2 SDK SE v1.2, releases).

▶ Applets from the *Java Tutorial* - These applets are part of the official Java tutorial.

Other Applets

AudioItem
BouncingHeads
Bubbles
Bullets

Java Glossary
Hangman
ImageLoop
ImageTest
ScrollingImages

TumblingDuke
UnderConstruction
WordMatch
XeoMenu

Applet Archive

Abacus
Cannon
Crossword
Dining Philosophers

Escher
LED
Neon Sign
Nuclear Plant

Pythagoras
Star Field
System Info
Voltage

Applets at Work

▶ To read about some fascinating examples of applets at work, see the following features on our Web site:

Applet Power!
Personal Applet Power
Applets Power the Client

Figure 5-8   Sun's *Java* applet page

that there is the very real possibility that security violations can occur. To counter this possibility, a special security model, called the Java sandbox, has been developed. Details of how the sandbox implements security are beyond the scope of this textbook. Generally, though, the **Java sandbox** confines Java applet actions to a set of rules defined by the security model. These rules apply to all untrusted Java applets. **Untrusted applets** are Java applets that have not been proven to be secure. When Java applets are run within the constraints of the security sandbox, they do not have access to security-compromising code in the system. For example, Java applets obeying the sandbox rules cannot perform file input, output, or delete operations. This prevents secrecy (disclosure) and integrity (deletion or modification) violations. (See **Java sandbox** in the Online Companion.) Java application programs, unlike Java applets, run outside a Web browser and can perform *any* action on your computer—including malevolent acts.

Java applets that are loaded from a local file system are trusted. They do not run within the constraints of the Java applet sandbox. **Trusted applets** have full access to system resources on the client computer. They are "trusted" to not do any damage. **Signed Java applets** contain embedded digital signatures from a trusted third party, which are proof of the identity of the source of the applet. If the applet is signed, then it can be let "out of the sandbox" to use the full system resources. The notion is that if you know who produced the applet and you trust that party, then you have recourse if the applet actually causes damage. Theoretically, malevolent applets are always produced anonymously. See **Security and signed applets** in the Online Companion for further details.

JavaScript is a scripting language developed by Netscape to enable Web page designers to build active content. Supported by popular browsers, JavaScript shares many of the structures of the full Java language. When you download a Web page with embedded JavaScript code, it executes on your (client) computer. Like other active content vehicles, JavaScript can invoke privacy and integrity attacks by executing code that destroys your hard disk, disclosing the e-mail stored in your mailboxes, or sending sensitive information to a particular Web server on the Internet. Other secrecy violations could include recording the URLs of Web pages you visit and capturing information you enter into any CGI Web forms. For example, if you enter credit card numbers while trying to reserve a rental car, a malevolent JavaScript program could copy the credit card number. Because such a violation occurs on the client computer, it doesn't matter whether a secure communications link is established to a commerce server. The culprit is on the client machine—your system—which is outside the security perimeter of the Web site. JavaScript programs, unlike Java applets, do not operate under the restrictions of the Java sandbox security model.

In 1999, *The New York Times* revealed that RealNetworks, the maker of the very popular RealPlayer multimedia Web browser plug-in, had been surreptitiously gathering information from its users. Easily downloaded and installed from the Internet, RealPlayer was recording user information such as the RealPlayer user's name, e-mail address, country, ZIP code, computer operating system, and a number of other details. RealPlayer used the Internet connection to send back to RealNetworks the information that it had gathered. Computer expert Richard Smith (the person credited with tracking down the Melissa virus author) discovered the RealNetworks privacy violation. Soon after the discovery, and after a great deal of public embarrassment, RealNetworks issued a statement that a software **patch**, a small piece of code designed to correct a software bug, was available for all current users. The patch would prevent the RealNetwork software from collecting and transmitting user information.

A JavaScript program cannot commence execution on its own, unlike Java programs or Java applets. To launch an ill-intentioned JavaScript program, you must somehow explicitly start the program yourself. A simple way for malicious intenders to accomplish this is to disguise the nasty program as a retirement calculator whose button you must click to see (supposedly) your retirement income potential. Once you click the button, the JavaScript program launches and does its evil deed—another example of a Trojan horse.

ActiveX Controls

ActiveX is an object, called a control, that contains (programmers say "encapsulates") programs and properties that Web designers place on Web pages to perform particular tasks. ActiveX components spring from many programming languages, such as C++ or Visual Basic. However, unlike Java or JavaScript code, ActiveX controls run only on computers running Windows (95, 98, NT, or 2000) and only on browsers that support them. Once ActiveX code is completed, programmers put it inside an ActiveX envelope, compile the control, and place it on a Web page. When a Web browser downloads a Web page containing an embedded ActiveX control, the control is executed on the client computer. Shockwave, an animation and entertainment control plug-in for Web browsers, is available as an ActiveX control. Other examples include Web-enabled calendar controls and a tremendous number of Web games. See **ActiveX controls library** in the Online Companion for a comprehensive list of ActiveX controls.

The security danger with ActiveX controls is that once they are downloaded, they execute like any other program on your machine. They have full access to all system resources, including operating system code. This has very dangerous implications. A malevolent ActiveX control could reformat your hard disk, send out e-mail to all the people listed in your address book, or simply shut down your computer. Because ActiveX controls have full access to your computer, they can cause secrecy, integrity, or necessity violations. The actions of ActiveX controls cannot be halted once they begin execution. They can be managed, however. In the section called "Securing Client Computers," you will learn how to protect your computer by categorizing the ActiveX controls into trusted and untrusted groups. If you set up your browser security features just right (as described in Chapter 6), the browser will signal when you are about to download an ActiveX control. Figure 5-9 shows an example of the warning issued when Internet Explorer detects an ActiveX control.

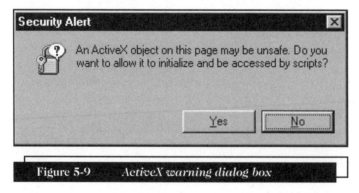

Figure 5-9   *ActiveX warning dialog box*

Graphics, Plug-Ins, and E-Mail Attachments

We mentioned earlier that graphics, browser plug-ins, and e-mail attachments can harbor executable content. Some graphics file formats have been specifically designed to contain instructions on how to render a graphic. That means that any Web page containing such a graphic could be a potential threat, because the code embedded in the graphic could cause harm to your computer. Similarly, browser plug-ins, which are programs that enhance the capabilities of browsers, handle Web content that a browser cannot handle. Plug-ins are normally beneficial and perform tasks for a browser, such as playing audio clips, displaying movies, or animating graphics. QuickTime, for example, is a plug-in that downloads and plays movies that are stored in a special format.

Many plug-ins perform their duties by executing commands buried within the media they are manipulating. This opens the door to the possibility that someone intent on doing harm to your computer could embed commands within a seemingly innocuous video or audio clip. The ill-intentioned commands hidden within the object that the plug-in is interpreting could damage your computer by erasing some (or all) of your files. Figure 5-10 shows **Netscape's Plug-ins** page, which contains category listings of plug-in software it offers for free downloading.

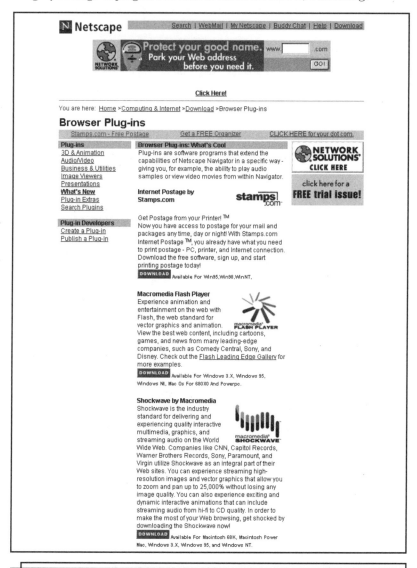

Figure 5-10   *Netscape's Plug-ins page*

The potential dangers lurking in e-mail attachments get a lot of news coverage and are the most familiar to the general population. E-mail attachments provide a convenient way to send nontext information over a text-only system—electronic mail. Attachments can contain word processing files, spreadsheets, databases, images, or virtually any other information you can imagine. When you receive attachments, most programs, including the popular browser e-mail programs, display the attachment by automatically executing an associated program; for example, the recipient's Excel program detaches an attached Excel workbook and displays it, or Word opens and displays a Word document. While this activity doesn't inherently cause damage, Word and Excel macro viruses *inside* the loaded documents and workbooks can damage your computer and reveal otherwise confidential information. A **virus** is software that attaches itself to another program and can cause damage when the host program is activated. **Worm** viruses, such as the Internet Worm described in this chapter's introduction, replicate themselves on other machines. Worms can spread quickly through the Internet. A **macro virus** is a type of virus that is coded as a small program, called a macro, and is embedded in a file. You have probably read about recent examples of e-mail attachment-borne virus attacks, including Happy99 Worm, Triplicate, and Chernobyl. **Symantec**, among other companies, keeps tabs on viruses and provides antivirus software.

E-mail attachments containing viruses and other malicious software are reported daily. Some of the most famous in recent years include the "ILOVEYOU" virus, also known as the "love bug," and its variants containing Visual Basic Script (VBScript) code. (**VBScript** is a programming language that runs on most microcomputers.) The "ILOVEYOU" virus has been attributed to a 23-year-old Filipino computer student. The virus spread through the Internet with amazing speed as an e-mail message, infecting the computer of anyone who opened the e-mail attachment and clogging e-mail systems with thousands of copies of the useless e-mail message. The virus spread quickly because it automatically sent itself to as many as 300 addresses stored in a computer's Microsoft Outlook address book. Besides replicating itself explosively through e-mail, the virus caused other harm, destroying digital music and photo files stored on the target computers. The "ILOVEYOU" virus also searched for other users' passwords and forwarded that information to the original perpetrator. Within days, the virus spread to more than 20 countries and caused an estimated $7 billion to $10 billion in damages—most of it in lost worker productivity.

In another case, a virus was sent as an e-mail attachment disguised as a text file attachment. It spread widely across the Internet. A subject line containing the words "Funny" or "Life-Stages" was the tip-off that the message might contain the virus. Written as a VBScript program, the virus clogged e-mail servers because of its ability to replicate and send itself via e-mail to others. The virus has been classified as a worm because of its ability to move through the Internet on its own. It did not damage any files or corrupt data, but it cost a great deal in lost productivity and labor charges needed to unclog e-mail servers and reinitialize them. CERT, described later in this chapter, reported that individual users received as many as 30 copies of the virus. Some system managers reported as many as 120,000 copies of the e-mail message and attached virus passing through individual mail servers. Click **Virus Information** for a comprehensive Web site containing descriptions of thousands of viruses.

The term **steganography** describes information (a command, for example) that is hidden within another piece of information. This information can be used for good or for bad purposes. Frequently, computer files contain redundant or insignificant information that can be replaced with other information. The latter resides in the background and is virtually undetectable. Steganography provides a way of hiding an encrypted file within another file so that a casual observer cannot detect that there is anything of importance in the containing file. Encrypting a file protects it from being read, and steganography makes it invisible. Steganography is comparable to hiding an encrypted microdot containing trade secrets by gluing it to the pupil of a person's eye in a portrait. It is simultaneously hidden and secured. The casual viewer sees a person. Closer inspection reveals a microdot. Several vendors provide software that implements **steganography**.

Communication Channel Threats

The Internet serves as the electronic chain linking a consumer (client) to an electronic commerce resource (commerce server). Now that you understand security threats to client machines, the next asset in the chain to consider is the communication channel connecting clients to servers—namely, the Internet.

The Internet is not at all secure. Though the Internet has its roots in a military network, that Defense Advanced Research Projects Agency (DARPA) network was built only to provide redundancy, not secure communications, in case one or more communications lines were cut. In other words, its original design goal was to provide several alternative paths on which to send critical military information. The military clearly planned to send sensitive information in an encrypted form so that any messages traveling over the network would remain secret and tamperproof. However, the security of messages traversing the network was provided by software that converted the messages into unintelligible strings of characters called cipher text. Today, the Internet remains unchanged from its original, insecure state. Messages on the Internet travel a random path from a source node to a destination node. A message passes through a number of intermediate computers on the network before reaching its final destination, and the path can vary each time a message is sent between the same source and destination points. It is impossible to guarantee that every computer on the Internet through which messages pass is safe, secure, and friendly. For all you know, a message sent from a merchant in Manchester, England, to a supplier in Cairo, Egypt, may have passed through a competitor's computer located in Beirut, Lebanon. Because you cannot control the path and do not know where your message packets have been, it is quite possible that some intermediary can read your messages, alter your messages, or even completely eliminate your messages from the Internet. That is, any e-mail message traveling on the Internet is subject to secrecy, integrity, and necessity violations. This section describes those problems in more detail. Chapter 6 describes several ways to defeat the security problems presented in this chapter.

This section discusses Internet channel security threats within the classifications of secrecy, integrity, and necessity. This organization provides a good structure for examining direct security threats to the Internet network itself.

Secrecy Threats

Secrecy is one of the highest profile security threats mentioned in articles and the popular media. Closely linked to secrecy is privacy, which receives a great deal of attention also. Hardly a day passes in which one doesn't read about a privacy infringement. Secrecy and privacy, though similar, are different issues. Secrecy is the prevention of unauthorized information disclosure. **Privacy** is the protection of individual rights to nondisclosure. (The **Privacy Council**, which helps businesses implement smart privacy and data practices, has created an extensive Web site surrounding privacy—covering both business and legal issues. Figure 5-11 shows its Privacy Library page for the United States.) Secrecy is a technical issue requiring sophisticated physical and logical mechanisms, whereas privacy protection is a legal matter. A classic example of the difference between secrecy and privacy is e-mail. A company may protect its e-mail messages against secrecy violations by using encryption (see Chapter 6). In encryption, a message is encoded into an unintelligible form that only the proper recipient can transcribe back into the original message. Secrecy countermeasures protect outgoing messages. E-mail privacy issues swirl around whether company supervisors should be permitted to randomly read employees' messages. Disputes in this area center around who owns the e-mail messages: the company or the employee who sent them. The focus in this section is on secrecy—keeping the bad guys from reading information they should not be reading.

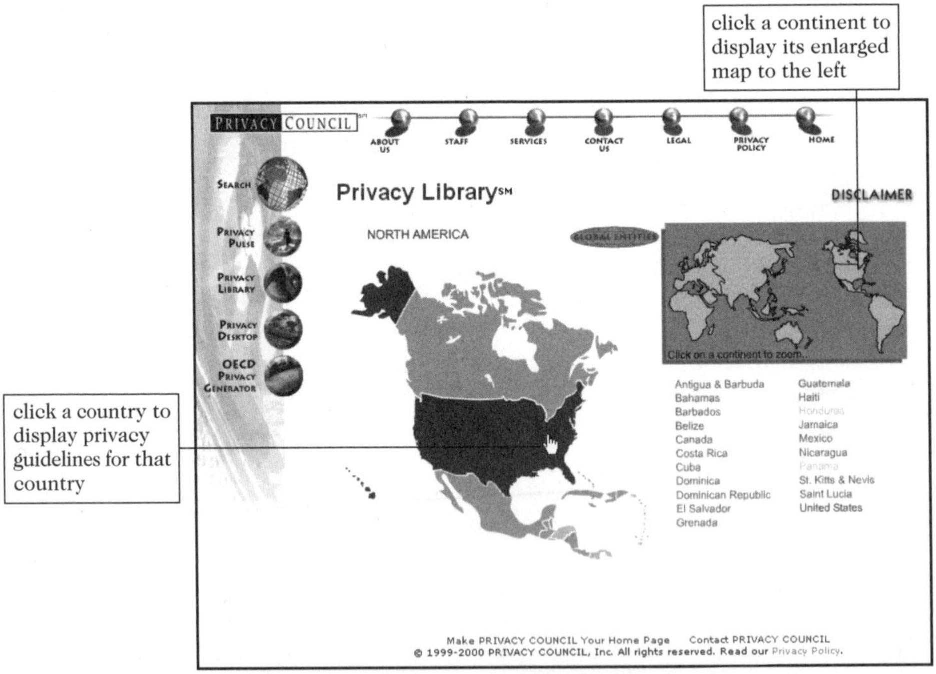

click a continent to display its enlarged map to the left

click a country to display privacy guidelines for that country

Figure 5-11   *Privacy Council's Privacy Library page*

You have already learned that a significant danger of conducting electronic commerce is theft of sensitive or personal information, including credit card numbers, names, addresses, and personal preferences. This can occur any time anyone fills out a form or submits credit card information over the Internet, because it is not difficult for an ill-intentioned person to record information packets (a secrecy violation) from the Internet for later examination. The same problems can occur in e-mail transmissions. Special software applications called **sniffer programs** provide the means to tap into the Internet and record information that passes through a particular computer (router) while traveling from its source to its destination. Using sniffer program is analogous to tapping a telephone line and recording a conversation. Sniffer programs can read e-mail messages as well as electronic commerce information.

Security experts periodically find electronic holes, called backdoors, in electronic commerce software. These can be left open accidentally by the software developer, or they can be left open intentionally. A **backdoor** allows anyone with knowledge of the existence of a backdoor or a system password to cause damage by observing transactions, deleting data, or stealing data. McMurtrey/Whitaker & Associates, owner of the Cart32 electronic commerce shopping cart software, recently confirmed that its Cart32 software contained a backdoor through which credit card numbers could be obtained by anyone with a backdoor password. The company quickly supplied a software fix, or patch, to eliminate the backdoor. Though the backdoor was a software mistake and not one created intentionally by a disgruntled employee, the consequences of a backdoor attack can be disastrous if credit card numbers are compromised, regardless of the origin of a backdoor.

Credit card number theft is an obvious problem, but proprietary corporate product information or prerelease data sheets mailed to corporate branches can be intercepted and passed along easily too. Often, corporate confidential information is even more valuable than a few credit cards, which usually have spending limits. Purloined corporate information can be worth millions of dollars.

Breaching secrecy on the Internet is not difficult. Here's an example of how you can inadvertently leak confidential information that an eavesdropper or another Web site server can retrieve afterwards. Suppose you log on to a Web site called www.anybiz.com that contains a form with text boxes for your name, address, and e-mail address. When you fill out those text boxes and click the submit button, the information is sent to the Web server for processing. One popular method of transmitting your data to a Web server is to collect your text box responses and place them at the end of the target server's URL (address). The captured data and the HTTP request to send the data to the server are then sent. So far, no violations have occurred. Suppose you change your mind, decide not to wait for a response from the anybiz.com server to which you have sent your information, and jump to another Web site, www.somecompany.com, instead. The server somecompany.com may choose to collect Web demographics and log the URL from which you just came. This can help the site manager determine how electronic commerce traffic has reached the site. By recording the previous anybiz.com URL, somecompany.com has breached secrecy by recording the confidential information you recently entered. This doesn't always occur, but it *can* occur.

You are continually revealing information about yourself when you use the Web. This includes your IP address (Internet address) and the browser you are using. This, too, is an example of a secrecy breach. There is at least one Web site that offers an "anonymous browser" service that hides personal information from sites that you visit. Called **Anonymizer**, the Web site provides a measure of secrecy as long as you use the site as your portal, which is the beginning site from which you visit other sites. Anonymizer acts as a firewall (see Chapter 6) and shields private information from leaking out. It does this by placing the Anonymizer address on the front end of any URL addresses you visit. This shield reveals to other Web sites information only about the Anonymizer Web site, not about you. For example, if you visit Amazon.com, Anonymizer would present this URL: http://www.anonymizer.com:8080/http:// www.amazon.com. Figure 5-12 shows Anonymizer's home page with a request to anonymously surf to Amazon.com's home page.

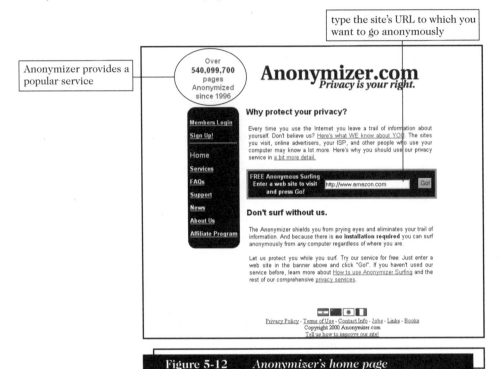

Figure 5-12   *Anonymizer's home page*

You can try out the Anonymizer service for free (the free version has a slight built-in delay). Click **Anonymizer** in the Online Companion, type a URL in the textbox (see Figure 5-12), and click the "Go" button. Then, click the "Surf for FREE" button on the next page that Anonymizer displays.

Integrity Threats

An integrity threat, also known as **active wiretapping**, exists when an unauthorized party can alter a message stream of information. Unprotected banking transactions, such as deposit amounts transmitted over the Internet, are subject to integrity violations. Of course, an integrity violation implies a secrecy violation, because an intruder who alters information can read and interpret that information. Unlike secrecy threats, where a viewer simply sees information he or she should not, integrity threats can cause a change in the actions a person or corporation takes, because a mission-critical transmission has been altered.

Cyber vandalism is an example of an integrity violation. **Cyber vandalism** is the electronic defacing of an existing Web site's page. The electronic equivalent of destroying property or placing graffiti on objects, cyber vandalism occurs whenever someone replaces a Web site's regular content with his or her own. Recently, there have been several reports of Web page defacing in which vandals have replaced business content with pornographic material and other offensive content.

Masquerading or **spoofing**—pretending to be someone you are not or representing a Web site as an original when it really is a fake—is one means of creating havoc on Web sites. Using a security hole in a Domain Name Server (DNS), perpetrators can substitute the address of their Web site in place of the real one to spoof Web site visitors. For instance, a hacker could create a fictitious Web site masquerading as www.widgetsinternational.com by exploiting a DNS security hole that substitutes his or her fake IP address for Widgets International's real IP address. All subsequent visits to Widgets International would be redirected to the fake Web site. There, the hacker could alter any orders to change the number of widgets ordered and redirect shipment of those products to another address. The integrity attack consists of altering an order and passing it to the real company's commerce server. The commerce server is unaware of the integrity attack and simply verifies the consumer's credit card number and passes the order on for fulfillment.

Integrity threats can alter vital financial, medical, or military information. You can imagine the impact if someone were to capture a message that credits a bank for $10 million dollars, and the credit then changes to a debit. Similarly, an integrity violation that alters an e-mailed résumé can affect the applicant's chances of being hired by the receiving company. Information alteration can have very serious consequences for businesses and people.

Necessity Threats

The purpose of a **necessity threat**, also known by other names such as a **delay, denial**, or **denial-of-service threat** (**DOS**), is to disrupt normal computer processing or to deny processing entirely. A computer that has experienced a necessity threat slows processing to an intolerably slow speed. For example, if the processing speed of a single ATM transaction slows from one or two seconds to 30 seconds, users will abandon ATMs entirely. Similarly, slowing any Internet service will drive customers to competitors' Web or commerce sites—possibly discouraging them from ever returning to the original commerce site. (Someone once said that an *Internet hour* is equivalent to 10 seconds or so of real time.) In other words, slowing processing can render a service unusable or unattractive. Clearly, an online newspaper that reports three-day-old news is worth very little to just about everyone.

Denial-of-service attacks remove information altogether or delete information from a transmission or file. One documented denial attack caused selected PCs that have Quicken, an accounting program, installed on every computer to divert money to a different bank account. The denial attack denied money from its rightful owners. In another famous denial-of-service attack in early 2000, a rash of attacks occurred against high-profile electronic commerce sites such as eBay. The attackers used zombie computers to send a flood of data packets to overwhelm various electronic commerce servers and choke access to them by legitimate customers. Prior to the attack, perpetrators located vulnerable computers and loaded them with the software that attacked the commerce sites. The Robert Morris Internet Worm attack is another famous example of a denial-of-service attack.

Server Threats

The server is the third link in the client-Internet-server trio embodying the electronic commerce path between the user and a commerce server. Servers have vulnerabilities that can be exploited by anyone determined to cause destruction or to acquire information illegally. One entry point is the Web server and its software. Other entry points are any backend programs containing data, such as a database and server on which it runs. Perhaps the most dangerous entry points are Common Gateway Interface (CGI) programs or utility programs residing on the server. While no system is completely safe, the commerce server administrator's job is to make sure that security policies are documented and considered in every part of the electronic commerce system.

Web Server Threats

Web server software, such as the software described in Chapter 3, is designed to deliver Web pages by responding to HTTP requests. While Web server software isn't inherently high-risk, it has been designed with Web service and convenience as the main design goals. The more complex the software is, the higher the probability that it contains coding errors (bugs) and that it contains security weaknesses that provide openings through which evil-doers can enter.

Web servers running on most machines, including UNIX-based computers, can be set up to run at various privilege levels. The highest privilege level provides the most flexibility and allows programs, including Web servers, to execute all machine instructions and to have unlimited access to any part of the system, including highly sensitive and privileged areas. Correspondingly, the lowest privilege levels provide a logical fence around an executing program, preventing it from running whole classes of machine instructions and disallowing it access to all but the least sensitive areas of computer storage. In security circles, the rule is to provide a program the lowest privilege level it needs to do its job. Clearly, a system administrator who sets up accounts and passwords for users needs, momentarily, a very high privilege level—called *superuser* in the UNIX world—to modify sensitive and valuable areas of the system. Setting up a Web server to run in high-privilege status can lead to a Web server threat. Most of the time, a Web server provides ordinary services and mundane tasks that can be accomplished with a very low privilege level. If a Web server runs at a high privilege level, a malevolent person trying to exploit a Web server may be able to do so and subsequently execute instructions in privileged mode.

A Web server can compromise secrecy if it keeps the default setting of automatic directory listings selected. The secrecy violation occurs when the contents of a server's folder names are revealed to a Web browser. This frequently happens and is caused when you enter a URL such as http://www.somecompany.com/FAQ/ and expect to see the default page in the FAQ directory. The default Web page that the server normally displays is named index.htm or index.html. If that file is not in the directory, the Web server frequently will display all the folder names in the directory. Then, you can click folder names at random and visit folders that might otherwise be off limits. Figure 5-13 shows an example of displayed folder names.

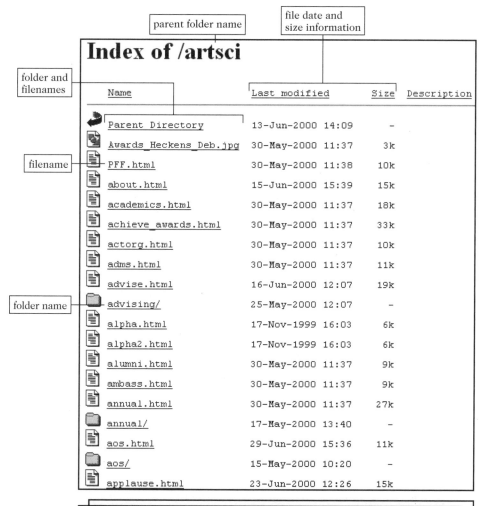

parent folder name

file date and size information

Index of /artsci

folder and filenames

Name	Last modified	Size	Description
Parent Directory	13-Jun-2000 14:09	-	
Awards_Heckens_Deb.jpg	30-May-2000 11:37	3k	
PFF.html	30-May-2000 11:38	10k	
about.html	15-Jun-2000 15:39	15k	
academics.html	30-May-2000 11:37	18k	
achieve_awards.html	30-May-2000 11:37	33k	
actorg.html	30-May-2000 11:37	10k	
adms.html	30-May-2000 11:37	11k	
advise.html	16-Jun-2000 12:07	19k	
advising/	25-May-2000 12:07	-	
alpha.html	17-Nov-1999 16:03	6k	
alpha2.html	17-Nov-1999 16:03	6k	
alumni.html	30-May-2000 11:37	9k	
ambass.html	30-May-2000 11:37	9k	
annual.html	30-May-2000 11:37	27k	
annual/	17-May-2000 13:40	-	
aos.html	29-Jun-2000 15:36	11k	
aos/	15-May-2000 10:20	-	
applause.html	23-Jun-2000 12:26	15k	

filename

folder name

183

Figure 5-13 *Displaying folder and filenames with a Web browser*

Administrators of other sites, such as Microsoft's, are careful to turn off the folder name display feature. If you attempt to browse a folder where protections prevent browsing, the Web server issues a warning message, such as "You cannot browse this directory."

Web servers can compromise security by requiring you to enter a username and password. The act of entering the username to allow admittance to a particular part of the Web space is not in itself a secrecy or privacy violation. However, the confidential username and password may be subsequently revealed when you visit multiple pages within the same Web server's protected or premium content area. The reason this can occur is that some servers require that you reestablish your username and password for each page you visit in the premium content area because the Web is stateless—it cannot remember what happened during the last transaction. The most convenient way to remember a username and password is to store the user's confidential information in a cookie on his or her computer. That way, the Web server can request confirmation of the data by requesting that the computer send a cookie. Trouble occurs because a cookie's information might be transmitted in an insecure way and copied by an eavesdropper. While cookies are not inherently unsafe, a Web server must not request that a cookie be transmitted unprotected.

A Server Side Include (SSI) is a small program embedded in a Web page that is executed by the server (sometimes called a servlet). (SSIs were described in Chapter 3.) Any time a program is executed on a server and the program comes from an unknown or untrusted source—a user's Web page, for example—there is always the possibility that the SSI may request some execution that is illegal. The embedded SSI code could be an operating system directive that requests that the password file be displayed or sent back to a particular location. The **W3C Threat Document** (listed in the Online Companion) provides additional information and answers frequently asked questions about server security.

The File Transfer Protocol (FTP) program, described in Chapter 2, while not a Web server, is part of the complete Web server package. The FTP program can reveal threats to the Web server's integrity. One possibility for unauthorized information disclosure occurs when there are no protection mechanisms on the folders that an FTP user can browse. For instance, suppose a regular business client has an account on another business's computer and can periodically upload data to the business partner's computer. Using an FTP client program, a system administrator can log on to the business partner's computer, upload data, and then proceed to open and display the contents of other folders on the Web server computer. This is not difficult to do if protections are missing. With a Web client program, you can double-click the parent directory folder to move up the folder hierarchy, double-click some other folder, such as the other company's privileged folder, and then download any information you see there. This security breach is possible simply because the company has forgotten to restrict its business partner's browsing capabilities to a single folder.

One of the most sensitive files on a Web server, if it exists, holds Web server username and password pairs. If that file is compromised, anyone can electronically break into privileged areas while masquerading as someone else. A masquerader can obtain usernames and passwords if that information about users is readily available and not encrypted. Most Web servers provide secure storage of user authentication information. But it is up to the Web server administrator to ensure that the Web server is instructed to always apply protection mechanisms to the data. (Chapter 6 describes protection mechanisms for authentication information.)

The passwords that users select can be a threat. Users sometimes select passwords that are easily guessed because they are their mother's maiden name, the name of one of their children, a telephone number, or some easily obtained identification number such as a social security number. So-called dictionary attack programs cycle through an electronic dictionary, trying every word in the book as a password. There are simple remedies for this attack, but users' passwords, once broken, may provide a covert opening for illegal entry into a server that can remain undetected for a very long time.

Database Threats

Electronic commerce systems store user data and retrieve product information from databases connected to the Web server. Besides storing product information, databases connected to the Web contain valuable and private information that could irreparably damage a company if it were disclosed or altered. Most modern, large-scale database systems use extensive database security features that rely on usernames and passwords.

Once a user is authenticated to a database, selected portions of the database are visible to that user. Security is enforced in databases through the use of privileges, which are stored in the database. However, some databases either store username/password pairs in a nonsecure way, or they fail to enforce security altogether and rely on the Web server to enforce security. If someone obtains user authentication information, then he or she can masquerade as a legitimate database user and reveal or download private and valuable information. Trojan horse programs hidden within the database system can also reveal information by downgrading the system (releasing sensitive information to a less protected area of the database that a broader population of users can peruse). When information is downgraded, all users have access—including potential intruders.

A large number of papers and Web sites discuss database security at length. Examine links in the Online Companion for examples of database security considerations. The **Database threats resource center** link in the Online Companion describes threats specific to database systems and contains a large collection of white papers discussing security solutions. Database security requires careful attention from a good database administrator. Figure 5-14 shows a Web page describing Oracle's database security features. (Oracle powers many of the large, well-known electronic commerce sites, including Amazon.com, Disney Store Online, eBay, and E*Trade.)

Introduction to Oracle Advanced Security

This chapter introduces the Oracle Advanced Security option encryption, **checksumming**, and authentication features. These features are available to network products using Net8, including Oracle8i, Designer 2000, Developer 2000, and any other Oracle or third-party products that support Net8.

Topics covered in this chapter:

- About the Oracle Advanced Security Option

- Architecture of the Oracle Advanced Security Option

- Secure Data Transfer Across Network Protocol Boundaries

- System Requirements

- Oracle Configuration for Network Authentication

- Oracle Products Not Yet Supported

About the Oracle Advanced Security Option

The Oracle Advanced Security option (formerly Secure Network Services and Oracle Advanced Networking Option) provides a comprehensive suite of security features to protect enterprise networks and securely extend corporate networks to the Internet. The Oracle Advanced Security option provides a single source of integration with network encryption and authentication solutions, single sign-on services, and security protocols. By integrating industry standards, it delivers unparalleled security to the Oracle network and beyond.

Network Security in a Distributed Environment

Figure 5-14   *Oracle security features page*

Common Gateway Interface Threats

Recall that a Common Gateway Interface (CGI) implements the transfer of information from a Web server to another program, such as a database program. CGIs and the programs to which they transfer data provide active content to Web pages. For example, a Web page might contain a list box asking you to fill in the name of your favorite professional sports team. Once you submit your choice, CGI programs process the information and look up the latest scores for the sports team you designated, place the scores into a Web page, and send the generated page back to your client browser.

Because CGIs are programs, they present a security threat if misused. Just like Web servers, CGI scripts can be set up to run with their privileges set to high—unconstrained. Defective or malicious CGIs with free access to system resources are capable of disabling the system, calling privileged (and dangerous) base system programs that delete files, or viewing confidential customer information, including usernames and passwords. When programmers discover inadequacies or errors in CGI programs, they rewrite and replace them. Older, retired CGIs that aren't erased can provide openings into the system that have been long forgotten by the systems development staff. And, because CGI programs or scripts can reside just about anywhere on the Web server (that is, in any folder or directory), they are hard to track down and manage. However, anyone who is determined enough can track down replaced CGI scripts, examine them, ascertain their weaknesses, and exploit those weaknesses to gain access to a Web server and its resources. Unlike JavaScript, CGI scripts do not run inside a protective security perimeter, or sandbox.

Other Programming Threats

Another serious Web server attack can come from programs executed by the server. Java or C++ programs that are passed to Web servers by a client or that reside on a server frequently make use of a buffer. A **buffer** is an area of memory set aside to hold data read from a file or database. A buffer is necessary whenever any input or output operation takes place, because a computer can process file information much faster than the information can be read from input devices or written to output devices. So a buffer serves as a "landing zone" for incoming or outgoing data. Database information about to be processed, for example, is gathered in a buffer so that either the entire collection or a large quantity of it is in the computer's memory. Then, the data are made available to the processing unit for manipulation and analysis. The problem with buffers is that programs filling them can go awry and overfill the buffer, spilling the excess data outside the designated buffer memory area. Usually, this occurs because the program contains an error or bug that causes the overflow. Sometimes, however, the mistake is intentional. In either case, buffer overflows can have moderate to very serious security consequences.

Anyone who has programmed probably has experienced the consequences of a buffer overflow or a runaway code segment that causes data or instructions to overwrite an out-of-bounds area in memory. The normal result of such a programming error is that the program halts with an exception—an error condition—and the process stops. Occasionally, the entire computer (PC or mainframe) halts (crashes). Intentional crashes caused by malevolent code are deliberate denial attacks. The Internet Worm attack was, in part, such a program. It caused an overflow condition that eventually consumed all resources until the host machine could no longer function.

A somewhat more insidious version of a buffer overflow attack writes *instructions* into critical memory locations so that when the intruder program has completed its work of overwriting buffers, the Web server resumes execution by loading internal registers with the address of the main attacking program's code. This type of attack can open the Web server to severe damage because the resumed program—which is now the attacker program—may regain control at a very high privilege or super user level. This opens up just about every file to disclosure and destruction by the marauding program.

Figure 5-15 shows a graphical representation of data being read from a file. The data flow into a memory buffer and then into a system area that is called the save area. A **save area** is where programs store critical information, such as the contents of all central processing unit registers and partial results of a program's computations just before control is passed to another program. When control returns to the original program, the save area contents are reloaded back into CPU registers, and control returns to the next instruction in the program. In an attack, however, control returns to the attacking program, not the benign program that gave up control. The **Buffer overflow attacks** links in the Online Companion describe details about two different Web servers' buffer vulnerabilities.

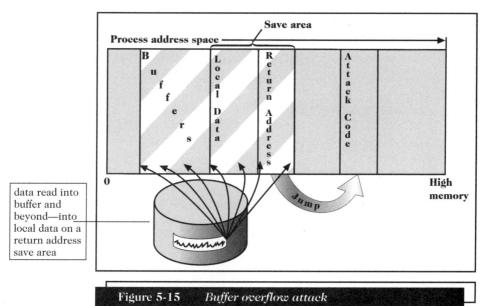

Figure 5-15   *Buffer overflow attack*

A similar attack, one in which excessive data are sent to a server, can occur on mail servers. Called a **mail bomb**, the attack occurs when hundreds or even thousands of people each send a message to a particular address. The accumulated mail received by the target of the mail bomb exceeds the allowed e-mail size limit and can cause e-mail systems to jam or malfunction. While it is fairly easy to track the people responsible for the attack, it is debilitating nonetheless. Mail bombs may seem similar to spam, but they are just the opposite. Spamming occurs when one person or organization sends a *single* message to thousands of people, and is more of a nuisance than a security threat.

CERT—COMPUTER EMERGENCY RESPONSE TEAM

Over a decade ago, a group of researchers quickly met to study and eliminate the infamous Internet Worm attack. The National Computer Security Center, part of the National Security Agency, initiated a series of meetings to figure out how to respond to future security breaks that might affect thousands of people. Soon after that meeting of security experts, DARPA created the **Computer Emergency Response Team**

(CERT®) Coordination Center. DARPA chose **Carnegie Mellon University** in Pittsburgh as the center's home. CERT members were responsible for setting up an effective and quick communications infrastructure among security experts so that future security breakouts could be avoided or quickly exterminated.

In the first 10 years of its existence, CERT has responded to over 14,000 security events and incidents occurring within the U.S. government and in the private sector. CERT continues its mission today and provides a wealth of information to help Internet users and companies building commerce sites become more aware of security risks. CERT posts **CERT alerts** to inform the Internet community at large about recent security events. The **CERT attack advisory** section of the Online Companion contains valuable information about the Domain Name Server (DNS) attack previously discussed in this chapter. Chapter 6, which describes remedies to the threats presented in this chapter, provides information about CERT-identified security risks and how to recognize and prevent them. Figure 5-16 shows a CERT alert page.

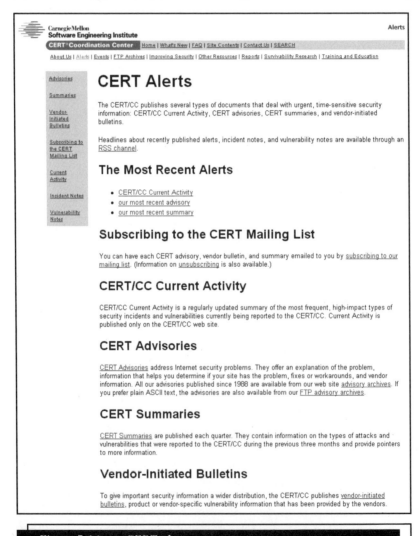

Figure 5-16   CERT alerts

Summary

In this chapter, you learned about the importance of electronic commerce security. Attacks against electronic commerce systems can disclose or manipulate proprietary information. Any commerce security policy must address secrecy, integrity, necessity, and intellectual property rights. Threats to commerce can occur anywhere in the commerce chain, beginning with a client computer and ending with the commerce and back office servers. News accounts of virus attacks have kept Web users aware of the ubiquitous security risks to client computers. More subtle threats are delivered as client-side applets. Java, JavaScript, and ActiveX controls are examples of programs and scripts that run on client machines and that have the potential to breach security.

Communication channels in general and the Internet in particular are especially vulnerable to attacks. The Internet is a vast network and because no one can control the nodes through which Internet traffic passes, there are ever-present dangers of unauthorized disclosure of private information, alteration of critical business documents, and theft and loss of important commerce messages. The Internet Worm of 1988 is a classic example of a security threat that used the Internet as a vehicle to travel around the world and infect thousands of computers within minutes.

Commerce servers are susceptible to security breaches in many of the same ways as are client machines. Worse, though, is that client security breaks can occur to any client that connects to the affected server. Common Gateway Interface (CGI) programs that run on servers have the potential to damage databases, abnormally terminate server software, and subtly change proprietary information. Attacks can come from within the server in the form of programs, or they can come from outside the server. One external attack occurs when a message overflows a server's internal storage region and overwrites crucial information. Overwritten information is replaced with either data or instructions that cause other programs on the server to execute.

CERT, the Computer Emergency Response Team, was formed to address security outbreaks by linking key scientific experts. When large security outbreaks occur, the member scientists can converge and discuss methods to locate and eliminate the electronic attacker. The stakes are high. Without adequate security safeguards in place in electronic commerce clients and servers, electronic commerce cannot last very long. Effective security policies along with adequate detection and enforcement are the only ways to safeguard electronic communications and commerce transactions across the electronic frontier.

Key Terms

Active Content	Denial Threat
Active Wiretapping	Domain Name Ownership Change
ActiveX	Eavesdropper
Applet	Fair Use
Backdoor	Integrity
Buffer	Intellectual Property
Computer Security	Java Sandbox
Copyright	Logical Security
Countermeasure	Macro Virus
Cyber Vandalism	Mail Bomb
Cybersquatting	Masquerading
Delay Threat	Name Changing
Denial of Service Threat (DOS)	Name Stealing

Necessity
Necessity Threat
Patch
Physical Security
Privacy
Ripper
Ripping
Save Area
Secrecy
Security Policy
Signed Java Applet

Sniffer Program
Spoofing
Steganography
Threat
Trojan Horse
Trusted Applet
Untrusted Applet
VBScript
Virus
Worm
Zombie

Review Questions

1. Explain why Web sites use cookies. What problem do cookies solve? What does a cookie contain? How large are cookies? Where are they stored? Use the Online Companion to help do your research.

2. What is steganography? What does steganography have to do with security? Use the Online Companion to research the question. Write at least 100 words about steganography.

3. List some of the Web server security risks. See if you can find a Web site where you can list the names of the directories on the site and print them.

4. What security risks does the Internet pose? Are the risks mostly related to secrecy, or can a message's integrity be compromised also?

5. Why are programs such as CGI scripts and Java programs that run on client machines or on a Web server considered security threats? Explain, in a general way, how programs could breach security. Do Java script programs pose an equally serious security risk?

Exercises

1. Brought Back Bugs is a used Volkswagen dealer in Lincoln, Nebraska. The dealership has hired you to create a Web site for it. One of its requirements is that the site should display a few banner advertisements showing the week's specials. You decide that active content is what's needed to provide the rotating advertisement. Your immediate task is to investigate Java, JavaScript, Jscript, and Java applets—you need to learn everything there is to know about Java-related alternatives. Using the Online Companion and Web search engines, research Java applications and write a two-page paper about what you have discovered about Java, JavaScript, and the other Java namesakes. You might want to start by figuring out the history of Java. Who invented it? What are the advantages and disadvantages of Java over ActiveX controls? Who publishes JavaScript and Jscript? Are electronic commerce customers restricted to a particular browser to run Java-enhanced Web pages? Be sure to list the URLs for any Web sites that you visit and that are particularly helpful in your search.

2. Suppose you suspect that your commerce server may be under attack from at least one malevolent source. Describe in a general way what types of threats are possible on a commerce server. Consult outside sources on the Internet to help you answer this question. What are the three categories of security threats and how might they apply to your organization's Web server? Answer these two separate questions in 100–200 words.

3. Write a 300-word paper about the CERT organization. Include information about when it was founded, what groups or people are members, and where it is headquartered. Include in your discussion at least three current security alerts, specifying the name of the virus or attack program, the date the alert was posted, and two sentences about each reported security alert. Use the Online Companion, any of several Internet search engines, and the CERT home page to help you locate information. Use at least one of the sources in the references at the end of this chapter and include a citation to that reference.

For Further Study and Research

Ahuja, V. 1997. *Secure Commerce on the Internet.* Boston: AP Professional.

Alexander, S. 2000. "Viruses, Worms, Trojan Horses and Zombies," *Computerworld*, 34(18), May 1, 74.

Applegate, L. et. al. 1996. "Electronic Commerce: Building Blocks of New Business Opportunity," *Journal of Organizational Computing and Electronic Commerce*, 6(1), June, 1–10.

Bhimani, A. 1996. "Securing the Commercial Internet," *Communications of the ACM*, 39(6), June, 29–35.

Ferguson, P. and D. Seine. 1998. "Network Ingress Filtering: Defeating Denial of Service Attacks Which Employ IP Source Address Spoofing," *Network Working Group, RFC 2267*, January.

Gardner, D. 1998. "E-mail Bug Stirs Up a Scare," *Infoworld*, August 3, 20.

Harrison, A. 2000. "Denial-of-Service Victims Share Lessons Learned," *Computerworld*, 34(25), June 19, 8.

Isenberg, D. 2000. "Many Trademarks, But Just One Domain Name," *Internet World*, July 1, 86.

Kerstetter, J. 1998. "Human Side of Hacking," *PC Week*, 15(31), August 3, 6.

Kushner, D. 1999. "The Domain Name Game," *PC Magazine*, November 12. Available online at: (http://www.zdnet.com/pcmag/stories/reviews/0,6755,2385324,00.html).

Lawson, N. and J. Garris. 1999. "Plug Your Company's Common Security Holes," *PC Magazine*, 18(10), May 25, 217–218.

Lim, F. 1998. *ActiveX and the Internet.* El Granada: Scott/Jones, Inc.

Machrone, B. 2000. "Always More Threats," *PC Magazine*, 19(13), July, 101.

McClure, S. 1998. "E & Y Teaches the Fine Art of Hacking at Your Site," *Infoworld*, July 27, 88.

McGraw, G. and E. Felton. 1999. *Securing Java: Getting Down to Business with Mobile Code.* New York: John Wiley & Sons.

Merkow, M., J. Breithaupt, and K. Wheeler. 1998. *Building SET Applications for Secure Transactions.* New York: John Wiley & Sons.

Morris, R. and K. Thompson. 1979. "UNIX Password Security," *Communications of the ACM*, 22(11), November, 594–597.

Neuman, B. and T. Tso. 1994. "An Authentication Service for Computer Networks," *IEEE Communications*, 32(9), September, 33–38.

Null, C. 2000. "Name Grab," *PC Computing*, 13(4), April, 40–42.

Oppliger, R. 1997. "Internet Security: Firewalls and Beyond," *Communications of the ACM*, 40(5), May, 92–102.

Parker, D. 1983. *Fighting Computer Crime.* New York: Charles Scribner's Sons.

Radcliff, D. 2000. "Domain name game," *Computerworld*, 34(24), June 12, 71.

Rockwell, B. 1998. *Using the Web to Compete in a Global Marketplace*. New York: John Wiley & Sons.

Smith, J. 2000. "'Love bug' prompts new Philippine law," *USA Today*, June 14. Available online at: (http://www.usatoday.com/life/cyber/ccarch/ccjoe021.htm).

Spafford, E. 1988. "The Internet Worm Program: An Analysis," *Purdue University Computer Science Department Technical Report CSD-TR-823*.

Stein, L. 2000. "Napster: Asking for Trouble. Getting It." *Web Techniques*, 5(5), May, 12–15.

Vijayan, J. 2000. "Possible S & P Security Holes Reveal Risks of E-Commerce," *Computerworld*, 34(22), May 29, 6.

Vijayan, J. 2000. "Analysts: Better to Be Safe Than Sorry With Viruses," *Computerworld*, 34(26), June 26, 20.

Vijayan, J. 2000. "Joke Virus Spreads Fast, Clogs Servers," *Computerworld*, 34(26), June 26, 20.

Weaver, J. 2000. "The Name Game," *PC Computing*, 13(4), April, 108–114.

192

IMPLEMENTING SECURITY FOR ELECTRONIC COMMERCE

INTRODUCTION

Jim Lockhart noticed the problem late Tuesday evening. Software engineers working on several different projects came to him in quick succession complaining that their computers had "died" and that files were gone. When Jim questioned each engineer, several of them indicated that they kept an electronic mail client program open on the Windows 2000 desktop, but that they read mail sporadically. All the engineers had the same story: They were reading their e-mail, and when they opened an e-mail attachment, their computers began misbehaving. The first symptoms were that the files they had been working on were reduced to zero bytes in length.

The Worm.ExploreZip virus had struck Jim's company, CSD. Both a worm and a virus, Worm.ExploreZip hides in the body of an e-mail message and e-mails itself out as an attachment named zipped_files.exe. Whenever a user sends an e-mail message to a computer infected with the Worm.ExploreZip virus, the addressee receives an e-mail containing the damaging payload. In the message body carrying the attachment is this seemingly innocent text: "Hi (Recipient Name)! I received your email and I shall send you a reply ASAP. Until then, take a look at the attached zipped docs. Bye." Once the unwary recipient opens the attachment, the worm begins doing its malevolent work by searching the computer's hard drive and destroying selected file types. The user doesn't realize that he or she has unleashed the destructive program until it is too late. It does make a difference whether Jim labels it a virus or a worm. Each requires a different cure. The lesson here is that all users must continuously use all the security tools at their disposal. Potentially huge resources, including time and money, are at stake.

Protecting electronic assets that comprise electronic commerce systems is not an option but a necessity if commerce is to thrive. The electronic world will always have to deal with viruses, worms, Trojan horses, eavesdroppers, and destructive programs whose goals are to disrupt, delay, or deny communications and information flow between consumers and producers. Billions of dollars are at stake, and security protection must continually be developed to provide consumers with confidence in the online systems with which they interact and through which they conduct business. This chapter describes security measures that protect client computers, the Internet over which commerce information flows, and the commerce server.

LEARNING OBJECTIVES

In this chapter, you will learn about:
- What security measures can reduce or eliminate intellectual property theft
- How to secure client computers from attack by viruses and by ill-intentioned programs and scripts downloaded in Web pages
- How to authenticate users to servers and authenticate servers
- What protection mechanisms are available to secure information sent between a client and server so that the information is not disclosed
- How to secure message integrity, preventing another program from altering information as it travels on the Internet
- What safeguards are available to enable commerce servers to authenticate users
- How firewalls can protect intranets and corporate servers against being attacked through the Internet
- What role the Secure Socket Layer, Secure HTTP, and secure electronic transaction protocols play in protecting electronic commerce

PROTECTING ELECTRONIC COMMERCE ASSETS

Regardless of whether companies are doing business over the Internet or face to face, security is an extremely serious issue. Customers engaging in business-to-consumer commerce and businesses engaging in business-to-business commerce need to feel confident that their transactions are secure from prying eyes and safe from alteration. Today, the volume of business sales conducted online is huge and will grow even larger over the coming years. Some conventional retail and wholesale venues that existed prior to the advent of electronic commerce may even disappear in selected market segments.

Thirty years ago, security meant physical security: alarmed doors and windows, guards, security badges to admit people to sensitive areas, surveillance cameras, and so on. Back then, interactions between people and computers were limited to dumb terminals connected directly to large mainframe computers. There were no other connections to computers. Computer security at that time meant dealing with the few people who had access to terminals. Anyone wishing to run programs did so by submitting programs in the form of decks of punched cards fed into card readers. People reclaimed their card decks and the output results—usually pages of fan-folded, green-bar piles of paper—from the input/output clerk. Security was pretty simple.

Both the audience of computer users and the methods to access computing resources have increased tremendously since those early years of computing. Millions of people now have access to computing power over both private and public networks that connect millions of computers. It is not a simple matter to determine who is using a computing resource, because the user could be located in Cape Town, South Africa, but using a computer in Berkeley, California. A whole new series of security tools and methods have evolved and are employed today to protect electronic assets. The transmission of valuable information such as electronic receipts, purchase orders, credit card numbers, and order confirmations has drastically changed the way security is viewed and has introduced new electronic and automatic methods to deal with security threats.

Data security measures date back to the time of the Roman Empire, when Julius Caesar coded information to prevent enemies from reading secret war and defense plans. Modern electronic security methods trace their roots to the defense sector also. The U.S. Department of Defense was the main driving force behind both early security requirements and more recent advances as well. Fewer than 20 years ago, the Defense Department formed a committee to develop computer security guidelines for handling classified information on computers. The result of that committee's work was *Trusted Computer System Evaluation Criteria*, known in defense circles simply as the "Orange Book," because its cover was orange. It spelled out rules for mandatory access control—the separation of confidential, secret, and top secret information—and established criteria for certification levels for computers ranging from D (not trusted to handle multiple levels of classified documents at once) to A1 (the most trustworthy level).

While that work was groundbreaking in defining security terms, conditions, and tests for security, it did not address how to handle electronic commerce computer security. Nonetheless, that early security work has been beneficial because it spawned commerce security research, which resulted in commercially applicable and practical security solutions. The early work also provided a much-needed formal approach to security. For example, security experts have learned that you cannot hope to produce secure commerce systems unless there is a written security policy in place. That policy must spell out what assets are to be protected, what is needed to protect those assets, an analysis of the likelihood of the threats, and the rules to be enforced to protect those assets. The security policy must be regularly reviewed and revised as threat conditions change. Without a written policy, it is difficult to implement any security at all.

Both defense and commercial security guidelines state that you must protect assets from unauthorized disclosure, modification, or destruction. However, military security policy differs from commercial policy because military applications stress separation of levels of security. Corporate information is usually classified as either "public" or "company confidential." The typical security policy concerning confidential company information is straightforward: Do not reveal company confidential information to anyone outside the company.

As mentioned in Chapter 5, a security policy should protect a system's privacy, integrity, and availability (necessity) and authenticate users. When you recast these goals for electronic commerce, they become those shown in Figure 6-1. Dr. Eugene Spafford, professor of computer sciences at Purdue University and computer security expert, provides some insight about the importance of securely conducting electronic commerce. In an interview with the *Purdue University Perspective*, he said, "The protection of information is a major concern as it relates to national defense, commerce, and even our private lives. It also is a business with enormous growth potential."[1] Clearly, robust security is vital to the health and continued growth of electronic commerce. (See **CERIAS**—pronounced "serious"—in the Online Companion for more information about the world's first comprehensive information security center.)

Requirement	Meaning
Secrecy	Prevent unauthorized persons from reading messages and business plans, obtaining credit card numbers, or deriving other confidential information.
Integrity	Wrap information in a digital envelope so that the computer can automatically detect modified messages.
Availability	Provide delivery assurance for each message segment so that messages or message segments cannot be lost undetectably.
Key management	Provide secure distribution and management of keys needed to provide secure communications.
Nonrepudiation	Provide undeniable, end-to-end proof of each message's origin and recipient.
Authentication	Securely identify clients and servers with digital signatures and certificates.

Figure 6-1   *Minimum requirements for secure electronic commerce*

This chapter examines security by looking at how to protect assets, beginning with the client. First, we will examine protections available for client computers. Then, we will investigate security for the carrier—the Internet. Finally, we will examine server security.

PROTECTING INTELLECTUAL PROPERTY AND PRIVACY ONLINE

Protecting digital intellectual property poses problems that are different from traditional intellectual property security. Traditional intellectual properties, such as written works, art, and music, are protected by national and, in some cases, international laws.

---

[1] "Lilly supports information security initiatives," 1999. *Purdue University Perspective*, 26(2), Spring, 13.

Digital intellectual properties, including art, logos, and music posted on Web sites, are also protected by laws. While those laws act as a deterrent, they do not prevent violations from occurring, and they do not provide a means to reliably trace the path taken by the violators to acquire the intellectual property. The real dilemma for digital property is how to display and make available intellectual property on the Web while simultaneously protecting those copyrighted works. While absolute intellectual property protection so far has proven elusive, there are measures you can take to provide some level of protection and accountability for copyrights held on digital works.

Protecting Intellectual Property

In the United States, Congress has been trying to deal legislatively with digital copyright issues. Recently, the U.S. Department of Justice launched the **Computer Crime and Intellectual Property Section** to provide information and updates on hacking, software piracy, and the latest security information as well as the latest information on cyber crime prosecutions. Part of that site is devoted to **Protecting Intellectual Property** and provides valuable information about both intellectual property attacks and countermeasures you can employ to protect intellectual assets (see Figure 6-2).

Figure 6-2   *Intellectual Property Section Web page*

An international body, the **World Intellectual Property Organization** (**WIPO**) has been trying to promote and oversee digital copyright issues internationally. Meanwhile, a few companies are providing the first and second generations of products that provide a measure of protection for digital copyright holders. While the volume of case law dealing with copyright issues is still light and the field is still young, early indications are that copyright laws in the United States, at least, apply to the Internet and other digital media. The **Information Technology Association of America**

(**ITAA**), which is a trade organization representing U.S. information technology, has written a comprehensive paper on protecting copyrighted digital information. In that paper, **Intellectual Property Protection in Cyberspace**, the author discusses the current problems in digital copyright protection and presents some solutions. The author describes three solutions:

- Host name blocking
- Packet filtering
- Proxy servers

All three approaches illustrate how an Internet service provider might try to block access to an entire offending site through IP blocking, packet filtering, or the use of a proxy server to filter requests. (See the Online Companion for a link to the full paper containing further details.) However, none of these approaches is really effective in preventing theft or providing identification of property obtained without the copyright holder's permission.

Several methods show promise in the battle to protect digital works. Examples of these methods include software metering, digital watermarks, and digital envelopes (sometimes called message authentication codes). Of course, the degree of protection provided varies widely, and none of these methodologies is foolproof, but they do provide some protection. New and improved methods are continually being developed. One promising technique employs steganography to create a digital watermark. The watermark is a digital code or stream embedded undetectably in a digital image or audio file. It can be encrypted to protect its contents, or simply hidden among the bits—digital information—comprising the image or recording. **Verance Corporation** is a company that provides, among other products, digital audio watermarking systems to protect audio files on the Internet. Its systems identify, authenticate, and protect intellectual property. Verance's ARIS MusiCode system enables recording artists to monitor, identify, and control the use of their digital recordings. The audio watermarks do not alter the audio fidelity of the recordings in which they are embedded. The Verance SoniCode product provides verification and authentication tools. SoniCode was originally developed by ARIS Technologies, which is now owned by Verance Corporation. SoniCode can ensure that telephonic conversations haven't been altered. The same is true for audiovisual transcripts and depositions. **Blue Spike** produces a competitive watermarking system called **Giovanni**. The technology uses encryption keys (discussed later in this chapter) to generate a watermark that can be hidden in an artist's digital audio or video work. Like the SoniCode system, the Giovanni watermark authenticates the copyright and provides copy control. **Copy control** is an electronic mechanism for limiting the number of copies that one can make of a digital work.

A group of more than 180 companies and organizations devoted to providing protection for intellectual property—digital music in this case—is the **Secure Digital Music Initiative** (**SDMI**) organization. Its members represent information technology and consumer electronics companies, security technology firms, Internet service providers, and the music recording industry. SDMI's charter is to develop open, public technology specifications that protect the playing, storing, and distributing of digital music for the express purpose of promoting the new market—the Internet—for digital music. When complete, the new technology specified by the SDMI's technical oversight will, as the organization states in its Web site, "…provide consumers with convenient online access to digital distribution of music, enable copyright protection for artists' works, and promote the development of new music-related businesses."

Digimarc Corporation is another company providing watermark protection systems and software. Its products embed a watermark that allows any works protected by its Digimarc system to be tracked across the Web. In addition, the watermark can link viewers to commerce sites and databases. It can also control software and playback devices. Finally, the imperceptible watermark contains copyright information and links to the image's creator. That enables nonrepudiation of a work's authorship and facilitates electronic purchase and licensing of the work. One of Digimarc's most fascinating products is called **MediaBridge**. MediaBridge technology places a small, imperceptible digital watermark into magazine ads and other printed material. Although you cannot see the watermark, a Digimarc-approved Web camera can. When the PC camera recognizes the code, it signals your Web browser to launch and open a specific Web page. Digimarc provides free software that you download and install. The concept provides a lot of interesting electronic commerce ideas. For example, a print ad in *Wired* for an automobile could contain a Digimarc watermark. By holding the print ad up to a digital camera attached to your PC, the watermark automatically leads magazine readers to a related Web site selling the product pictured in the magazine. (Click **MediaBridge Participating Magazines** in the Online Companion for a list of magazines containing MediaBridge watermarks.)

SoftLock.com is a company that produces, among many other products, a technology allowing authors and publishers to lock files containing digital information for sale on the Web. Using SoftLock, authors and other copyright holders can allow purchasers to unlock the files they purchased. This provides a secure means to post files that can be downloaded, because those files do not function without a purchased key unlocking them. Figure 6-3 shows SoftLock's home page.

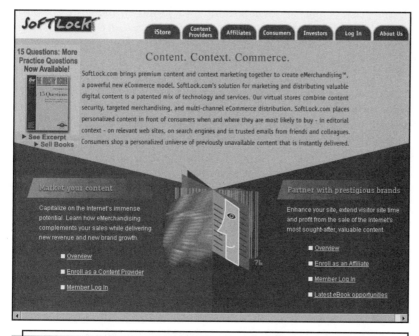

Figure 6-3   *SoftLock.com home page*

Protecting Privacy

Cookies were introduced in Chapter 4. These small pieces of text are stored on your computer and contain private information that is not encrypted. That means that anyone could read and interpret a cookie to gather and record the information it contains. This information includes credit card data, passwords, and login information. Because most cookies are like tickets that allow admission to Web sites that the client computer has previously visited, a cookie can provide entrance to anyone who has the cookie in his or her possession. While cookies typically do not harm client machines directly, they can cause damage.

Cookies are designed to solve a problem inherent with Web servers—saving information about a Web user from one session to another. There are two kinds of cookies: **session cookies**, which exist until you shut down your browser, and **persistent cookies**, which can exist indefinitely. An electronic commerce site may use both kinds of cookies. For example, a session cookie might contain information about a particular shopping session, whereas a persistent cookie contains information to help the Web site recognize you the next time you shop there. Each time your browser moves to a different part of a merchant's Web site, the merchant's server asks your computer to send back the cookie that the server previously stored on your computer.

The privacy problem exists because *any* machine that supplies any part of a Web page you visit can send your computer a cookie for storage and subsequent retrieval. Advertisers and Web trend analysis companies can learn a great deal about your browsing habits from retrieving cookies stored on your computer. For example, if you visit two different Web pages on different sites that display banner ads from the same company, that company can determine from its stored cookies that you have visited the two Web pages. Information the company may have acquired from one site (perhaps you filled out a form on one site on which a company has an advertising banner) can be used at the other site you visited that contains an advertising banner from the same company—a clear but subtle privacy violation.

Sometimes an image from an advertiser is imperceptible to the naked eye. A tiny graphic that invisibly adorns a Web page is called a **Web bug**. Only a few pixels in size, a Web bug's only purpose is to provide a way for a Web site to place cookies on your computer. (The Internet advertising community refers to Web bugs as "clear GIFs" or "1-by-1GIFs.")

WebSideStory, a San Diego-based company, provides software that analyzes Internet traffic data and provides reports to Web sites about who visits their site and what sites the visitors came from. WebSideStory's **HitBox** software technology remotely, securely, and anonymously collects and warehouses data from Web site visitors. To see the cookies that are stored on your machine, click **Cookie Demonstration** in the Protecting Intellectual Property section of the Online Companion (if you don't think that your computer has any cookies on it, you're in for a surprise). After the WebSideStory Privacy Center page opens, locate and then click the Edit Your Profile link to display the contents of a cookie that WebSideStory has created on your computer. It is surprising how much the Web trend analysis companies know about your machine, isn't it? Figure 6-4 shows an example of the information stored in a cookie collected by a Web site—information specific to the computer that browsed the Web site.

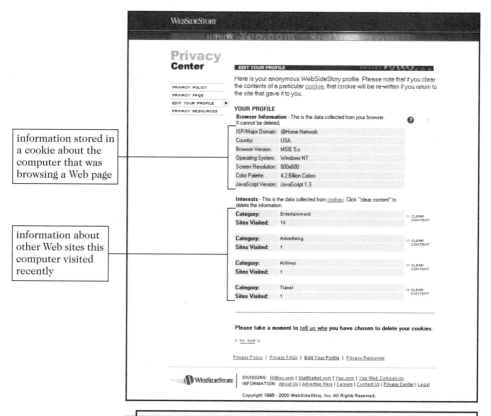

information stored in a cookie about the computer that was browsing a Web page

information about other Web sites this computer visited recently

Figure 6-4 *Private information stored in a cookie on your computer*

The most foolproof way to protect yourself from revealing private information or being tracked by cookies is to disable cookies entirely. If you are using Netscape Navigator, select Edit, Preferences, and then select the Advanced item in the left column. Then, click the Disable Cookies option button. In Microsoft Internet Explorer, select Tools, Internet Options, and then select the Disable option for both session and stored cookies. The problem with that approach—denying all cookies—is that you lose the "good" cookies along with the others. You will have to enter more information each time you revisit a Web site. The extra work may be worth the added privacy, however.

A better approach is to use one of the many third-party programs that prevent cookie storage selectively. Known as **cookie blockers**, some of the cookie monitoring programs, such as **WebWasher**, plug into your browser and allow you to disable cookies from advertising banners. Other cookie blockers allow you to filter cookies by Internet (IP) address, allowing in the "good" cookies and denying storage to all others. **Junkbuster Proxy** is an advertising cookie blocker. Junkbuster Proxy blocks selected cookies by domain name and allows your browser to send back to a Web server vanilla wafers. **Vanilla wafers** are cookies that your browser creates that contain little or no personal information. Other cookie managers include **AdSubtract**, **Cookie Cruncher**, **Cookie Crusher**, **Cookie Pal**, **Cookie-Server**, **IEClean** and **NSClean**, and **Window Washer**.

PROTECTING CLIENT COMPUTERS

Client computers, usually PCs, must be protected from threats that originate from software and data on the Internet that are downloaded to the client computer. Chapter 5 provides a detailed account of client computer threats, but here's a quick recap to refresh your memory: Ordinary Web pages that are delivered to your computer in response to your browser's request are static displays of information and are completely harmless (regardless of whether they happen to offend anyone). Active contents, delivered over the Internet in dynamic Web pages, are not harmless. They can be one of the most serious threats to client computers.

Recall that active content consists of programs that are embedded in Web pages and that provide dynamic content and appearance for Web pages. Most of the time, the programs perform their appointed duties and nothing more; they are quite benign. Occasionally, however, threats masquerade as harmless active content but cause damage when they are executed on your computer. Programs that make Web pages dynamic are usually written in Java or JavaScript. The other popular active content tools are ActiveX controls. Besides threats from programs inside Web pages, downloaded graphics, browser plug-ins, and e-mail attachments can harbor threats that could harm client computers when the hidden programs are activated.

Another threat to client computers is a malevolent server site masquerading as a legitimate Web site. There have been an alarming number of cases of misplaced trust in which users and their client computers were duped into revealing information to illicit Web sites, receiving nothing in return. This is really a client computer security concern, because it is the responsibility of the client computer to know its server—the commerce site it is dealing with—to avoid being tricked. The next sections discuss built-in protection mechanisms designed to prevent or greatly reduce the probability of the preceding client threats.

Monitoring Active Content

The Netscape Navigator and Microsoft Internet Explorer browsers are equipped to recognize when they are about to download Web pages containing active content. When your browser downloads Web pages and runs programs embedded in them, it gives you a chance to confirm that the programs are from a source you know and trust. The way that both of these popular browsers ensure security varies slightly, so they are presented separately in this section. You can choose to read more about both browsers' security features, or simply read about the security features of the browser that you use. But first, you will learn about digital certificates, which are essential in providing assurance to clients and servers that the participant is authenticated. How can your Web browser identify a program's source reliably—how do you know you can trust a site? The answer lies in digital certificates, which are described next.

Digital Certificates

A **digital certificate**, also known as a **digital ID**, is an attachment to an e-mail message or a program embedded in a Web page that verifies that a user or Web site is who it claims to be. In addition, the digital certificate contains a means to send an encrypted message—encoded so others cannot read it—to the entity that sent the original Web page or e-mail message. In the case of a downloaded program containing a digital certificate, the encrypted message identifies the software publisher (ensuring that the identity of the software publisher matches the certificate) and indicates whether the certificate is still valid. When a message or Web page contains an attached digital certificate, it is a **signed** message or code. Signed code or messages serve the same function as a photo on a driver's license or passport. They provide proof that the holder is the person identified by the certificate. Just like a passport, a certificate does not imply anything about either the utility or quality of the downloaded program. What a certificate *does* supply is a level of assurance that the software is the genuine article and not a forgery. The idea behind certificates is that if you trust the software developer, the certificate provides you with assurance that the signed software came from that trusted developer.

Digital certificates are used for many different types of online transactions, including electronic commerce, electronic mail, and electronic funds transfers. A digital ID verifies a Web site to a shopper and, optionally, identifies a shopper to a Web site. A Web site's digital certificate is a shopper's assurance that the Web site is the "real" store and not an imitator, because digital certificates are built so they cannot be forged easily. Web browsers or e-mail programs automatically exchange digital certificates invisibly when they are requested to validate the identity of each party involved in a transaction.

What does a digital certificate contain? Figure 6-5 shows the general structure of a digital certificate. You will learn in the sections that follow how and when certificates are exchanged order to provide assurance between clients and servers. Figure 6-6 displays the digital certificate owned by Amazon.com. Whenever your browser indicates that it has established secure communications with a Web site—when a lock appears in the browser's status line—you can double-click the lock to display the Web site's digital certificate.

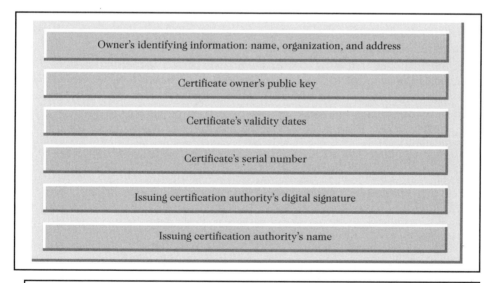

| Owner's identifying information: name, organization, and address |
| Certificate owner's public key |
| Certificate's validity dates |
| Certificate's serial number |
| Issuing certification authority's digital signature |
| Issuing certification authority's name |

Figure 6-5   *Structure of a digital certificate*

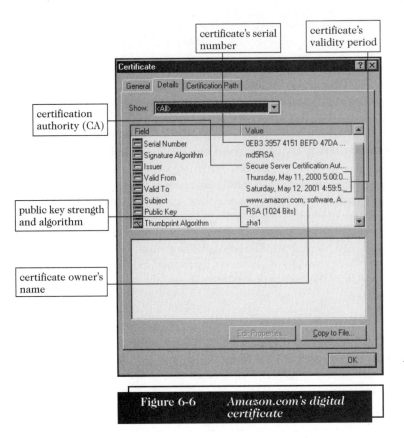

certificate's serial number

certificate's validity period

certification authority (CA)

public key strength and algorithm

certificate owner's name

Figure 6-6   Amazon.com's digital certificate

The software publisher is not the same as the entity that signs the certificate. The certificate is only an approval of the code and does not indicate who authored it. Companies that sign the software first obtain a software publisher certificate from one of a few dozen primary or secondary certification authorities. A **certification authority (CA)** issues a digital certificate to an organization or individual. If you compare a digital certificate to a passport, then a CA is analogous to the U.S. State Department, which is the agency that issues passports. The State Department requires anyone wanting a passport to supply several forms of proof of identity along with a photo. Similarly, a CA requires entities applying for digital certificates to supply appropriate proof of identity. Once the CA is satisfied, it issues a certificate. Then, the CA signs the certificate—its stamp of approval is affixed—in the form of a public encryption key, which "unlocks" the certificate for anyone who receives the certificate attached to the publisher's code. A **key** is simply a number—usually a long binary number—that is used with the encryption algorithm to "lock" the characters of the message you want to protect so that they are undecipherable (unless you know the key, of course). All things being equal, longer keys provide significantly better protection than shorter keys. In effect, the CA is guaranteeing that the individual or organization that presents the certificate is who it claims to be. There are only a small number of CAs. One of the oldest and best known is **VeriSign**. The certificates it issues are as trustworthy as VeriSign itself. Other CAs are listed in the Online Companion. Figure 6-7 shows VeriSign's home page.

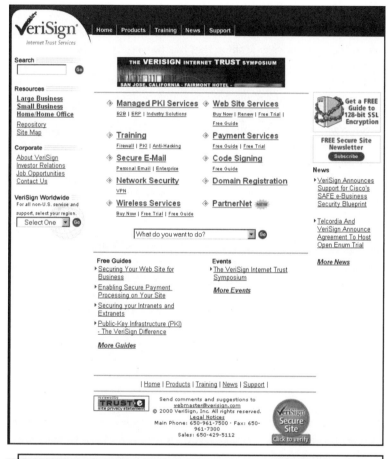

Figure 6-7   *VeriSign, a well-known certification authority*

Identification requirements vary from one certification authority to another. One CA may require a driver's license for individuals' certificates, while others may require a notarized form or fingerprints. CAs usually publish their identification requirements so that any Web user or site accepting certificates from each CA understands how stringent the CA's validation procedures are. Certificates are classified as low, medium, or high assurance, based largely on the identification requirements imposed on certificate seekers.

VeriSign provides certificate issuing and revocation services and offers several classes of certificates—from Class 1 through Class 4—that are differentiated by their assurance level, which is the confidence level one can assume based on the process the CA uses to verify the owner's identity. Class 1 certificates are the lowest level and bind e-mail addresses and associated public keys. Class 4 certificates apply to servers and the server organizations. Requirements for Class 4 certificates are significantly greater than those for Class 1. VeriSign's Class 4 certificate, for example, offers assurance of the individual's identity and of that person's relationship to the specified company or organization.

Like physical forms of identification, digital certificates expire after a period of time (often, one year). This built-in limit provides protection for both users and businesses. Limited-duration certificates guarantee that businesses and individuals must submit their credentials for reevaluation periodically. Besides becoming invalid when their time is expired, certificates can be revoked. If the CA determines that a corporation has a history of delivering bad or malicious code, it can unilaterally refuse to issue new certificates and revoke all existing certificates. You can click a hyperlink to view the corporation's timestamp. The time limit automatically eliminates the problem of a digital certificate outlasting the individual or business it identifies. This is important because a large number of dot-com businesses have gone out of business lately.

The next sections describe the security features built into the two most popular Web browsers, Microsoft Internet Explorer and Netscape Navigator.

Microsoft Internet Explorer

Microsoft Internet Explorer provides client-side protection right inside the browser. Internet Explorer also reacts to ActiveX and Java-based active content. Internet Explorer uses Microsoft Authenticode technology to verify the identity of downloaded active contents, which are programs. Authenticode can check for two important items from a downloaded ActiveX control: who has signed the code and whether the code has been modified since it was signed. Authenticode technology verifies that the program has a valid certificate. However, it does not prevent a misbehaving program from being downloaded and run on your computer. That is, Authenticode technology can only verify that XYZ Corporation, which you trust, has signed the code. If a publisher has not attached a certificate to the active content, you can set up Internet Explorer so that the Web page's code is not downloaded at all. Unfortunately, Authenticode cannot guarantee that XYZ's Java or ActiveX control will perform flawlessly. That responsibility is yours, and you must decide whether you trust active content from individual companies, or so-called "zones."

Here's how Authenticode works: When you download a page containing a certificate and active content, Authenticode detaches the certificate (sometimes called the signature block), verifies the identity of the certification authority, verifies that the content is from the publisher, and ensures that the program is unaltered from its original state. A list of trusted CAs is built into Internet Explorer along with their public keys, and Authenticode scans the list to locate a match with the CA supplying the certificate. If the public key in the list matches the public key in the certificate, the CA is known to be a genuine one. The CA's public key is used to unlock the certificate and within the unlocked certificate is the software publisher's signed digest— a summary of the certificate itself. If the signed digest proves that the software publisher signed the downloaded code, the certificate is displayed. That display assures you that the supplier is valid.

Figure 6-8 shows a security warning and certification validation dialog box. Authenticode determines that the active content has a signed, valid certificate. In this case, Beatnik, Inc., is the publisher, VeriSign is the CA, and the software being downloaded is the Beatnik Player Web browser plug-in. If you were to download a Web page containing active content that was not signed, the dialog box would indicate that there is no valid certificate. Whether Internet Explorer displays a security warning depends on how you have configured security on your browser.

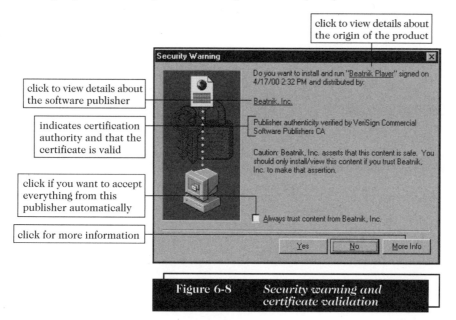

click to view details about the origin of the product

click to view details about the software publisher

indicates certification authority and that the certificate is valid

click if you want to accept everything from this publisher automatically

click for more information

Figure 6-8   *Security warning and certificate validation*

You can specify different security settings that determine how Internet Explorer handles the programs and files it downloads, depending on the source of the files being downloaded. Microsoft Internet Explorer divides the Internet into zones (categories). That way, you can classify particular Web sites into one of the zones and then assign security levels appropriate for each zone, or group of Web sites. There are four zones: *Internet, local intranet, trusted sites*, and *restricted sites*. The Internet zone is anything that is not on your computer, not on an intranet, or not assigned to any other zone. The local intranet zone typically contains Web sites that don't require a proxy server (any sites on the Connections tab [see Figure 6-9], for example), the internal corporate network to which your client machine is attached, and other local intranet sites. The trusted sites zone contains sites you trust. You know you can safely download content from those sites without security worries because they are trustworthy. The restricted sites zone contains Web sites you do not trust. They are not necessarily awful sites bent on destruction. They may simply be sites with which you are not yet familiar. Figure 6-9 shows that you can assign a security level from the options of Low, Medium-Low, Medium, and High. If you want to fine-tune those designations, click the Custom Level... button. Figure 6-9 also shows the four Internet Explorer zones and the dialog box that allows you to customize security for a level.

You could select Low for the intranet zone setting if you are confident that anything you download within the corporate intranet is safe. Figure 6-10 summarizes the default behavior of the four security settings.

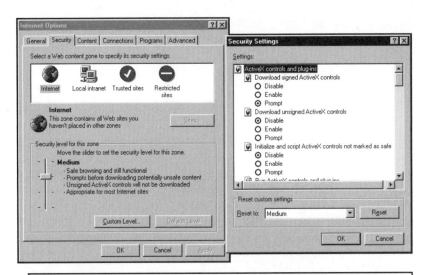

Figure 6-9   *Internet Explorer zones and their security levels*

Security level	Default security protection provided
High	Provides the safest way to browse, but less functional. Disables less secure features. Disables cookies.
Medium	Safe browsing is functional. Prompts before downloading potentially unsafe content. Does not download unsigned ActiveX components.
Medium-Low	Downloads everything without prompts. Runs most content without prompts. Does not download unsigned ActiveX components.
Low	Supplies minimal safeguards and warnings. Downloads and runs most content without prompts. Can run all active content.

Figure 6-10   *Internet Explorer security zone default settings*

Authenticode technology boils down to a yes/no decision on who and what you trust. While you can fine-tune your security settings, the protections are still a choice between running or not running active code. Nothing in Authenticode provides ongoing monitoring of code *during* its execution. So, seemingly safe code that Authenticode permits into your computer can still malfunction—due either to a programming mistake or an intentional act. In other words, once you pass judgment on the trustworthiness of a site, zone, or vendor, you leave the door wide open to security breaches when you permit downloaded content. By examining the CERT advisories posted on the Internet (see Chapter 5), it is clear that most alerts are in response to errors discovered by the developers themselves. That means, of course, that most damage to computers is caused by programming errors that slipped by because the software was not tested as thoroughly as it should have been.

Netscape Navigator

Netscape Navigator allows you to control whether active content is downloaded to your computer. If you decide to allow Netscape Navigator to download active content, you can view the signature attached to Java and JavaScript controls (ActiveX controls do not execute with Netscape Navigator). You set security in the Preferences dialog box. To display this dialog box, select Edit, Preferences. Then, click Advanced in the left panel of the Preferences dialog box. The right panel displays your security settings, as shown in Figure 6-11. You can choose to enable or disable Java and JavaScript. Also in the Preferences dialog box, you can decide what to do about cookies. Three option buttons establish how to handle cookies: You can choose to unconditionally accept all cookies, select cookies that are sent back to the server, or disallow cookies completely. Independently, you can select a check box to receive a warning before accepting cookies.

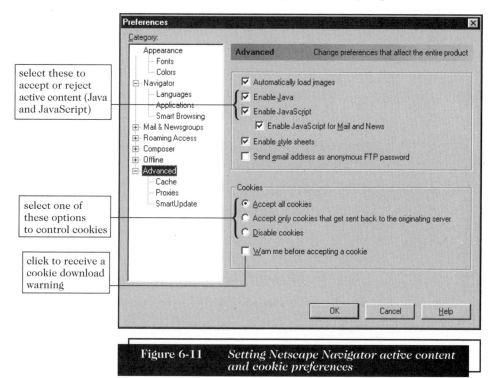

Figure 6-11 *Setting Netscape Navigator active content and cookie preferences*

If you choose to allow Java or JavaScript active content, you will always receive an alert from Netscape Navigator. The alert indicates whether the active content is signed and allows you to view the attached certificate (if available) to determine whether to grant or deny permission to download the active content. Figure 6-12 shows a Netscape Navigator alert from attempting to download a plug-in from Headspace, Inc. Notice that Netscape Navigator judges the risk to be high. Clicking the Details button on the security alert displays more information about the current download request. Clicking the Grant button allows the download process to proceed. Clicking the Deny button denies access, and the Java applet or JavaScript is not downloaded. You can examine the vendor's certificate attached to the active content by clicking the Certificate button (see

Figure 6-12). Figure 6-13 shows the Headspace certificate that is attached to the plug-in installation program.

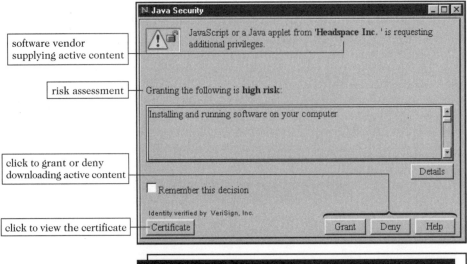

Figure 6-12   *A typical Netscape Navigator Java security alert*

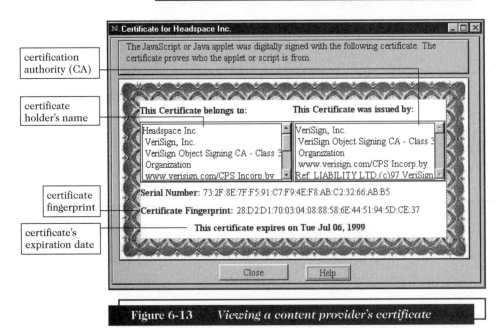

Figure 6-13   *Viewing a content provider's certificate*

Notice that the content provider's certificate has a serial number unique to the certificate and a signature (the list of numbers and letters following the Certificate Fingerprint label). The certificate has a limited lifetime. Certificates are often renewed annually, though some certificates last more than a year.

Using Antivirus Software

No client computer defense would be complete without antivirus software. While it isn't the purpose of this textbook to examine in detail antivirus software or endorse one product over another, it is important that you include this valuable class of protection in any security plan. Antivirus software only protects your computer from viruses that are already downloaded to your computer. So, the antivirus software is a defense strategy.

One of the most likely places to find viruses is in electronic mail attachments. Some e-mail systems, such as Yahoo! Mail, let you scan attachments using antivirus software before downloading e-mail. Many of the most infamous viruses—including the ILOVEYOU virus and its imitators—arrived as electronic mail attachments. When e-mail recipients opened the attachments, the viruses launched their attacks. Any antivirus program worth considering must also be able to scan e-mail attachments and warn of the embedded viruses. Some companies are using a novel approach to protect client computers from e-mail viruses that disable electronic mail programs (such as Outlook Express, Netscape Messenger, and Eudora) on client machines.

These companies that outlaw the use of e-mail client programs have turned to **application service providers** (**ASPs**)—Web-based sites that provide applications such as spreadsheets, human resources management, or e-mail to companies for a fee—to supply their e-mail services. Using ASPs instead of installing and maintaining e-mail systems and other software packages can save a tremendous amount of time and money. Many larger organizations are happy to turn to ASPs so the companies can focus on their core business functions. ASPs such as San Francisco-based **Critical Path** or New York-based **MessageClick** have helped to eliminate the e-mail virus problems because the ASP detected and purged viruses before e-mail reached corporate client computers.

Regardless of which vendor's software you choose, it is only effective if you continually keep the antivirus data files current. The data files contain virus-identifying information that is used to detect viruses on your computer. Because people generate new viruses by the hundreds every month, you must be vigilant and update your antivirus data files regularly so that the newest viruses are recognized and eliminated. The Online Companion contains links to the leading antivirus software companies.

Calling in Computer Forensics Experts

There is a small group of firms, heartily endorsed by corporations and security organizations, whose job is to *break into* client computers. No, these aren't nefarious individuals intent on doing physical damage to client computers. Called **computer forensics experts**, these computer sleuths are hired to probe PCs and locate information that can be used in legal proceedings. The field of **computer forensics** is responsible for the collection, preservation, and analysis of computer-related evidence.

Joseph Schorr of the Newhouse News Service reported the case of a naïve individual whose computer was turned against him to prove a case. Schorr described how a computer forensics company, **Data Discovery**, combed through a company computer used by the disgruntled former employee. Accused of stealing equipment, the former employee thought he had erased all incriminating files and e-mail messages from his computer. Learning that his computer was about to be seized under court order so it could be examined, he attempted to install a program that was supposed to erase the hard drive and eliminate all traces of any information that he thought could be used against him. When the computer forensics company later examined the computer, its experts were able to restore and print incriminating files and even locate a few files squirreled away in unlikely folders on the former employee's computer. These same electronic sleuths also could tell that the employee was attempting to hide the evidence against himself, because the date stamp on the disk-cleaning software was recent. The employee did not practice secure computing, thankfully, and was caught.

The field of computer forensics is small, but growing. Other companies specializing in this interesting work include **Berryhill Computer Forensics**, **Computer Forensics Inc.**, and **DIBS Computer Forensics**. Figure 6-14 shows the Berryhill Computer Forensics Web page outlining the company's services.

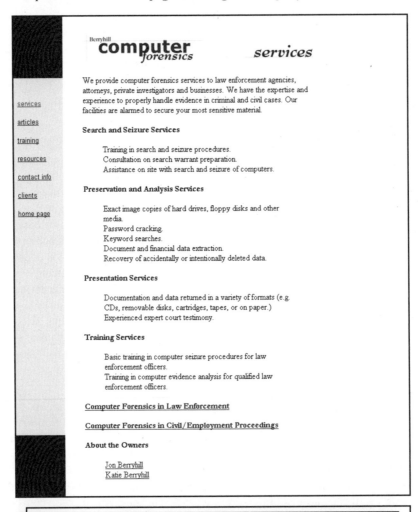

Figure 6-14   Berryhill Computer Forensic's home page

PROTECTING ELECTRONIC COMMERCE CHANNELS

Protecting the electronic commerce channels is by far the most visible segment of computer security. Hardly a day passes without a newspaper or magazine article detailing attacks on the Internet or descriptions of youthful attackers gaining entrance to a computer system by way of an insecure communications channel, such as intranets, extranets, or the Internet. Consequently, a great deal of attention has been given to protecting assets while they are in transit between client computers and remote servers.

Providing commerce channel security means providing channel secrecy, guaranteeing message integrity, and ensuring channel availability. In addition, a complete security plan includes authentication—ensuring that those using computers are who they say they are. Because user authentication is a security measure used to protect commerce servers and not commerce channels, it is described in the "Protecting the Commerce Server" section later in the chapter. In the next section, you will learn how authentication is part of the protocols that provide security services (bearing in mind that the details about authentication procedures are presented later). You will now examine each of these security services for commerce channels, beginning with transaction privacy.

Providing Transaction Privacy

Since you cannot prevent eavesdroppers from snooping on the Internet, businesses must use techniques that prevent eavesdroppers from reading Internet messages that they intercept. Sending a message over the Internet is like sending a postcard through the mail: It will probably reach its destination, but everyone involved with delivering it can read the message. The only way to prevent snoopers from copying your credit card number, for example, is to encrypt it before you send it out over the Internet. Encrypting electronic mail or Internet commerce transactions is like writing a message on the postcard in a language only you and the recipient understand. No one else in the entire world understands that language, so even though other people might intercept the message, it will make no sense to them unless they are the intended recipients.

Encryption

Encryption is the coding of information by using a mathematically based program and a secret key to produce a string of characters that is unintelligible. The science that studies encryption is called **cryptography**, which comes from a combination of the two Greek words *krupto* and *grafh,* which mean "secret" and "writing," respectively. That is, cryptography is the science of hiding messages so that only the sender and receiver can read them.

Cryptography is not related to steganography, which makes text invisible to the naked eye. Cryptography doesn't attempt to hide text; it converts text to strings that are visible but that do not appear to have any meaning. A string of these unintelligible characters is made up of combinations of bits, many of which correspond to alphabetic or numeric characters, that create a message that seems to be a random assemblage.

The program that transforms text, called **clear text**, into **cipher text** (the random assemblage of bits) is called an **encryption program**. Messages are encrypted just before they are sent over a network or the Internet. Upon arrival, each message is decoded, or **decrypted**, using a **decryption program**—a type of encryption-reversing procedure. Encryption programs, and the logic behind them, called **encryption algorithms**, are considered so vitally important to preserving security within the United States that the National Security Agency has control over their dissemination. Some encryption algorithms are considered so important that the U.S. government has banned publication of details about them. Currently, it is illegal to export some of these encryption algorithms. This affects several U.S. companies who supply encryption software or software containing encryption software. Web pages containing software whose distribution is restricted contain warnings about U.S. export laws. The Freedom Forum Online contains a number of articles on lawsuits and legislation surrounding encryption export laws. Critics consider publication restrictions a freedom of speech issue. If you are interested in reading more about the latest arguments in the ongoing debates over freedom of speech and export law, search the **Freedom Forum Online** using the keyword "encryption" as the search term.

One interesting and necessary property of encryption programs, or algorithms, is that someone can know the details of the encryption program and still not be able to decipher the encrypted message without the key used in the process of encoding the message. The resistance of an encrypted message to attack attempts is directly dependent on the size, in terms of bits, of the key used in the encryption procedure. A 40-bit key is considered minimal, whereas longer keys, such as 128-bit keys, provide much more secure encryption. A sufficiently long key can make the security of messages unbreakable.

The type of key and associated encryption program used to "lock" a message or otherwise manipulate a message subdivides encryption into three functions:

- Hash coding
- Asymmetric encryption
- Symmetric encryption

Hash coding is a process that uses a **hash algorithm** to calculate a number, called a **hash value**, from a message of any length. It is a fingerprint for the message, because it is almost certain to be unique for each message. Due to the design of good quality hash algorithms, the probability of two different messages resulting in the same hash value, which would create a **collision**, is extremely small. Hash coding is a particularly convenient way to tell whether a message has been altered in transit, because its original hash value and the hash value computed by the receiver will not match after a message is altered.

Asymmetric encryption, or **public-key encryption**, encodes messages by using two mathematically related numeric keys. In 1977, Ronald Rivest, Adi Shamir, and Leonard Adleman invented the RSA Public Key Cryptosystem while they were professors at MIT. Their invention revolutionized the way sensitive information is exchanged. In their system, one key of the pair, called a **public key**, is freely distributed to the public at large—to anyone interested in communicating securely with the holder of both keys. The public key is used to encrypt messages. The second key—called a **private key**—belongs to the key owner, who carefully keeps the key secret. The owner uses the private key to decrypt messages sent to him or her. Here is an overview of how the encryption system works: If Herb wants to send a message to Allison, then he obtains Allison's public key from any of several well-known public places. Then, he encrypts his message to Allison using her public key. Once the message is encrypted, only Allison can read the message by decrypting it with her secret key. Because the keys are unique, only one secret key can open the message encrypted with a corresponding public key, and vice versa. Reversing the process, Allison can send a private message to Herb using Herb's public key to encrypt the message. When he receives Allison's message, Herb uses his super-secret private key to decrypt the message and then read it. If they are sending e-mail to one another, the message is secret only while in transit. Once a message is downloaded from the mail server and decoded, it is stored in plain text on the recipient's machine for all to view.

Symmetric encryption, also known as **private-key encryption**, encodes a message by using a single numeric key, such as 456839420783, to encode and decode data. Because the same key is used, both the message sender and the message receiver must know the key. Encoding and decoding messages using symmetric encryption is very fast and efficient. However, the key must be guarded. If the key is made public, then all previous messages are vulnerable, and both the sender and receiver must use new keys for future communications. It is difficult to securely distribute new keys to authorized parties. The catch is that to transmit *anything* privately, it must be encrypted. This includes the new, secret key. Another big problem with private keys is that they do not scale well in large environments such as the Internet. There must be a private key for each pair of users on the Internet who want to share information privately. That's a huge number of key-pair combinations, and is similar to a telephone system of private lines without switching stations. Enabling 12 people to have a private key pair between all pairs (or private telephone lines between each pair), would require 66 private keys. In general, for N individual Internet clients, you would need approximately $\frac{1}{2}N^2$ private key pairs!

In secure environments such as the defense sector, using private-key encryption is simpler, and it is, in fact, the prevalent method to encode sensitive data. Distribution of classified information and encryption keys is straightforward in the defense sector. It requires guards (two-person control) and secret transportation plans. The **Data Encryption Standard** (**DES**) is an encryption standard adopted by the U.S. government for encrypting sensitive or commercial information. It is the most widely used private-key encryption system. However, the DES private key size is increased periodically, because individuals are using increasingly fast computers to break messages encoded with shorter keys. Not long ago, for example, the Electronic Frontier Foundation's Deep Crack key breaker used 100,000 PCs on the Internet to break a DES-encrypted test message in under 23 hours. (See **Cracking the 56-bit DES system** in the Online Companion.)

Today, the U.S. government uses a more robust version of the Data Encryption Standard, called **Triple Data Encryption Standard** (**3DES**). Triple DES offers good protection—it cannot be cracked even with today's supercomputers—and will do so for several years to come. However, the U.S. government's **National Institute of Standards and Technology** (**NIST**) has been developing a new encryption standard designed to keep government information secure. The new standard is called the Advanced Encryption Standard (AES), but NIST has not yet decided which of several possible algorithms (programs) it will use to implement the new standard. When implemented, the AES standard will supposedly be stronger and more efficient than the Triple Data Encryption Standard. The requirements for the AES state the following:

- AES will be available worldwide and royalty-free.
- Algorithms that implement AES must use symmetric-key (private-key) cryptography.
- Algorithms must support private-key sizes of 128, 192, and 256 bits.
- The selected encryption algorithm must be unclassified and publicly disclosed.

Visit **AES** in the Online Companion for more information about the forthcoming standard.

Public-key systems provide several advantages over private-key encryption methods. First, the combination of keys required to provide private messages between enormous numbers of people is small. If N people want to share secret information with one another, then only N unique public key pairs are required—far fewer than an equivalent private-key system. Second, key distribution is not a problem. Each person's public key can be posted in the subway if necessary and does not require any special handling to distribute. Third, public-key systems make implementation of digital signatures possible. This means that an electronic document can be signed and sent to any recipient with nonrepudiation. That is, with public-key techniques, it is not possible for anyone other than the signer to have electronically produced the signature; in addition, the signer cannot later deny electronically signing the electronic document. (You will learn more about digital signatures in the "Ensuring Transaction Integrity" section later in the chapter.)

Public-key systems have disadvantages. One disadvantage is that public-key encryption and decryption are significantly slower than private-key systems. This extra time can add up quickly as you and your customers conduct commerce on the Internet. Public-key systems aren't meant to replace private-key systems. They complement each other. As you learn more, you will find that public-key systems are used to transmit private keys to Internet participants so that additional, more-efficient communications can occur in a secure Internet session. Figure 6-15 shows a graphical representation of the hashing, private-key, and public-key encryption methods: Figure 6-15a shows hash coding, Figure 6-15b depicts private-key encryption, and Figure 6-15c illustrates public-key encryption in which Herb is sending a private message to Allison.

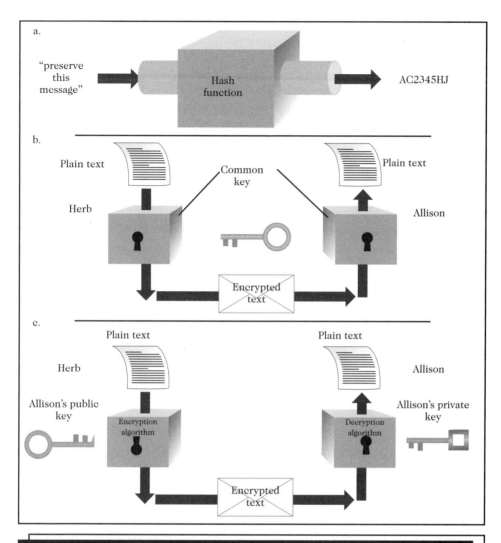

a.

"preserve this message" → Hash function → AC2345HJ

b.

Plain text   Common key   Plain text

Herb   Allison

Encrypted text

c.

Plain text   Plain text

Herb   Allison

Allison's public key   Encryption algorithm   Decryption algorithm   Allison's private key

Encrypted text

Figure 6-15   *(a) Hash coding, (b) private-key, and (c) public-key encryption*

Encryption Algorithms and Standards

There are several encryption, or cipher, algorithms that can be used with secure commerce servers. The U.S. government approves the use of several of these inside the United States, whereas others—typically weaker algorithms—are approved for use outside the United States. Secure commerce servers usually can accommodate most, if not all, of these different algorithms, because they must be able to communicate with browsers. To accommodate various browsers and different versions of those browsers, servers must offer a small array of ciphers. The brevity of this textbook precludes spending much time on each of the more visible and important algorithms, but some are briefly described in Figure 6-16.

The encryption names in the left column of Figure 6-16 are not arranged in order of importance. They are merely in alphabetical order. Explaining what each of the named algorithms does is beyond the scope of this textbook. They are listed here so you can become familiar with their names, and a few of them will be mentioned later in this chapter. Any secure server or browser uses one or more of these algorithms when it encodes information. You will learn later how to interrogate secure servers to determine which of the preceding encryption algorithms they support. Algorithms in Figure 6-16 are of three different types, indicated in the Type column. You have already learned two of the types: private-key and public-key. Several different algorithms, or methods, are available for public-key encryption and Figure 6-16 lists several different private-key algorithms.

Algorithm	Type	Comments
AES	Private key	Just a standard; promises to be stronger than 3DES
Blowfish	Private key	Block cipher; developed by Bruce Schneier
DES	Private key	Block cipher; developed in the 1970s
ECC	Public key	
IDEA	Private key	Block (considered the best algorithm available)
LUC	Public key	
MD2	Digest (Hash)	Considered dead; developed by RSA Security, Inc.
MD4	Digest (Hash)	Considered insecure; 128-bit digest; developed by RSA Security, Inc.
MD5	Digest (Hash)	Fair to good security; 128-bit digest; developed by RSA Security, Inc.
RC2	Private key	Block cipher; developed by RSA Security, Inc.
RC4	Private key	Stream cipher
RC5	Private key	Block cipher
RC6	Private key	Block cipher
RSA	Public key	Block cipher; developed by RSA Security, Inc.
SHA1	Digest (Hash)	Replaces SHA algorithm; 160-bit hash value
Skipjack	Private key	Block cipher
Triple DES	Private key	Encrypt-decrypt-encrypt sequence with three keys

Figure 6-16   *Significant encryption algorithms and standards*

Why are there more than one of each algorithm? Isn't one algorithm satisfactory for all security work? The answer is that the different algorithms have different strengths, and some algorithms are older and have proven to be inadequate for modern uses and high-speed central processing units. A third type listed in the figure is called Digest (Hash). Digest algorithms, which are hash code algorithms, do not encrypt information at all. Instead, they compute a fixed-length number from an entire message. The fixed-length number, often 128 bits long, is a signature that summarizes the message's contents. The term "digest" is used in this context because the computed numbers are digests—summaries—of an entire message. They are the signature of the message, or the message signature. Message signatures assure message recipients that the message is unaltered, because the received message should generate the same digest number as the original one. If it doesn't, then the recipient knows someone has altered the original message.

The series of algorithms called MD2, MD4, and MD5 in Figure 6-16 are message digest (thus the abbreviation MD) algorithms. MD2 was once considered a good function. Today, it is all but dead because much better digest functions are available (MD5, for example). If you are interested in learning more details about any particular algorithm, you can use a search engine and locate complete specifications on any of the algorithms shown in Figure 6-16. Another option is to use the Online Companion links to begin your research.

Secure Sockets Layer Protocol

The **Secure Sockets Layer** (**SSL**) system from Netscape Communications and the Secure HyperText Transfer Protocol (S-HTTP) from CommerceNet are two protocols that provide secure information transfer through the Internet. SSL and S-HTTP allow both the client and server computers to manage encryption and decryption activities between each other during a secure Web session.

SSL and S-HTTP have very different goals. Whereas SSL secures connections between two computers, S-HTTP sends *individual* messages securely. Encryption of outgoing messages and decryption of incoming messages happen automatically and transparently with both SSL and S-HTTP. SSL works at the transport layer of the multilayer Internet protocol set, and S-HTTP works at the application layer—the top layer.

SSL provides a security handshake in which the client and server computers exchange a brief burst of messages. In those messages, they agree upon the level of security they will use to exchange digital certificates and perform other tasks. Each computer unfailingly identifies the other. It is not a problem if the client does not have a certificate, because the client is the one who is sending sensitive information. On the other hand, the server with whom the client is doing business ought to have a valid certificate. Otherwise, you (the client) cannot be certain the commerce site is actually who it says it is. After identification, the SSL encrypts and decrypts information flowing between the two computers. This means that information in both the HTTP request and any HTTP responses are encrypted. Encrypted information includes the URL the client is requesting, any forms containing information the user has completed (which might include a credit card number), and HTTP access authorization data such as usernames and passwords. In short, *all* communication between SSL-enabled clients and servers is encoded. When SSL encodes everything flowing between the client and server, an eavesdropper will receive only unintelligible information.

Because SSL resides on top of the TCP/IP layer of the Internet protocol suite, SSL can secure many different types of communications between computers in addition to HTTP. For example, SSL can secure FTP sessions, enabling private downloading and uploading of sensitive documents, spreadsheets, and other electronic data. SSL can secure Telnet sessions in which remote computer users can log on to corporate host machines and send their passwords and usernames. The protocol that implements SSL is HTTPS. By preceding the URL with the protocol name HTTPS, you are signifying that you would like to establish a secure connection with the remote server. For example, if you were to enter the protocol and URL https://www.amazon.com, you would immediately establish a secure link with Amazon.com. The locked padlock in the status bar verifies a secure connection.

Secure Sockets Layer comes in two strengths: 40-bit and 128-bit. The designations indicate the length of the private session key generated by every encrypted transaction. A **session key** is a key used by an encryption algorithm to create the cipher text from plain text during a single secure session. The longer the key, the more resistant the encryption is to attack. You can tell when your browser has entered into an SSL session because the lock appearing in both the Internet Explorer and Netscape Navigator status bars is closed (locked). Otherwise, the lock appears to be open. Once the session is ended, the session key is discarded permanently and not reused for subsequent secure sessions.

Here is how SSL works with an exchange between a client and an electronic commerce server site: Remember that SSL has to authenticate the commerce site (at least) and encrypt any transmissions between the two computers. When a client browser first lands on a server's secure Web site, the server sends a hello request to the browser (client). The browser responds with a client hello. The exchange of these greetings, or the handshake, allows the two computers to determine the compression and encryption standards that they both support.

Next, the browser asks the server for a digital certificate—akin to asking for photo identification: "Prove to me that you are www.gateway.com." In response, the server sends to the browser a certificate signed by a recognized certification authority. The browser checks the digital signature on the server certificate (see Figure 6-13) against the public key of the CA stored within the browser. Once the CA's public key is verified, the endorsement is verified. That action authenticates the commerce server.

Both the client and server agree that their exchanges should be kept secure because they involve transmitting over the Internet credit card numbers, invoice numbers, and verification codes. To implement secrecy, SSL uses public-key (asymmetric) encryption and private-key (symmetric) encryption. While public-key encryption is very handy, it is very slow compared to private-key encryption. That's why SSL uses private-key encryption for nearly all its secure communications. But how do the client and server share a private key with each other without exposing it to an eavesdropper? The answer is that the browser generates a private key for both to share. Then the browser encrypts the private key it has generated using the server's public key. The server's public key is stored in the digital certificate that the server sent to the browser during the authentication step. Once the key is encrypted, the browser sends it to the server. The server, in turn, decrypts the message with its private key and exposes the shared private key. From this point on, public-key encryption is no longer used. Instead, only private-key encryption is used. All messages sent between the client and the server are encrypted with the shared private key, also known as the session key. When the session ends, the session key is discarded. A new connection between a client and a secure server starts the entire process all over again, beginning with the "hello browser," "hello server" exchange. Depending on what they agree upon, the client and server can use a 40-bit encryption or a 128-bit encryption. The algorithm may be DES, triple DES, or the RAS encryption algorithm. Whichever combination is used, both client and server have agreed beforehand on the encryption "language" they will use. Figure 6-17 illustrates the SSL handshaking that occurs before a client and server exchange private-key encoded business information for the remainder of the secure session.

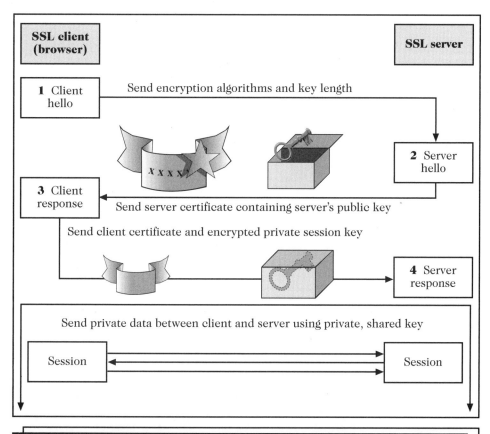

Figure 6-17 *Establishing an SSL session*

You can determine details about an SSL site by visiting Netcraft's Web Server Survey site. Netcraft's surveys were first mentioned in Chapter 3, where you learned that if you enter a URL at Netcraft's site, Netcraft would return details about the Web server software. The same is true for SSL-enabled servers, which run different Web server software than sites that are not SSL-enabled.

To discover what software and encryption algorithms a commerce site supports, click the Online Companion link **What's that SSL site running?** to start the query process. Then, enter a URL and click the Examine button to return information about the commerce. Frequently SSL queries take longer to process than the Netcraft Web server query for sites not running SSL—as much as 45 seconds or so. Figure 6-18 shows an example of the output returned, which is information about Sun Microsystems' SSL-enabled Web server. At the top of the query results page is a link called **What does it all mean?**. You can click that link for an explanation of each field of the returned SSL query. **The Netcraft Secure Server Survey** contains interesting information about certification authorities and the number of sites certi-fied by them. You will have to settle for slightly out-of-date information unless you want to purchase a current report, which costs approximately $1800.

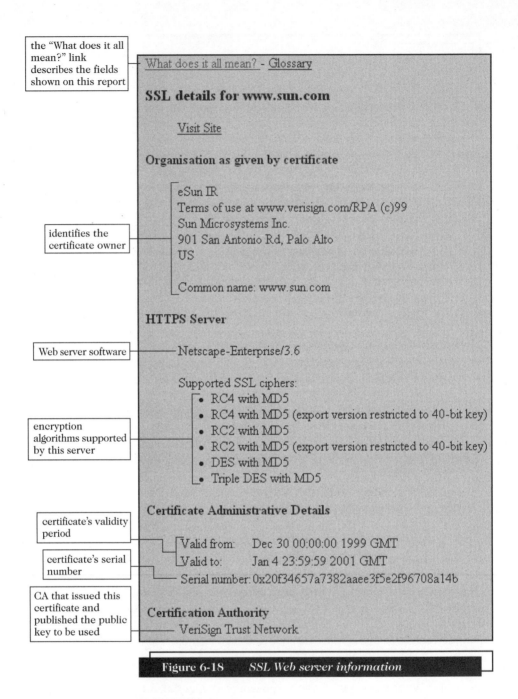

Figure 6-18   SSL Web server information

Secure HTTP (S-HTTP) Protocol

Secure HTTP (S-HTTP) is an extension to HTTP that provides a number of security features, including client and server authentication, spontaneous encryption, and request/response nonrepudiation. Developed by the **CommerceNet Consortium**, the protocol operates at the topmost layer of the protocol suite—the application layer. It provides symmetric encryption for maintaining secret communications, public-key encryption (from RSA Security, Inc.) to establish client/server authentication, and **message digests** (summaries of messages as small integer numbers) for data integrity. (Data integrity is described in the next section, "Ensuring Transaction Integrity.") Interestingly, the client or the server can use S-HTTP techniques separately. That is, a client browser may require security through the use of a private (symmetric) key, whereas the server may require client authentication by using public-key techniques.

The details of S-HTTP security are conducted during the initial negotiation session between the client and server. Either the client or the server can specify that a particular security feature be required, optional, or refused. When one party stipulates that a particular security feature be required, the client or server will continue the connection only if the other party (client or server) agrees to enforce the specified security. Otherwise, no secure connection is established. Suppose the client browser specifies that encryption is required to render all communications secret. This means that the transactions of a high-fashion clothing designer purchasing silk from a Far East textile house will remain confidential. Eavesdropping competitors cannot learn which fabrics are featured next season. On the other hand, the textile mill may insist that integrity be enforced so that quantities and prices quoted to the purchaser remain intact. In addition, the textile mill may want assurances that the purchaser is who he or she claims to be, and not an imposter. Called nonrepudiation (see Chapter 9), this security property provides positive confirmation of an offer by a client and makes it impossible for the client to deny ever having made the offer. It is, in effect, a secure digital signature (defined and described in the next section).

S-HTTP differs from SSL in the way it establishes a secure session. Whereas SSL carries out a client/server handshake exchange to set up a secure communication, S-HTTP sets up security details with special packet headers that are exchanged in S-HTTP. The headers define the type of security techniques, including the use of private-key encryption, server authentication, client authentication, and message integrity. Header exchanges also stipulate which specific algorithms each side supports, whether the client or the server (or both) supports the algorithm, and whether the security technique (for example, secrecy) is required, optional, or refused. Once the client and server have agreed to security implementations enforced between them, all subsequent messages between them during that session are wrapped in a secure container, sometimes called an envelope. A **secure envelope** encapsulates a message and provides secrecy, integrity, and client/server authentication. It is a complete package, in other words. With it, all messages traveling on the network or Internet are encrypted so that they cannot be read. Messages cannot be *undetectably* altered because integrity mechanisms provide a detection code that signals that a message has been altered. Clients and servers are authenticated with digital certificates issued by a recognized certification authority. The secure envelope embodies all of these security features.

You have learned how encryption provides message secrecy and confidentiality, and you have learned how digital certificates serve to authenticate a server to a client, and vice versa. However, implementing message integrity may be a mystery to you still. What methods can you use to ensure that an interloper does not change a message that states "Buy 25 shares of Microsoft" to "Buy 25,000 shares of Microsoft"? The answers are in the next section.

Ensuring Transaction Integrity: Hash Functions and Digital Signatures

Electronic commerce ultimately involves a client browser sending payment information, order information, and payment instructions to the commerce server, and the commerce server responding with an electronic confirmation of the order details. If an Internet interloper alters any of the order information in transit, there can be harmful consequences. For instance, the perpetrator could alter the shipment address or quantity so that he or she receives the merchandise instead of the original customer. This is an example of an **integrity violation**, which occurs whenever a message is altered while in transit between the sender and receiver.

While it is difficult and expensive to *prevent* a perpetrator from altering a message, there are security techniques that allow the receiver to *detect* when a message has been altered. When the receiver—a commerce server, for example—receives a damaged message, the receiver simply asks the sender to retransmit the message. Apart from being annoying, a damaged message harms no one as long as both parties are aware of the alteration. The damage occurs when unauthorized message changes are transparent to the message's receiver.

Hash Functions

A combination of techniques creates messages that are both tamperproof and authenticated. Additionally, those techniques provide the property of nonrepudiation—making it impossible for the message's creator to claim that the message was not his or hers and that he or she did not send it. To eliminate fraud and abuse caused by commerce messages being altered, two separate algorithms are applied to a message. First, a hash algorithm (see Figure 6-16) is applied to the message. Hash algorithms are **one-way functions**, meaning that there is no way to transform the hash value back to the original message. That is fine, because a hash value is only compared with another hash value to see if there is a match—the original, prehash values are never compared with one another. (The UNIX operating system uses a hash function to encrypt passwords when they are created. Then, when a user logs on to a system, his or her encrypted hash code is compared to the stored hash code to check for a match. Original passwords aren't stored and are impossible to generate.)

MD5 is an example of a hash algorithm used extensively in electronic commerce. **Ron Rivest** (mentioned earlier in the "Providing Transaction Privacy" section) developed the MD5 algorithm, among several others. A hash algorithm has these characteristics: It uses no secret key, the message digest it produces cannot be inverted to produce the original information, the algorithm and information about how it works are publicly available, and hash collisions are nearly impossible.

Once the hash function computes a message's hash value, that value is appended to the message. Suppose the message is a purchase order containing the customer's address and payment information. When the merchant receives the purchase order and attached message digest, he or she calculates a message digest for the message (exclusive of the attached message digest). If the message digest value that the merchant calculates matches the message digest attached to the message, the merchant then knows the message is unaltered—that is, no interloper altered the amount or the shipping address information. Had an Internet snoop altered the information, then the merchant's software would compute a message digest that is different from the message digest that the client calculated and sent along with the purchase order.

Digital Signatures

There is another problem, however. Because the hash algorithm is public and (by design) widely known, anyone could intercept a purchase order, alter the shipping address and quantity ordered, re-create the message digest, and send the message and new message digest on to the merchant. Upon receipt, the merchant would calculate the message digest and confirm that the two message digests match. The merchant is fooled into concluding that the message is unadulterated and genuine. To prevent this type of fraud, the sender encrypts the message digest using his or her private key.

An encrypted message digest (message hash value) is called a **digital signature**. A purchase order accompanied by the digital signature provides the merchant positive identification of the sender and assures the merchant that the message was not altered. How does a signature provide message integrity, nonrepudiation, and client authentication? Since the message digest is encrypted using public-key techniques, only the owner of the public/private key pair could have encrypted the message digest. Thus, when the merchant decrypts the message with the user's public key and subsequently calculates a matching message digest value, the result is proof that the sender is authentic. Furthermore, matching hash values prove that only the sender could have authored the message (nonrepudiation), because only his or her private key would yield an encrypted message that could be successfully decrypted by an associated public key. This solves the spoofing problem. If necessary, both parties can agree to provide transaction secrecy in addition to the integrity, nonrepudiation, and authentication that the digital signature provides. Simply encrypting the entire string—digital signature and message—guarantees message secrecy. Used together, public-key encryption, message digests, and digital signatures provide quality security for Internet transactions. Figure 6-19 illustrates how a digital signature and a message are created and sent.

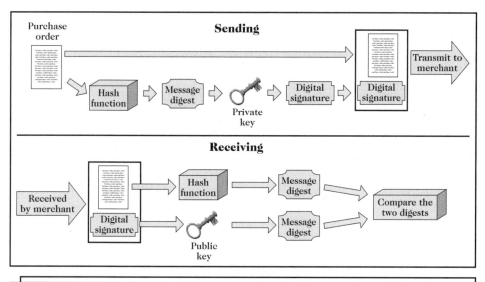

Figure 6-19   *Sending and receiving a signed message*

During the summer of 2000, in the same place where the U.S. Declaration of Independence and the U.S. Constitution were signed (Congress Hall, which stands next door to Independence Hall in Philadelphia), U.S. President Bill Clinton signed a bill giving digital signatures the same legal status as "wet ink," or traditional signatures created the old-fashioned way with ink and paper. "Under this landmark legislation, online contracts will now have the same legal force as equivalent paper contracts," President Clinton said at the signing ceremony. President Clinton first signed the paper version of the new digital signature legislation with an ink-filled pen. Then, he signed the electronic version of the bill with a smart card (see Chapter 7) containing his digital signature. After doing so, the name "Bill Clinton" appeared on the screen under the text of the new law entitled *Electronic Signatures in Global and National Commerce Act* (E-Sign Act). Under this law, both businesses and consumers doing business in the United States will be able to use digital signatures in place of ink signatures. This will save time and money for both businesses and consumers, because it will preclude the time-consuming process of sending physical documents to others and waiting for the documents to be signed and returned. Instead, people can electronically sign all sorts of legal documents, such as online car lease agreements, electronic loan papers, and purchase orders. Click **Electronic Signature Legislation** for more information.

Guaranteeing Transaction Delivery

A denial or delay of service attack removes or absorbs resources. You read about these types of security attacks in the "Protecting Client Computers" section of this chapter. In such an attack, Java programs could download with a Web page and then proceed to tie up your processor until the mouse freezes and your computer no longer reacts to keyboard keystrokes. The same sort of attack could occur on the commerce channel—a network or the Internet. One way to deny service is to flood the Internet with a large number of packets in order to crash the server or to slow it down to levels of service that are unacceptable to everyone attempting to do business. One defense, perhaps the most effective one, for that type of irrational behavior is the threat of punishment. Some attacks have caused a crash of the operating system itself. In other cases, the Web server has crashed and thus temporarily removed the commerce server. Denial attacks can also mean removal of Internet packets, causing them to disappear altogether. If this happens frequently to a particular commerce site, shoppers will start to avoid that site.

Neither encryption nor digital signatures protect information packets from theft or slowdown. However, the Transmission Control Protocol (TCP) half of the TCP/IP pair is responsible for end-to-end control of packets. When it reassembles packets at the destination in the correct order, it handles all the details when packets do not appear. Among TCP's duties are to request that the client computer resend data when packets seem to be missing. That is, no special computer security protocol beyond TCP/IP is required as a countermeasure against denial attacks. The TCP/IP protocol builds checks into the data so that it can tell when data packets are altered, inadvertently or otherwise. For more details about the TCP/IP protocol, you should consult one of the several good textbooks available on the subject. Also look in the Online Companion for links to Internet sites that discuss TCP/IP.

PROTECTING THE COMMERCE SERVER

The security protections described in previous sections centered around protecting the client computer and protecting commerce transactions on the Internet or commerce channel. While a great deal of discussion in the security world focuses on protecting the Internet from hackers and intruders, there is very little discussion about protecting the very heart and soul of electronic commerce—the electronic commerce server and associated servers connected to it.

The commerce server, along with the Web server, responds to requests from Web browsers through the HTTP protocol and CGI scripts. Several pieces of software comprise the commerce server software suite, and all the software packages must be considered both individually and as a whole when you develop a security model. Software comprising a typical Web commerce server includes an FTP server, a Web server, a mail server, a remote login server, and operating systems on host machines. An FTP server facilitates delivery of soft goods to consumers. E-mail servers service electronic mail to and from the corporation. Web servers manage Web requests from customers. A remote login server allows field personnel to remotely log on to the corporate computer to perform a variety of tasks. Of course, each piece of hardware in the commerce server collection has an underlying operating system that carries out the very fundamental services that all the commerce software programs request. A detailed presentation of security protections for all these related parts of the commerce server is beyond the scope of this chapter and this textbook. However, this section does present an overview of several important security solutions to the server threats that were outlined in Chapter 5. The next sections discuss security solutions for commerce servers.

Access Control and Authentication

Access control and authentication refers to controlling who and what has access to the commerce server. Authentication, as you already know, is verification of the identity of the entity wanting access to the computer—principally through digital certificates. Just as users can authenticate servers with whom they are interacting, servers can authenticate individual users. When a server requires positive identification of a client computer and its user, it requests that the client send a certificate. The server can authenticate a user in several ways. First, the certificate represents the user's admittance voucher. If the server cannot decrypt the user's digital signature contained in the certificate using the user's public key, then the certificate did not come from the true owner. Otherwise, the server is certain that the certificate came from the owner. This procedure prevents fraudulent certificates of "admission" to a secure server.

Second, the server checks the timestamp on the certificate to ensure that the certificate has not expired (see Figure 6-18). A server will reject an expired certificate and provide no further service. Expired certificates that are stolen or discarded insecurely become available for an imposter to scoop out of the digital trash basket.

Third, a server can use a callback system in which the user's client computer address and name are checked against a list of usernames and assigned client computer addresses. Such a system works especially well in the confines of an intranet where usernames and client computers are closely controlled and systematically assigned. On the Internet, a callback system is more difficult to manage, particularly if client users are mobile and work from different locations. In any case, you can see how certificates issued by trusted CAs play a central role in authenticating client machines and their users. Certificates provide attribution—irrefutable evidence—if a security breach occurs.

Usernames and passwords have been used for decades to provide some element of protection for servers through attribution and quasisecure identification. You use passwords every day for access to your e-mail server, to Telnet to the university or corporate computer, and to log on to subscription services, such as E*Trade, on the Internet. To authenticate users using usernames and passwords, the server must acquire and store a database containing rightful users' usernames and passwords. The system always allows users to be added or deleted, and it provides a facility (usually) to change passwords. Most modern systems, especially those springing up on the Internet, provide a user-hint style of memory refresher in case you forget your password. Usually you can retrieve a forgotten password by requesting that the server mail you your password, though this requires that you have an e-mail account outside the system to which you are requesting access. Otherwise, mailing your password to an account you cannot access is a classic Catch-22[2] situation.

Many Web server systems store usernames and passwords in a flat file, foregoing the time and expense needed to set up a full-scale database merely to store names. (A **flat file** is a single file containing all the data—unlike a database in which data are stored in several related files that are linked together.) It is unlikely that large commerce sites keep your username/password combination in a flat file. You can bet that large sites employ industrial-strength databases to handle their catalog of products. It makes sense to use that same database to store user preferences as well as usernames and passwords. To do otherwise would cause large delays with thousands of users attempting to log on simultaneously.

Regardless of where login information is stored, the most popular and safest way to store passwords—a method used by UNIX systems—is to store usernames in clear text and encrypt passwords. When you or a system creates a new username and password combination, the password is encrypted using a one-way encryption algorithm—the same type of algorithm as those that produce the message digests that you read about previously. With the clear text username and encrypted password stored, the system can validate users, when they log on, by checking the username they enter against the list of usernames stored in the database. The password that a user enters when he or she logs on to a system is encrypted. Then the resulting encrypted password from the user is checked against the encrypted password stored in the database. If the two encrypted versions of the password match for the given user, the login is accepted. That's why even a system administrator cannot figure out a forgotten password on UNIX operating systems. Instead, the administrator simply assigns a temporary password that you can change to another password of your own choosing.

As you know, passwords are not foolproof, and anyone intent on stealing a password can usually figure out a way to do so. Usernames and passwords are just one weapon in the security battle, however. Figure 6-20 shows an example of a login box for accessing subscription-only areas of *The Wall Street Journal*'s Web site. Notice that you can save

---

[2]The term Catch-22 comes from the Joseph Heller novel of the same name.

your own username and password as a cookie on your own machine. That allows you to access subscription areas of the site without entering your username and password on subsequent site visits. The trouble with that system of cookies is that the information is stored on your computer in clear text. If the cookie contains login and password information, then that information is visible to anyone who cares to print the cookies stored on your computer.

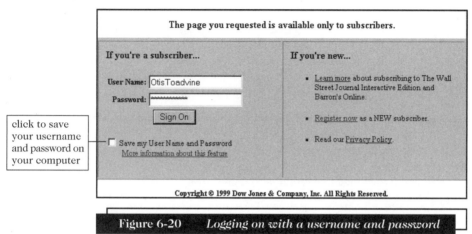

click to save your username and password on your computer

Figure 6-20   Logging on with a username and password

Web servers often provide access control list security to restrict file access to selected users. An **access control list** (**ACL**) is a list or database of files and other resources and the usernames of people who can access the files and other resources. Each file has its own access control list. When a client computer requests Web server access to a file or document that has been configured to require an access check, the Web server checks the resource's ACL file to determine if the user is allowed to access that file. This system is especially convenient to restrict access of files on an intranet server so that individuals can only access selected files on a need-to-know basis. The Web server can exercise fine control over resources by further subdividing file access into the subactivities of read, write, or execute. For example, some users may be permitted to read the corporate employee handbook but not allowed to update or write to the file. Only the human resources (HR) manager would have write access to the employee handbook, and that access privilege is stored along with the HR manager's ID and password in an ACL.

Operating System Controls

Most operating systems (except those running small computers) have a username and password user authentication system in place. That system provides the security substructure for Web servers residing on the host computer on which the operating system runs. The UNIX operating system (and its variants) is the operating system that runs the majority of the Web server platforms today. UNIX contains several native protection mechanisms that prevent unauthorized disclosure and that enforce integrity at a file level. There are many different implementations of UNIX, including AIX, Irix, Linux, HP-UX, SCO, Solaris, SunOS, and Ultrix. Each one is a particular vendor's implementation of the original UNIX created by AT&T's Bell Labs (now called AT&T Labs) in 1969. Access control lists and username/password protections are probably the best known of the UNIX security features. For further details about operating system (UNIX, Windows 2000, and so on) security features, you should consult some of the literature that is available on this topic. The Online Companion has several links to information also.

Firewalls

A **firewall** is a computer and software combination that is installed at the entry point of a networked system. The firewall provides a defense, sometimes the first line of defense, between a network to be protected and the Internet or other network that could pose a threat. All corporate access to and from the Internet flows through firewalls. The network and computers being protected are *inside* the firewall, and any other network is *outside*. Firewalls are computers that have the following characteristics:

- All traffic from inside to outside and from outside to inside the network must pass through it.
- Only authorized traffic, as defined by the local security policy, is allowed to pass through it.
- The firewall itself is immune to penetration.

Those networks inside the firewall are often called **trusted**, whereas networks outside the firewall are called **untrusted**. Acting as a filter, firewalls permit selected messages to flow into and out of the protected network. For example, one security policy a firewall might enforce is to allow all HTTP (Web) traffic to pass back and forth but to disallow FTP or Telnet requests either into or out of the protected network. Ideally, firewall protection should prevent access to networks inside the firewall by unauthorized users and thus prevent access to sensitive information. Simultaneously, a firewall should not obstruct legitimate users. Authorized employees outside the firewall ought to have access to firewall-protected networks and data files. Firewalls can separate corporate networks from one another and prevent personnel in one division from accessing information from another division of the same company. Using firewalls to segment a corporate network into secure zones serves as a coarse, need-to-know filter. (Be sure to check out the firewall links in the Online Companion.)

In the protocol stack, firewalls operate at the application layer. They also can operate at the network and transport layers. (Consult any networking textbook for more information on the OSI model and its protocol stack or see Chapter 2.) Far-flung, disparate corporate sites comprising the corporation must all have a firewall at each external connection to the Internet. Such a system ensures an unbroken security perimeter that is effective for the entire corporation. In addition, each firewall in the corporation must follow the same security policy. Otherwise, one firewall might permit one type of transaction to flow into the corporate network that another excludes. The result is an unwanted access that is permitted throughout the corporation because one firewall left a small security door open to the entire enterprise network.

Firewalls are, or should be, stripped of any unnecessary software. For example, a newly purchased UNIX-based computer comes with a plethora of software that provides a complete computing environment. All software shipped with the computer must be examined and removed if it does not serve a purpose other than support of

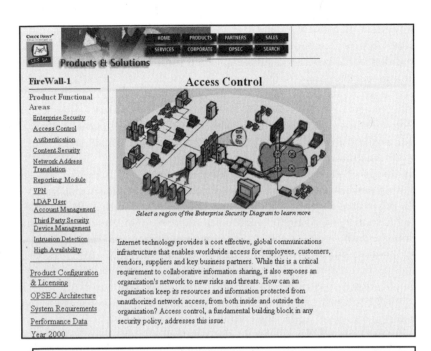

Figure 6-21   *Check Point Software's FireWall-1 Web page*

Summary

This chapter presents security solutions to the security threats described in Chapter 5. Several techniques are available and are being developed to protect intellectual property. The three general assets to protect are client computers, electronic commerce channels, and the commerce server. Key security provisions in each of these parts of the customer-Internet-commerce server linkage are secrecy, integrity, and available service. Encryption provides secrecy, and several forms of encryption are available. They are either private-key or public-key techniques. While public-key encryption eliminates the problem of sharing a secret key, it is much slower than private-key encryption. Private-key encryption is used during most commerce sessions because it is fast and efficient. Integrity protections ensure that transactions—messages

and commerce transactions—are not altered. Digital certificates provide both integrity controls and user authentication. A trusted third party or its trusted subordinate organizations, known as certification authorities, provide digital certificates to users and organizations. Several Internet protocols, including Secure Sockets Layer and Secure HTTP, provide secure Internet transmission capabilities. Finally, the commerce server, like the client computer, must be protected. Protections for the server include access control and authentication provided by username and password login procedures and client certificates. Firewalls provide a hardware solution that separates the trusted inside computer networks and clients from the untrusted outside networks, including other divisions of a company's enterprise network system and the Internet.

the operating system. Because the firewall computer is used only as a firewall and not as a general-purpose computing machine, only essential operating system software and firewall-specific protection software should remain on the computer. Having fewer software programs on the system should reduce the chances for malevolent software security breaches. Access to a firewall should be restricted to a console physically connected directly to the firewall machine. Otherwise, you must provide for remote administration of the firewall, which opens up the possibility of a break in the firewall by an imposter remotely accessing the firewall along the same path that an administrator would use.

Firewalls are classified into categories. These categories are packet filters, gateway servers, and proxy servers. **Packet-filter firewalls** examine all data flowing back and forth between the trusted network (within the firewall) and the Internet. Packet filtering examines the source and destination addresses and ports of incoming packets and denies or permits entrance to the packets based on a preprogrammed set of rules.

Gateway servers are firewalls that filter traffic based on the application they request. Gateway servers limit access to specific applications such as Telnet, FTP, and HTTP. Application gateways arbitrate traffic between the inside network and the outside network. In contrast to a packet filter technique, an application-level firewall filters requests and logs them at the application level rather that at the lower IP level. A gateway firewall provides a central point where all requests can be classified, logged, and later analyzed. An example is a gateway-level policy that permits incoming FTP requests but blocks outgoing FTP requests. That policy prevents employees inside a firewall from downloading potentially dangerous programs from the outside.

Proxy servers are firewalls that communicate with the Internet on the private network's behalf (like a proxy vote at a stockholder's meeting where one individual votes on your behalf). When you configure a browser to use a proxy, the firewall passes your browser request to the Internet. When the Internet sends back a response, the proxy server relays it back to the browser. Proxy servers are also (and were originally) used to serve as a huge cache for Web pages. The @Home network, for example, uses proxy servers for its cable modem subscribers to store frequently accessed pages. The proxy server returns pages from the cache before fetching them from the Internet. Figure 6-21 shows an informative Web page from **Check Point Software Technologies**, which produces a firewall product called FireWall-1.

Key Terms

Access Control List (ACL)
Application Service Provider (ASP)
Asymmetric Encryption
Certification Authority (CA)
Cipher Text
Clear Text
Collision
Computer Forensics
Computer Forensics Expert
Cookie Blocker
Copy Control
Cryptography
Data Encryption Standard (DES)
Decrypted
Decryption Program
Digital Certificate
Digital ID
Digital Signature
Encryption
Encryption Algorithm
Encryption Program
Firewall
Flat File
Gateway Server

Hash Algorithm
Hash Coding
Hash Value
Integrity Violation
Key
Message Digest
One-Way Function
Packet-Filter Firewall
Persistent Cookie
Private Key
Private-Key Encryption
Proxy Server
Public Key
Public-Key Encryption
Secure Envelope
Session Cookie
Session Key
Signed (message or code)
Symmetric Encryption
Triple Data Encryption Standard (3DES)
Trusted (network)
Untrusted (network)
Vanilla Wafer
Web Bug

Review Questions

1. What is watermarking? In general terms, how does it work, and what does it protect? Provide an answer of 200 words or fewer.

2. Using your Web browser, research and briefly describe the contents of a digital certificate. Write 100 words and include a diagram, if necessary. Look, first, at **VeriSign**'s site for documents and research reports.

3. Write a 200-word paper outlining the field of computer forensics. Locate and describe very briefly at least one case where computer forensic experts used their talents to help the legal system. Be sure to list at least three sites where readers can look for more information about the field. Start by using the links in the Online Companion.

4. Briefly explain the difference between public-key encryption and private-key encryption. List advantages and disadvantages of each encryption method. Are there any restrictions on the use of certain strong algorithms? Explain.

5. Write 250 words about hash algorithms. How do they work? How do they differ from public-key or private-key encryption? Using the Online Companion and Internet search engines, list the names of five hash algorithms and describe their differences in a few words. For example, provide answers to these questions: how long is the hash value, and who proposed or invented the particular technique?

Exercises

1. Julius Caesar supposedly used secret codes, which are known today as Caesar ciphers. There are 26 such ciphers in the English alphabet. In the simplest, A is replaced with B, B is replaced with C, and so on, up to Z, which is replaced with A. This type of code is called a rotate-one Caesar cipher because it rotates the alphabet one place. A rotate-two cipher replaces A with C, B with D, and so on until Z, which is replaced with B. The following line is encrypted using a simple Caesar rotation cipher.

```
Mjqqt Hfjxfw. Mtb nx dtzw
hnvmjw? Xyfd fbfd kwtr ymj
Xjsfyj ytifd.
```

In the preceding cipher text, the letters have been changed but spaces and punctuation marks have not. Capitalization has been preserved. See if you can decipher the preceding message using a simple rotate-n cipher where the value of n is less than 10. For your assignment, write the cipher text first, and then below it, write the plain text decryption of the message. Good luck! The problem isn't too difficult.

2. Obtain your own temporary certificate from VeriSign. Click the VeriSign link in the Online Companion and locate the link on VeriSign's Web pages that ask you to try out a 30-day certificate. Once you receive the certificate, see if you can print out your public key. (Make sure you *do not* print out your private key.) Alternatively, you can obtain a free *permanent* certificate from Thawte rather than VeriSign to complete this exercise. Click the **Thawte** link under Exercise 2 in the Online Companion.

3. Your colleague Andrea Flemington is responsible for security of your company's electronic commerce Web site. Her manager has sent her, rather unexpectedly, to the one of the company's divisions in Sydney. Before leaving, she left a note on your desk asking you to help her with a brief analysis of other electronic commerce sites. In her note, she requested that you conduct a small survey of 10 Web sites that employ the Secure Sockets Layer protocol to protect their transactions. Andrea indicated that you could select six sites yourself, but she wanted you to be sure to include these sites in the survey: www.amazon.com, www.beyond.com, www.fedex.com, and www.gateway.com. Print out the query results from each of the 10 sites for reference as you write a report about your research. Create a table containing 11 rows and five columns. In the leftmost column of the table, list the name (or abbreviation) of each company (leave the cell in row one blank). Place in the first row of the table, beginning in cell 2, the following labels: HTTPS Server, SSL Ciphers, Validity Period, and CA's Name. Fill in each row with the appropriate information. Here is an example of the table:

	HTTPS Server	SSL Ciphers	Validity Period	CA's Name
Sun Microsystems	Netscape Enterprise 3.0	RC4 with MD5 RC2 with MD5 DES with MD5 Triple DES	January 1, 2001 to January 1, 2002	VeriSign

HTTP normally accesses a Web server through port 80 or 8080. FTP uses port 21, Telnet uses port 23, and SMTP uses port 25. What port number does SSL (HTTPS) appear to use? Finally, indicate in a separate paragraph the names of the top three certification authorities and the percentage of the certified sites for which they are responsible. You are allowed to use out-of-date information as long as your discussion includes the URL of the page containing the statistics.

For Further Study and Research

Aucsmith, D. (ed.) 1998. *Proceedings of the Second International Workshop on Information Hiding*. Portland, OR, April 14–17, Springer Verlang, 219–239.

Baltazar, H. 2000. "Hacker Attacks Welcomed," *eWeek*, 17(26), June 26, 30–34.

Berinato, S. 2000. "Expiration of RSA patents opens up Net security," *eWeek*, 17(31), August 25. Available online at: (http://www.zdnet.com/zdnn/stories/news/0,4586,2620278,00.html).

Berinato, S. 2000. "Surprise! RSA releases crypto patent ahead of schedule," *eWeek*, 17(32), September 6. Available online at: (http://www.zdnet.com/eweek/stories/general/0,11011,2624678,00.html).

Betts, M. 2000. "Digital Signatures Law to Speed Online B-to-B Deals," *Computerworld*, 34(26), June 26, 8.

Burke, R. and M. Fatmi. 1998. *MCSE Guide to TCP/IP*. Cambridge, MA: Course Technology.

Busby, M. 1999. *Demystifying TCP/IP*. Plano, TX: Wordware Publishing.

"Crash-Proof Web Sites," 2000. *PC Magazine*, 19(7), March 17. Available online at: (http://www.zdnet.com/pcmag/stories/reviews/0,6755,2457531,00.html).

Curtin, M. 2000. "On Guard: Fortifying Your Site Against Attack," *Web Techniques*, April, 46–50.

"Data Encryption Standard (DES)," 1993. *Federal Information Processing Standards Publication 46–2*, December 30.

DoD Directive 5215.1 CSC-STD-001-83. 1983. *Department of Defense Trusted Computer System Evaluation Criteria* (the "Orange Book"), Washington, DC.

"Enduring Web Site Tricks, VPN Phobia," 2000. *Computerworld*, 34(17), April 24, 70.

Gardner, E. 2000. "ZixIt Corp.," *Internet World*, July 1, 28.

Garfinkel, S. and G. Spafford. 1996. *Practical UNIX & Internet Security*. Cambridge, MA: O'Reilly & Associates.

Glass, B. 2000. "Keeping Your Private Information Private," *PC Magazine*, 19(11), June 6, 118–130.

Greenberg, P. 2000. "U.S. Lawmakers Reach E-Signature Accord," *E-Commerce Times*, June 9. Available online at: (http://www.ecommercetimes.com/news/articles2000/000609-9.shtml).

Harrison, A. 2000. "Advanced Encryption Standard," *Computerworld*, 34(22), May 29, 57.

Harrison, A. 2000. "Basically Uncrackable," *Computerworld*, 34(25), June 19, 82.

Isenberg, D. 2000. "Enforcing E-Contracts," *Internet World*, August 1, 62.

Katzenheisser, S. and F. Petitcolas (eds.). 1999. *Information Hiding Techniques for Steganography and Digital Watermarking*, Norwood. MA: Artech House.

Keen, P. 2000. "Designing Privacy for Your E-Business, *PC Magazine*, 19(11), June 6, 132.

Kutter, M. 1998. "Watermarking resistance to translation, rotation, and scaling," *Proceedings of SPIE: Multimedia Systems and Applications*, Vol. 3528, Boston, November 1–6, 423–431.

Lehman, D. 2000. "Privacy Policies Missing On 77% of Web Sites," *Computerworld*, 34(16), April 17, 103.

Merkow, M., J. Breithaupt, and K. Wheeler. 1998. *Building SET Applications for Secure Transactions*. New York: John Wiley and Sons.

Nash, K. 2000. "Fine-Tuned Security," *Computerworld*, 34(20), May 15, 76.

Petitcolas, F. and R. J. Anderson. 1999. "Evaluation of copyright marking systems," *Proceedings of the 1999 International Conference on Multimedia Computing and Systems (ICMCS '99)*, June 7–11.

Pleas, K. 1999. "Certificates, Keys, and Security," *PC Magazine*, 18(8), April 20, 227–230.

"Plug Those Leaky Privacy Policies," 2000. *Internet World*, September 15, 20.

Popovich, K. 2000. "Biometrics Gets Thumbs Up," *eWeek*, 17(24), June 12, 37.

Radcliff, D. 2000. "Digital Signatures," *Computerworld*, 34(15), April 10, 64.

Rivest, R. 1992. *The MD5 Message-Digest Algorithm*, IETF RFC 1321.

Rosencrance, L. 2000. "Microsoft, Others Roll Out Tools to Guard Online Privacy," *Computerworld*, 34(26), June 26, 8.

Rubin, A. 2000. "None of Your E-Business: Protecting Your Privacy in Cyberspace," *Web Techniques*, April, 55–58.

Schorr, J. 1999. "Computer forensics whiz is a hard-driving snoop," *San Diego Union-Tribune ComputerLink*, May 25, 2.

"WebAgain 1.0 erases graffiti," *eWeek*, 17(27), July 3, 57.

Zelnick, N. 2000. "Digital Signatures Won't Get Such a Quick Sign-on," *Internet World*, June 15, 20.

236

ELECTRONIC PAYMENT SYSTEMS

INTRODUCTION

Back in the fall of 1998, Mark Fenwick did what any good son would do: He sent his mother a birthday card. Unlike other years, Mark sent the card in time to arrive on her birthday. What was even more different from previous years was that he didn't leave his residence or his office desk to do it. No, that year Mark navigated to the **Greeting-cards.com** store and bought an electronic birthday card online. He paid for it using his **Mondex** smart card, a plastic card with an embedded chip containing electronic cash. Wells Fargo employees like Mark can pay for online purchases by swiping their smart cards through a special card reader. This transfers money from their card to the merchant. Mark and his coworkers at **Wells Fargo Bank** have been championing and testing smart cards since mid-1998. The bank hopes that Mondex will catch on in a big way for electronic commerce and take some of the market share from other payment methods, such as traditional credit cards. Wells Fargo and its smart card testers hope to persuade customers, who are reluctant to use their credit cards on the Internet, to switch to the new digital cards. If the trials go well, Wells Fargo plans to market the cards to a wider audience. Whether the smart card will become an economic success in the United States remains to be seen. It

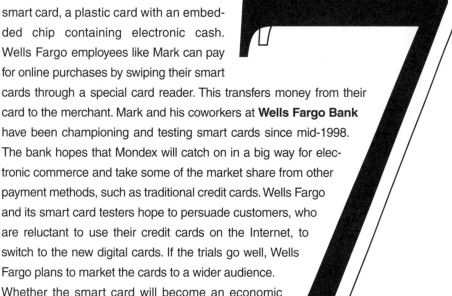

is used regularly in Europe, and that may provide the critical mass to keep it alive. In either case, it is a theoretically interesting idea and one of several described in this chapter. One thing is certain: The Internet is rapidly improving the way people work and shop, and the Internet economy undoubtedly will become the single largest part of the global economy's infrastructure. Before very long, the Internet will be viewed as a dominant vehicle for conducting business, not just a tool for finding information and exchanging e-mail messages.

Huge amounts of money are tendered every hour on the Internet. Money is deposited and managed electronically by online brokerages that individually manage assets worth more than $375 million. Each day, Dell takes in millions of dollars on its Web site. In addition, Americans alone will spend more than $41 billion online in 2002, according to a *PC Magazine* article. Systems are in place and being improved to handle these huge and growing transaction volumes. It is likely some payment systems may die a premature death. Those systems that suffer an early death do so for a variety of reasons. Some are ahead of their time; others are simply not perceived as being as safe or as convenient as their inventors thought they would be. Whatever the reason, some systems described here will be wild successes and others will fade away. It is too early to separate the winners from the losers. This chapter describes existing electronic payment systems and presents some historical background on a few other systems—not all of which have survived.

LEARNING OBJECTIVES

In this chapter, you will learn about:
- Distinct methods to collect payments from customers
- The history and near-term future for electronic cash
- The implementation of electronic cash systems
- Electronic wallets and how they work
- The role of smart cards in electronic commerce
- Credit and debit card processing for electronic commerce transactions
- The most popular electronic payment systems and those that show promise of gaining acceptance
- The SET protocol and how it protects credit card transactions

THE BASICS OF ELECTRONIC PAYMENT SYSTEMS

Chapter 3 described Web-based tools for electronic commerce—Web browser software. Millions of Web pages on the Internet provide a forum for discussing issues and a way to make information available in a global setting. Chapter 4 described electronic commerce software. The largest distinction between traditional Web servers and Web commerce servers is the concept of money: handling payments over the Internet. Electronic commerce involves the exchange of some form of money for soft or hard goods and services.

Implementation of electronic payment systems is in its infancy and still evolving. The technical, economic, cultural, and legal components of electronic payment systems are not fully understood. As a result, there are a number of competing proposals and implementations of electronic payment systems. One thing is clear to everyone involved in electronic payments: The electronic payment method is far cheaper than the dead-tree method of mailing out paper invoices and then later processing received payments. David Samuel, vice president of customer care at BEC Energy's Boston Edison (a company with more than 640,000 customers), says that electronic billing and payment systems are a win-win situation: They are convenient for customers and save the company a lot of money. Estimates indicate that the cost of billing one person varies between $1 and $1.50. Sending bills and receiving payments over the Internet promise to drop the billing/paying cost to an average of 50 cents per bill. The total savings is huge when you multiply the unit cost by the number of customers who could use electronic payment. In the case of BEC Energy, the company could save over 50 cents per customer—more than $320,000 every billing cycle. If that does not convince you of the value of electronic payment, then think about environmental conservation. John Dodge, a columnist for *The Wall Street Journal Interactive Edition*, wrote "GTE sends out 53.5 million bills annually, consuming 1.6 million pounds of paper. That's 2073 trees."

There are currently three basic ways to pay for your purchases: cash, check, or credit card. Electronic cash distribution and payment can be handled by wallets, smart cards, or proprietary, limited-use scrip (Flooz is an example). **Scrip** is digital cash minted by a small number of third-party organizations. Checks, for the purpose of the discussions in this book, do not refer to conventional paper checks that you can use to pay for Internet products after the seller is sure your check has cleared your bank; they are encrypted representations that resemble electronic cash. Electronic checks, which are not described in detail in this book, represent a small but growing percentage of online payment transactions. Credit and debit cards are by far the most popular form of electronic payment in the business-to-consumer market. A recent survey indicates that over 80% of Internet purchases are paid for with credit (or debit) cards. Debit cards, which draw directly from your bank account in the same way that checks do, are also used for online transaction processing.

When your customers arrive at your store's electronic checkout counter, you want to offer them payment options that are safe, convenient, and widely accepted. The key is to ascertain which choices work the best for your company and for your customers. You will have to make some choices and choose the best one or two solutions for your situation and budget. This chapter will provide you with the know-how to make those decisions.

Four technologies appear in this chapter. They are electronic cash, software wallets, smart cards, and credit/debit cards. Each has unique properties, costs, advantages, and disadvantages. Some represent solutions that are already popular and widely accepted, while others are only now catching on and have an unclear future. Still others do not appear to hold the promise of wide acceptance that they did only a few years ago. Regardless of how popular a particular payment solution is or whether it is faltering, you will learn what has worked and what appears not to be working. Knowing the upsides and downsides of each method will help you understand electronic commerce in general and electronic payments in particular so that you can make wise decisions.

All the preceding electronic payment methods work well for Web commerce sites. Business-to-business transactions almost always use different ways to invoice and pay for materials and services. According to the **GartnerGroup**, 12 billion business bills are generated in the United States each year, and 12% of those bills are paid electronically. Only 9% of all business-to-business electronic payments are made using credit cards.

The payment method consumers prefer varies slightly based on the venue. For auction sites—a venue whose operation and rules are different from a typical online store (see Chapter 10)—consumers pay for their purchases with credit or debit cards 77% of the time, according to a recent survey by **BizRate.com**. (BizRate.com is both a portal and Web site performance rating firm that gathers its statistics from online surveys.)

Though this chapter focuses on the business-to-consumer model, it would be incomplete without mentioning business-to-business briefly. Typically, business trading partners use their own private VPN network (see Chapter 2), extranet network (see Chapter 2), and electronic data interchange (EDI) to exchange documents and electronic accounting documents with each other.

ELECTRONIC CASH

Banks that issue credit cards make money, in part, by charging merchants a processing fee ranging from 1.5% to 3% of the value of the transaction. Often, they impose a minimum fee of 20 cents per transaction, or more. A GartnerGroup survey, whose results were reported in *eWeek* magazine, found that some banks charge dot-coms more than their equivalent brick-and-mortar stores—up to $1 more per credit card transaction. The survey reported that a $100 transaction charged on a credit card cost $3.10, compared to a charge of $2.10 to process the same transaction for a brick-and-mortar retailer.

You probably have observed that stores accepting credit cards may require a minimum purchase amount of $10 or $15. Merchants impose a minimum purchase _____ ause the fees for small purchase amounts would erode their profits enor- _____ same is true for Internet purchases. Small purchases are not profitable _____ ts who accept only credit cards for payment. Is there an online market _____ rchases on the Internet—say, for purchases below $4? You can bet on it!

That's where electronic cash can shine. With very low fixed costs, electronic cash provides the promise of allowing users to spend, for example, 50 cents for an online Sunday newspaper or $1.55 to send an electronic greeting card.

According to the latest estimates, approximately 31% of the U.S. population and most of the world do not have credit cards. Some of the adult population cannot obtain credit cards due to minimum income requirements or past debt problems. Children and teens—eager purchasers representing a significant percentage of online buyers—are ineligible simply because they are too young. For both of these groups, electronic cash provides the solution to paying for online purchases—using both micropayments and larger payments.

Even though there have been many failures in the last few years in electronic cash introductions, electronic cash just refuses to die. Compaq and IBM are among several companies that think electronic cash schemes are in their infancy, and these companies envision a rosy future for such methods. Electronic cash is attractive in two arenas: in the sale of goods and services of less than $10 dollars—the lower threshold for credit card payments—and in the sale of higher-priced goods and services to those without access to credit cards.

Internet payments for items costing from a few cents to approximately $10 are called **micropayments**. Micropayment champions see lots of applications for such small transactions, such as paying 5 cents for an article reprint or 25 cents for a complicated literature search. However, micropayments do not appear to be catching on with merchants or consumers yet. (Stay tuned.) A GartnerGroup survey indicated that 13% of the 75 merchants surveyed said they do offer items for $10 or less, but none of the surveyed merchants use micropayment systems. Examples of companies that offer micropayment systems include **iPin**, **eCharge**, **Qpass**, **1ClickBrands**, and **PayPal.com** (PayPal is discussed later in this chapter).

Regardless of the sometimes confusing and seemingly opposing groups of electronic payment systems, all electronic payment schemes have some issues that must be satisfactorily resolved to allay consumers' fears and give them confidence in the methodology. Otherwise, consumers will abandon the technology, no matter how promising it may seem. Concerns about electronic payment methods include privacy and security, independence, portability, divisibility, and convenience. These issues are particularly important when considering electronic cash payment systems, which are discussed next.

Privacy and security questions are probably the most important issues that have to be addressed with any consumer. Fundamentally, consumers want to know, "Is my transaction vulnerable? Can the electronic currency be copied, reused, or forged?" If the answer to any of these questions is yes, then people will not use the system and it will soon cease to exist. Electronic security concerns were discussed in Chapter 5.

Electronic cash, one of the forms of payment described in this chapter, brings with it unique security problems. Electronic cash should have two important characteristics in common with real currency. First, it must be possible to spend electronic cash only once, just as with traditional currency. Second, electronic cash ought to be anonymous, just as hard currency is. That is, security procedures should be in place to guarantee that the entire electronic cash transaction occurs between two parties such that the recipient knows that the electronic currency being received is not counterfeit or being used in two different transactions. In addition, the consumer (and sometimes the seller) should be able to use electronic cash to avoid revealing who he or she is—for a variety of completely legitimate reasons. Anonymity also prevents the seller from collecting information about individual or group spending habits. Of course, credit card transactions should be private and secure also. But the credit card user realizes that he or she is giving up some measure of privacy by using a credit card. Figure 7-1 shows an example of the **Beenz.com** U.S. home page. Beenz is a kind of scrip that consumers can earn and exchange for goods and services from supporting commerce sites.

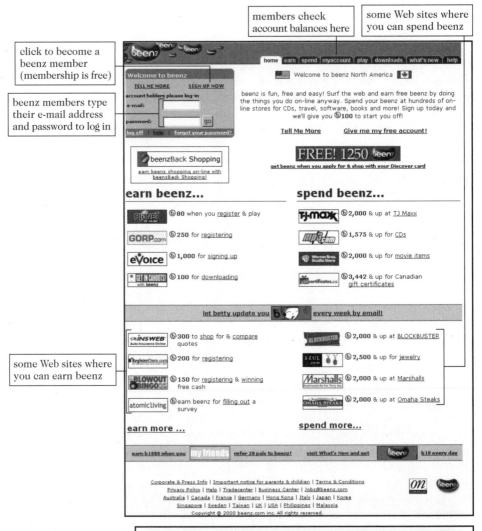

members check account balances here

some Web sites where you can spend beenz

click to become a beenz member (membership is free)

beenz members type their e-mail address and password to log in

some Web sites where you can earn beenz

Figure 7-1   *Beenz U.S. home page*

Electronic cash has the advantages of being independent, portable, and divisible. When electronic cash (also called e-cash and digital cash) is independent, it is unrelated to any network or storage device. That is, electronic cash is really not free-floating currency if its existence depends on a particular proprietary storage mechanism or "cash box" that is specially designed to hold one type of electronic cash. Electronic cash must be able to pass transparently across international borders and be automatically converted to the recipient country's currency. Electronic cash portability means that it must be freely transferable between any two parties in all forms of peer-to-peer transactions. In contrast, charge and credit cards do not possess this property of portability or transferability between every combination of two parties. In a credit card transaction, the credit card payment recipient must already have a merchant account established with a bank—a condition that is not required with electronic cash.

Divisibility is a property that distinguishes electronic cash from real currency. Divisibility determines the size of payment units. Both the number of different electronic cash units and their values can be defined independently of real currency. For example, participants in electronic cash transactions in the United States might decide that the smallest electronic cash unit that they want to deal with is $1. The next denomination might be $1.20, and so on. The denominations are up to the definers and are not limited to the typical breakdowns of a traditional cash system.

Perhaps the most important characteristic of cash is convenience. If electronic cash requires special hardware, software, or finely honed expertise, then it will not be convenient for people to use. Chances are good that people will cast a virtual no-confidence vote for any difficult-to-use electronic cash system and quickly cause the demise of the system.

Holding Electronic Cash: Online and Offline Cash

Two widely accepted approaches to holding cash exist today: online storage and offline storage. Online cash storage means that the consumer does not personally possess electronic cash. Instead, a trusted third party—an online bank—is involved in all transfers of electronic cash and holds the consumers' cash accounts. Online systems work by requiring merchants to contact the consumer's bank to receive payment for a consumer purchase, which helps prevent fraud by confirming that the consumer's cash is valid. This resembles the process of checking with a consumer's bank to ensure that a credit card is still valid and that the consumer's name matches the name on the credit card.

Offline cash storage is the virtual equivalent of money you keep in your wallet. The customer holds it, and no trusted third party is involved in the transaction. Protection against fraud is still a concern (you will see why later), so either hardware or software must prevent double or fraudulent spending. Smart cards, discussed later in this chapter, are the hardware solution to storing electronic money. Encrypted and thus tamperproof methods are the software solution (described later) to prevent double spending. **Double spending** is spending a particular piece of electronic cash twice by simply submitting the same electronic currency to two different vendors. By the time the same electronic currency clears the bank for a second time, it is too late to prevent the fraudulent act. Figure 7-2 shows the home products page of CyberCash, one of the best-known electronic cash sites on the Web.

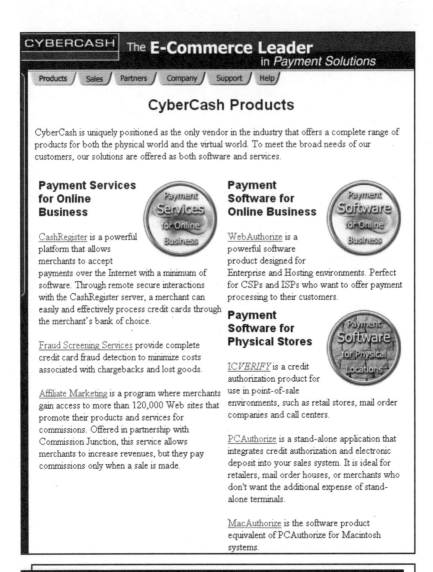

CYBERCASH The **E-Commerce Leader**
in *Payment Solutions*

Products / Sales / Partners / Company / Support / Help

CyberCash Products

CyberCash is uniquely positioned as the only vendor in the industry that offers a complete range of products for both the physical world and the virtual world. To meet the broad needs of our customers, our solutions are offered as both software and services.

Payment Services for Online Business

CashRegister is a powerful platform that allows merchants to accept payments over the Internet with a minimum of software. Through remote secure interactions with the CashRegister server, a merchant can easily and effectively process credit cards through the merchant's bank of choice.

Fraud Screening Services provide complete credit card fraud detection to minimize costs associated with chargebacks and lost goods.

Affiliate Marketing is a program where merchants gain access to more than 120,000 Web sites that promote their products and services for commissions. Offered in partnership with Commission Junction, this service allows merchants to increase revenues, but they pay commissions only when a sale is made.

Payment Software for Online Business

WebAuthorize is a powerful software product designed for Enterprise and Hosting environments. Perfect for CSPs and ISPs who want to offer payment processing to their customers.

Payment Software for Physical Stores

ICVERIFY is a credit authorization product for use in point-of-sale environments, such as retail stores, mail order companies and call centers.

PCAuthorize is a stand-alone application that integrates credit authorization and electronic deposit into your sales system. It is ideal for retailers, mail order houses, or merchants who don't want the additional expense of stand-alone terminals.

MacAuthorize is the software product equivalent of PCAuthorize for Macintosh systems.

Figure 7-2   *CyberCash, a pioneer in electronic cash*

The Advantages and Disadvantages of Electronic Cash

Billing for goods and services that customers have purchased is part of any business. Traditional billing methods in the brick-and-mortar paradigm are costly and involve generating invoices, stuffing envelopes, affixing proper postage to the envelopes, and sending the invoices to the customers. Meanwhile, the accounts payable department must keep track of incoming payments, post accounts in the database, and ensure that customer data are current.

Online stores have many of the same payment collection inefficiencies as their brick-and-mortar cousins. Most online customers use credit cards to pay for their purchases, while auction customers may also use conventional payments methods, including checks and money orders. Electronic cash systems, though less popular than other payment methods, do provide several advantages as well as some disadvantages unique to electronic cash.

For the most part, electronic cash transactions are more efficient (and therefore less costly) than other methods, and that should foster more business, which eventually means lower prices for consumers. Transferring electronic cash on the Internet costs less than processing credit card transactions. You have observed that conventional money exchange systems require banks, bank branches, clerks, automated teller machines, and a parallel electronic transaction system to manage, transfer, and dispense cash. That is expensive. On the other hand, electronic cash transfers occur across an existing infrastructure, the Internet, and through existing computer systems. So the fixed cost of hardware to handle electronic cash is nearly zero. Because the Internet spans the globe, the distance that an electronic transaction must travel does not affect cost. When considering moving physical cash and checks, distance and cost are proportional—the greater the distance that the currency has to go, the more it costs to move it. However, moving electronic currency from Detroit to San Francisco costs the same as moving it from San Francisco to Hong Kong. Everyone can use electronic cash. Merchants can pay other merchants in a business-to-business relationship, and consumers can pay each other. Electronic cash does not require that one party have any special authorization, as is required with credit card transactions.

Electronic cash has disadvantages, and they are significant. In the United States, for example, the idea of an Internet tax continues to be a hot topic. While not a distinct disadvantage, the taxation issue is merely the other side of the electronic commerce coin—the lack of taxation (sales taxes, for example, for out-of-state purchases) for Internet purchases is an advantage for both consumers and merchants. The concept of an Internet tax poses many problems and questions: Can a seller located in the United States charge and collect an Internet tax on items sold to a purchaser in Zimbabwe? Should Zimbabwe receive part of the tax? Worse, using electronic cash to pay any taxes provides no audit trail. In other words, electronic cash is just like real cash in that it cannot be easily traced.

Because true electronic cash is not traceable, another large problem arises: money laundering. Money laundering can easily occur through purchases of goods and services with electronic cash. Ill-gotten electronic cash is spent anonymously for like-valued goods. The goods are sold for real cash on the open market. Of course, goods can be purchased in another country, further complicating jurisdiction issues.

Just like its physical currency counterpart, electronic cash is susceptible to forgery. It is possible, though increasingly difficult, to create and spend forged electronic cash. (Like any other kind of Internet-based activity, without strong countermeasures, electronic forgeries can occur.) Beyond forgery, there are several potentially damaging digital economic factors. These factors have to do with the expansion of the money supply when banks loan electronic cash on consumer and merchant accounts in traditional bank accounts. These factors are beyond the scope of this textbook, but you should be aware that they exist. For more information, look at **Understanding the Digital Economy** and **The Economic and Social Impacts of Electronic Commerce** in the Online Companion.

Finally, electronic cash is, so far, a commercial flop. Merchants worldwide have been slow to adopt electronic cash as an acceptable payment system. Making electronic cash a popular alternative payment system requires wide acceptance and a solution to the pervasive problems of multiple electronic cash standards. Customers do not want to have to carry a dozen different brands of electronic cash to be able to purchase goods from a majority of the merchants that accept electronic cash. Establishing electronic cash as a popular payment method requires that a standard be developed for electronic cash disbursement and acceptance—a standard that individual vendors then implement for their individual electronic cash systems. Electronic cash from different vendors must be easily interchangeable so that customers can exchange one cash type for another when needed.

How Electronic Cash Works

To establish electronic cash, a consumer goes in person to open an account with a bank and shows some identification to establish identity. Whenever the consumer wants to withdraw electronic cash to make a purchase, he or she accesses the bank through the Internet and presents proof of identity—usually a digital certificate issued by a certification authority. After the bank verifies the consumer's identity (see Chapter 6), it issues the consumer a particular amount of electronic cash and deducts the same amount from the consumer's account. In addition, the bank may charge a small processing fee (proportionate to the amount of electronic cash issued). The consumer stores the electronic cash in a wallet (described later in this chapter) on his or her computer's hard disk, or on a special electronic card device called a smart card (also described later in this chapter).

Consumers can spend their electronic cash at electronic commerce sites that accept electronic cash for payment. Briefly, the consumer sends electronic cash (the logistics of this are described later in the chapter) to the merchant for the specified total cost of the goods or services. Then, the merchant validates the electronic cash (that is, the merchant determines that it is not forged and it belongs to the consumer). Only when the goods or services are shipped to the consumer can the merchant present the electronic cash to the issuing bank for deposit. The bank then credits the merchant's account for the transaction amount, minus a small service charge.

Throughout this process, the electronic cash must be protected from both theft and alteration. Additionally, the bank and merchant must be able to verify that the electronic cash belongs to the consumer who spent it. Electronic cash security is briefly discussed in the next section.

Providing Security for Electronic Cash

We have already mentioned one significant problem with electronic cash: its potential for double spending, which is apparent in the previous description of how electronic cash works. To prevent double spending, the main security feature is the threat of prosecution. Complex cryptographic algorithms are the keys to creating tamperproof electronic cash that can be traced back to its origins. A complicated two-part lock provides anonymous security that also signals when someone is attempting to double spend cash. When a second transaction occurs for the same electronic cash, a complicated process comes into play that reveals the identity of the original electronic cash holder. Otherwise, electronic cash that is used correctly maintains a user's anonymity. (See the Brands reference listed in the "For Further Study and Reference" section at the end of the chapter for more details about the double-lock system.) What's important is that a procedure is available both to protect the anonymity of electronic cash users and to simultaneously provide built-in safeguards to prevent double spending. Figure 7-3 shows a graphic representation of detecting double spending using a double-lock system.

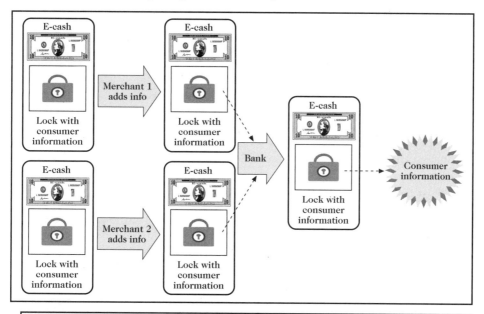

Figure 7-3   *Detecting double spending*

Double spending can neither be detected nor prevented with truly anonymous electronic cash. **Anonymous electronic cash** is electronic cash that, like bills and coins, cannot be traced back to the person who spent it. One way to be able to trace electronic cash (to prevent money laundering, for example) is to attach a serial number to each electronic cash transaction. That way, cash can be positively associated with a particular consumer. That does not solve the double spending problem, however. While a single issuing bank could detect if two deposits of the same electronic cash are about to occur, it is impossible to ascertain *who* is at fault—the consumer or the merchant. Of course, electronic cash that contains serial numbers is no longer anonymous, which is one reason to acquire electronic cash in the first place. Electronic cash containing serial numbers also raises a number of privacy issues, because merchants could use the serial numbers to track spending habits of consumers.

Creating truly anonymous electronic cash requires a bank to issue electronic cash with embedded serial numbers such that the bank can digitally sign the electronic cash while removing any association of the cash with a particular customer. The process begins when a consumer creates a random serial number that he or she sends to the bank issuing the electronic cash. The bank uses the consumer's random serial number along with the bank's digital signature and sends the random number, electronic cash, and digital signature as one package back to the user. When the user receives the electronic cash bundle, he or she extracts the original random serial number and keeps the bank's digital signature (see Chapter 6). The consumer can now spend the electronic cash, which is digitally signed by the bank. When the consumer spends the electronic cash and the merchant passes it along to the issuing bank, the bank validates the electronic cash because it contains the bank's digital signature. But, the bank cannot determine who the spender is. It only knows that the electronic cash is genuine. That's true anonymous cash. (See the Chaum 1992 reference listed in the "For Further Study and Reference" section for further details.)

Electronic Cash Systems

Electronic cash has not been nearly as successful in the United States as it has been in Europe and Japan. Compaq Computer's electronic cash technology allows users to use its NetCoin electronic cash to pay 45 cents for an hour's worth of playing a video game or 4 cents to download a piece of clip art. IBM is bullish on electronic cash and has posted software code for its IBM Micro Payments system (see **AlphaWorks**) so that software and Web site developers can try it and provide feedback. So, even though U.S. Web consumers at large haven't embraced electronic cash, it has plenty of supporters.

Many other attempts at establishing electronic cash systems have had lukewarm receptions by users. **CyberCash**, which created CyberCoin (its electronic cash and requisite software system to handle the cash) for micropayments, is not the huge success it was hoped to be. DigiCash is the acknowledged pioneer in the electronic cash field. Founded by David Chaum, DigiCash filed for bankruptcy. First Virtual has gotten out of online payments completely and has reinvented itself as **MessageMedia**. First Virtual tracked and recorded the transfer of information, products, and payments for accounting and billing. IBM's Micro Payments software is more for technically oriented folks than it is for consumers.

KDD Communications (KCOM) is the Internet subsidiary of Kokusai Denshin Denwa, which is Japan's largest global phone company. KCOM offers its own NetCoin electronic cash system and offers electronic cash through its NetCoin Center, which is supported by a software system called **MilliCent**. Shoppers can go to the NetCoin Center and fill their electronic wallets (see the next section on "Electronic Wallets") with electronic cash. Then, they can shop online for recipes, travel directories (10 cents each), or download MP3 music for less than a dollar per song. Soon, other content providers, such as Japanese newspapers, will provide access to their newspaper archives and charge a small fee to retrieve articles. There are even plans in Japan to open a donation site where browsers can donate electronic coins to a few organizations.

Reasons for past failures of electronic cash systems in the United States are not completely clear. Some industry observers blame the failure on the way that many electronic cash systems were implemented. Most of these systems required the user to download and install complicated client-side software that ran in conjunction with the browser. Also, there were a number of competing technologies, and therefore no standards were ever developed for the entire electronic cash system. The absence of electronic cash standards means that consumers are faced with choosing from an array of proprietary electronic cash alternatives—none of which are interoperable. **Interoperable software** will run transparently on a variety of hardware configurations and on different software systems.

Despite their rough start, not all U.S.-based electronic cash ventures have failed. Next, you will learn about some of the Internet companies that are currently offering electronic cash services and bill presentment and payment systems. The companies briefly described next are either already successful or on the road to success.

CheckFree

CheckFree, the largest online bill processor, provides online payment processing services to both large corporations and individual Internet users. CheckFree provides the infrastructure and software to permit users to pay all their bills with online electronic checks. Figure 7-4 shows a demonstration of a facsimile check in the process of being constructed and paid by an online consumer. To view the demonstration, click the **CheckFree** link in the Online Companion and then click the "see a demo" link to launch the JavaScript-driven demonstration. The demonstration walks you through the process of paying a bill online. Try it. Recently, CheckFree signed an agreement with Yahoo! to provide Yahoo! customers the option of paying directly from the Yahoo! site (see Figure 7-5).

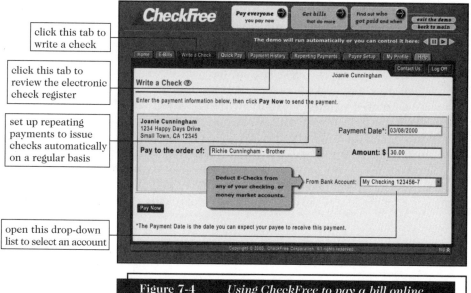

click this tab to write a check

click this tab to review the electronic check register

set up repeating payments to issue checks automatically on a regular basis

open this drop-down list to select an account

Figure 7-4   *Using CheckFree to pay a bill online*

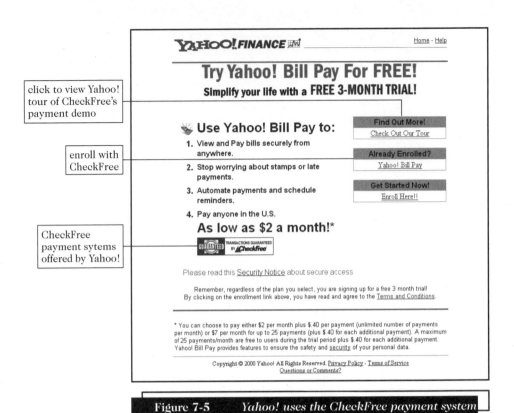

click to view Yahoo! tour of CheckFree's payment demo

enroll with CheckFree

CheckFree payment sytems offered by Yahoo!

Figure 7-5 *Yahoo! uses the CheckFree payment system*

Clickshare

Clickshare is an electronic cash system aimed at magazine and newspaper publishers. Clickshare's technology has occasionally been miscast as a micropayment-only system, such as MilliCent or IBM's Micro Payment. The ability to make micropayments is only one of Clickshare's features. Users with an Internet Service Provider (ISP) that supports Clickshare are automatically registered with Clickshare. When users click links leading to other sites that are registered with Clickshare, they can make purchases on those sites without having to register with Clickshare again. Clickshare keeps track of transactions and bills the user's ISP. The ISP, which already has an account relationship with the user, then bills the user for his or her purchases.

Another feature of Clickshare is that it tracks where a user travels on the Internet. This has significant value to advertisers and marketers who want to measure audience preferences. The micropayment capability is, according to the company, a by-product of the core functionality of tracking identified users.

Clickshare tracks users with the standard HTTP Web protocol and does not require cookies or software wallets. Clickshare claims to be the only company that can do this. (Click **Clickshare service diagram** for a diagram and explanation of how users are billed for the hyperlinks that they click.) Figure 7-6 shows Clickshare's home page.

Figure 7-6   *Clickshare's home page*

DigiCash (eCash)

As mentioned earlier, DigiCash was a trailblazer in the electronic cash field. David Chaum started the company in the Netherlands in order to use an encryption technology that the United States banned from export. Later, Chaum moved DigiCash to Silicon Valley. DigiCash made software products that allowed users to purchase goods and services on the Internet and to pay for them using anonymous electronic cash. The Mark Twain Bank of St. Louis was the only bank in the mid-1990s that used the DigiCash **eCash** system to allow consumers and businesses to exchange eCash for U.S. dollars. DigiCash reorganized and is now called eCash. ECash offers a wide variety of services, including a robust electronic cash payment system. Figure 7-7 shows the **eCash demonstration page**.

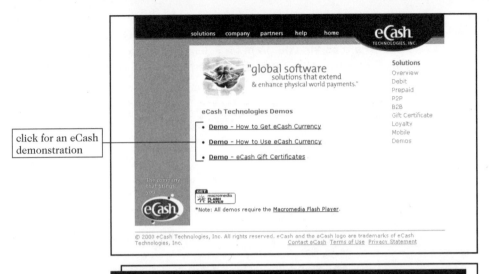

click for an eCash demonstration

Figure 7-7 *eCash demonstration page*

eCoin.Net

ECoins are electronic tokens issued by **eCoin.net**. Consumers can use the tokens to pay for online goods. ECoins provide a way to make micropayments online. The electronic cash is stored in an eCoin wallet on the consumer's computer. As with similar micropayment systems, you can purchase and download a single newspaper article for a few cents, browse a pay-per-view Web site, or download a single music track for 75 cents. Each merchant sets its own fees. Of course, a merchant's commerce site must be set up to handle eCoin electronic cash in order for the consumer to use this currency. A consumer can use eCoins by first installing the wallet software as a plug-in to his or her Web browser. Merchants that accept eCoins do not have to install special software. Instead, an eCoin-compatible commerce site creates special invoice tags in its HTML pages to enable the consumer's eCoin manager (the wallet installed in the user's browser).

The eCoin system uses a three-link chain consisting of a consumer, a merchant, and the eCoin server. The eCoin server operates as a broker that maintains and updates consumer and merchant accounts, accepts payment requests from the consumer's software, and computes invoices for the merchant site. The eCoin server runs on eCoin's own site. The eCoin system employs security features that prevent double spending. The eCoin structure makes consumers anonymous to merchants but not anonymous to the eCoin server. This is a conscious decision by eCoin.net to make all eCoin transactions traceable during the early phase of eCoin's operation. Figure 7-8 shows the eCoin home page. Click **eCoin.net** and other links on the eCoin.net home page to learn more about eCoins or to download an eCoin wallet.

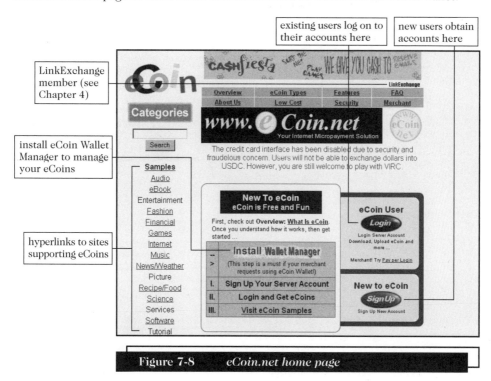

Figure 7-8   *eCoin.net home page*

InternetCash

Merchants who want to provide their online customers the option of using cash rather than credit cards to pay for purchases will want to consider **InternetCash**. It provides electronic currency that is very similar to traditional cash. Customers must first purchase an InternetCash card from stores such as Circle K, Sunglass Hut, or any of almost 10,000 brick-and-mortar stores. Similar to prepaid phone cards, the InternetCash cards come in denominations of $10, $20, $50, and $100. After purchasing a card, customers go online and activate their cards by entering a 20-digit code found on the back of the card and creating a PIN, or personal identification number. (A **personal identification number** is a random assemblage of digits, chosen by the customer, that serves as a password for monetary transactions.) Figure 7-9 shows the InternetCash activation display.

253

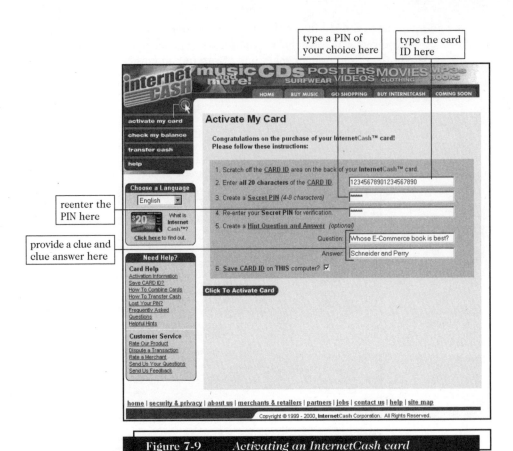

type a PIN of your choice here

type the card ID here

reenter the PIN here

provide a clue and clue answer here

Figure 7-9 *Activating an InternetCash card*

Once their card is activated, customers can pay for purchases using the InternetCash card at any site accepting them. Customers choose to use InternetCash at the electronic checkout counter, where an array of payment choices appears, including popular credit cards such as MasterCard and Visa. Selecting InternetCash as the payment choice causes a secure window to appear. If a customer selects InternetCash to pay for his or her purchases, the merchant sends the customer card number to InternetCash via the Internet. This causes the secure InternetCash payment window to appear on the purchaser's Web page. Once the customer has typed in his or her InternetCash PIN and the card is accepted, InternetCash's Web site creates a digital signature for the purchase. After verifying the payment information, the merchant submits the payment request to InternetCash for processing. InternetCash, in turn, verifies that (by checking the PIN) the customer has the right to use that account.

After verifying the customer's identity, InternetCash transfers the money to the merchant's account and sends the merchant a payment authorization number. The transaction's value is automatically deducted from the value of the card, which is stored at a central storehouse of card numbers and available cash balances. A customer's PIN automatically eliminates any customer nonrepudiation issues (see Chapter 5)—the customer cannot claim that he or she did not order the merchandise.

After the merchant receives confirmation that the customer's InternetCash payment was successful, the merchant then sends the customer an order confirmation number indicating that the sale has been fulfilled. On the merchant's server is the customer's digital signature (received from the customer when he or she placed the order) and a payment authorization number from InternetCash.

InternetCash is already part of the shopping cart packages of several commerce service providers, including Microsoft's Site Server, IBM's Net.Commerce, Intershop, and Mercantec's SoftCart, to name a few. Unfortunately, InternetCash is not interchangeable with other electronic cash types. InternetCash makes money by discounting monies paid to merchants from customer purchases. Customers do not pay a fee to use InternetCash, however. Because InternetCash values sit on a secure InternetCash server, customers do not have to worry about which client computer they are using to make InternetCash purchases, and InternetCash provides a convenient online purchase solution for teens and others who do not have access to credit cards.

MilliCent

MilliCent was developed by Digital, which is now part of **Compaq**. MilliCent is an electronic scrip system that does not issue one standard currency. Instead, each participating merchant creates and sells its own scrip to a broker at discount. Consumers register with one broker and buy broker scrip in bulk—generic scrip. Brokers receive payment from consumers in a variety of ways, but the most common way that consumers pay brokers is with a credit card. When a consumer locates a product on a merchant's Web site that he or she would like to purchase, the consumer converts the broker scrip into vendor-specific scrip. **Vendor-specific scrip** is scrip that a particular merchant will accept. The consumer's new scrip is stored in an electronic wallet on the consumer's computer. Paying for an item involves simply transferring a merchant's scrip, which is in a consumer's wallet, to the merchant in exchange for a purchased item. Merchants can then send a broker their own scrip and obtain a check. MilliCent, like other micropayment systems, provides a way for consumers to purchase items that have a very low value. A consumer can electronically pay a few cents up to a few dollars for items purchased on the Internet.

A broker is required in this setup for two reasons. First, the system of very small payments works because transactions can be aggregated so they are significant. That is, when a consumer buys a few dollars worth of scrip, there is little profit in the deal. But when several consumers purchase scrip, the aggregate amount—discounted so that the broker can make money—does make economic sense. The second reason for a broker is that it makes the entire system easier to use. A consumer has to deal with only one broker, not dozens, to satisfy his or her scrip needs for many merchants. The broker buys the scrip in bulk and does all the hard work of organizing the retail selling to consumers.

PayPal

PayPal (**PayPal.com**) is an electronic cash payment system that is a very popular way for consumers to pay for online purchases. Owned by X.com and touted as the world's first e-mail payment service, PayPal.com is a free service that earns a profit on the **float**, which is money that is deposited in PayPal accounts. PayPal is very popular with **eBay** auction customers—it is currently the number one electronic cash system used by eBay. PayPal eliminates the need to pay for online purchases by writing and mailing checks or using credit cards. PayPal allows consumers to send money instantly and securely to anyone with an e-mail address, including an online merchant. PayPal is a convenient way for auction bidders to pay for their purchases, and sellers like it because it eliminates the risks of accepting other types of online payments. PayPal transactions clear instantly so that the sender's account is reduced and the receiver's account is credited when the transaction occurs. Anyone with a PayPal account— online merchants or eBay auction participants alike—can withdraw cash from their PayPal accounts at any time. Figure 7-10 shows PayPal's home page.

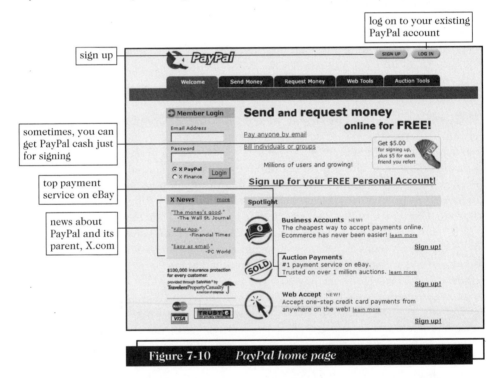

Figure 7-10   *PayPal home page*

To use PayPal, merchants and consumers first must register for a PayPal account. There is no minimum amount that a PayPal account must contain, and customers add money to their PayPal accounts by sending a check or using a credit card. Once members' payments are approved and deposited into their PayPal accounts, they can use their PayPal money to pay for purchases.

Merchants must have PayPal accounts to accept PayPal payments, but consumer-to-consumer markets such as eBay are more flexible. Using PayPal to pay for auction purchases is very popular. Figure 7-11 shows a partial list of over 22,000 items for which auction participants will accept PayPal cash. A consumer can use PayPal to pay a seller for purchases even if the seller does not have a PayPal account. When you use PayPal to pay for purchases from a seller or merchant who does not have a PayPal account, the PayPal service sends the seller or merchant an e-mail message indicating that a payment is waiting at the PayPal Web site. To collect PayPal cash, the seller or merchant who received the e-mail message must register and provide PayPal with payment instructions.

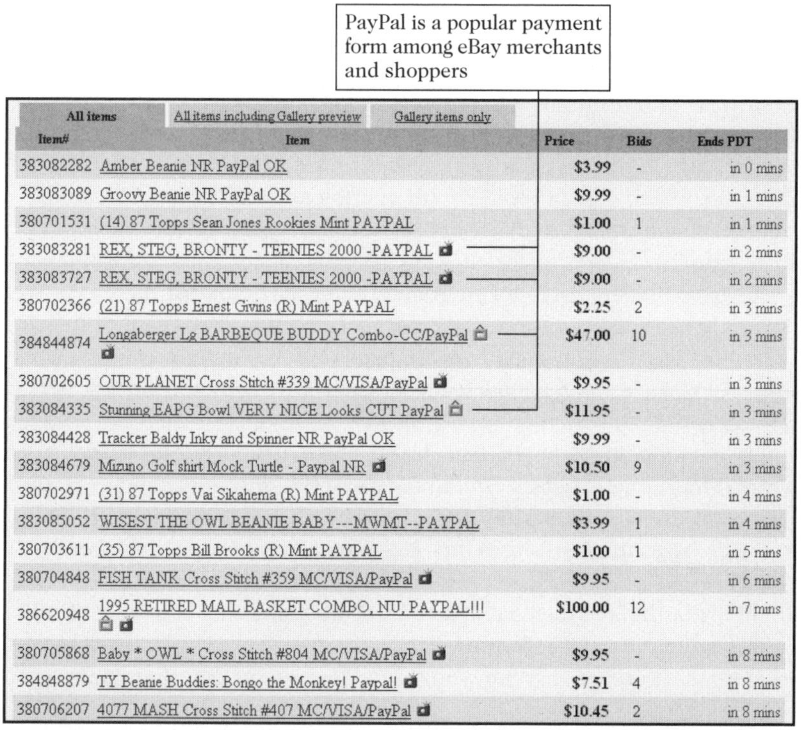

PayPal is a popular payment form among eBay merchants and shoppers

Item#	Item	Price	Bids	Ends PDT
	All items All items including Gallery preview Gallery items only			
383082282	Amber Beanie NR PayPal OK	$3.99	-	in 0 mins
383083089	Groovy Beanie NR PayPal OK	$9.99	-	in 1 mins
380701531	(14) 87 Topps Sean Jones Rookies Mint PAYPAL	$1.00	1	in 1 mins
383083281	REX, STEG, BRONTY - TEENIES 2000 -PAYPAL	$9.00	-	in 2 mins
383083727	REX, STEG, BRONTY - TEENIES 2000 -PAYPAL	$9.00	-	in 2 mins
380702366	(21) 87 Topps Ernest Givins (R) Mint PAYPAL	$2.25	2	in 3 mins
384844874	Longaberger Lg BARBEQUE BUDDY Combo-CC/PayPal	$47.00	10	in 3 mins
380702605	OUR PLANET Cross Stitch #339 MC/VISA/PayPal	$9.95	-	in 3 mins
383084335	Stunning EAPG Bowl VERY NICE Looks CUT PayPal	$11.95	-	in 3 mins
383084428	Tracker Baldy Inky and Spinner NR PayPal OK	$9.99	-	in 3 mins
383084679	Mizuno Golf shirt Mock Turtle - Paypal NR	$10.50	9	in 3 mins
380702971	(31) 87 Topps Vai Sikahema (R) Mint PAYPAL	$1.00	-	in 4 mins
383085052	WISEST THE OWL BEANIE BABY---MWMT--PAYPAL	$3.99	1	in 4 mins
380703611	(35) 87 Topps Bill Brooks (R) Mint PAYPAL	$1.00	1	in 5 mins
380704848	FISH TANK Cross Stitch #359 MC/VISA/PayPal	$9.95	-	in 6 mins
386620948	1995 RETIRED MAIL BASKET COMBO, NU, PAYPAL!!!	$100.00	12	in 7 mins
380705868	Baby * OWL * Cross Stitch #804 MC/VISA/PayPal	$9.95	-	in 8 mins
384848879	TY Beanie Buddies: Bongo the Monkey! Paypal!	$7.51	4	in 8 mins
380706207	4077 MASH Cross Stitch #407 MC/VISA/PayPal	$10.45	2	in 8 mins

Figure 7-11 *All of these eBay auctions accept PayPal*

Loyalty and Rewards Systems

Scrip is a form of electronic cash that is stored on your computer or on the server of the scrip vendor. You obtain scrip by depositing money at a scrip vendor's server. Scrip is the equivalent of a paper gift certificate. You can spend scrip just like cash to purchase an item at any merchant that accepts your particular brand of scrip. Currently, the number of online stores that accept various types of scrip is small but increasing.

Loyalty and rewards programs endorsed by various Web stores frequently use scrip to reward Web visitors for frequenting their stores. Also known as a rewards system, scrip from one program rarely is interchangeable with scrip from another. Frequently, Web sites award scrip to users who refer other customers to their Web sites or induce other customers to purchase items from scrip-awarding Web sites. Flooz and beenz are two of the most popular brands of scrip, and they have secured a large share of the scrip market. Flooz has become particularly visible because it is widely advertised on U.S. television. Whoopi Goldberg is the Flooz national spokeswoman. Many sites use scrip from both of these vendors, and scrip owners can send their scrip as gifts to anyone who has an e-mail address—an especially convenient alternative to paper gift certificates.

Beenz

Beenz is a brand of scrip that is marketed as a loyalty reward program for Internet consumers. It is a new way of attracting consumers to a Web site and rewarding them for visiting. **Beenz.com** has partnered with over 100 Web sites to offer its beenz reward points, which consumers earn for visiting and making purchases at Web sites that offer beenz. Web site members of the beenz program pay visitors in beenz scrip for visiting their sites or for clicking designated site links. Consumers collect beenz and later redeem them for merchandise at participating merchant Web sites. Beenz has only been in operation since March 1999, but already more than 25 million beenz transactions have earned users in excess of 370 million beenz scrip units. 290 million of those units have been spent. On average, one beenz is worth approximately one cent.

Beenz.com is not alone in the marketing loyalty program. Other industry giants such as MyPoints, Flooz, and America Online have started similar consumer reward programs to attract visitors to their sites. So many rewards programs are springing up that the RewardsPrograms.com site is devoted exclusively to tracking Internet rewards programs.

Beenz and Flooz both have the goal of attracting customers to Web sites. Unlike Flooz, beenz units are not available for purchase—they are rewards for consumers to use to purchase items. However, anyone who owns Flooz scrip, described next, can exchange Flooz for beenz.

Flooz

Flooz is scrip that you purchase using a credit card and either use for your own purchases or send to a recipient as a gift or payment. Flooz is redeemable for purchases at any merchant that accepts it. The Flooz exchange rate is one Flooz equals one dollar, and Flooz.com does not charge a service fee to purchase Flooz. You mail the scrip, along with an electronic greeting card (an **e-card**), to the recipient, who can spend his or her Flooz at online stores such as Proflowers.com, Nirvana Chocolates, Books.com, and Starbucks. Flooz earns a commission, which is paid by the merchant, when a consumer spends scrip at an online store. A merchant installs software that communicates with the Flooz.com servers and pays Flooz.com a commission. Consumers cannot exchange Flooz for cash; however, they can use it to pay for goods and services at participating merchants. Figure 7-12 shows the Flooz home page.

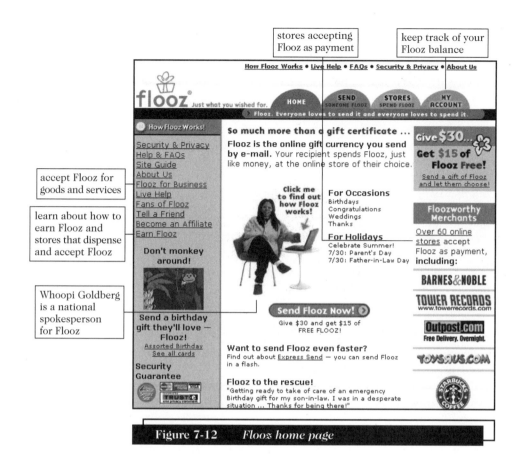

Figure 7-12 *Flooz home page*

The following annotations surround the figure:

- stores accepting Flooz as payment
- keep track of your Flooz balance
- accept Flooz for goods and services
- learn about how to earn Flooz and stores that dispense and accept Flooz
- Whoopi Goldberg is a national spokesperson for Flooz

ELECTRONIC WALLETS

As shoppers are becoming more enthusiastic about online shopping, they have begun to tire of repeatedly entering detailed shipping and payment information each time that they make an online purchase. Research repeatedly has shown that filling out forms ranks high on online customers' list of gripes about online shopping. That is one problem that electronic wallet technology is intended to solve. The other problem electronic wallets solve is providing a secure storage place for credit card data and electronic cash. Thus, an **electronic wallet**, serving a function similar to a physical wallet, holds credit cards, electronic cash, owner identification, and owner contact information and provides that information at an electronic commerce site's checkout counter. Occasionally, an electronic wallet contains an address book too. Electronic wallets make shopping more efficient. When consumers click on items to purchase, they can then click on their electronic wallet to order the item quickly. Though they do not yet do it, wallets could serve their owners by tracking purchases that they make and maintaining receipts for those purchases. Maintaining records of a consumer's purchasing habits is something that online giants such as Amazon.com have mastered, but an enhanced digital wallet could reverse that process and suggest where a consumer might find a lower price on an item that he or she purchases regularly.

Electronic wallets fall into two categories based on where they are stored. A **server-side electronic wallet** stores a customer's information on a remote server belonging to a particular merchant or (better yet) belonging to the wallet's publisher. The main weakness of server-side electronic wallets is that a client-server security breach could reveal thousands of users' personal information—including credit card numbers—to unauthorized parties. Typically, server-side electronic wallets employ strong security measures that minimize or eliminate the possibility of unauthorized disclosure. A **client-side electronic wallet** stores a consumer's information on the consumer's own computer. Storing an electronic wallet on the user's computer shifts the responsibility for maintaining security to the user. Because no user information is stored on a central server, there's no chance that an attack on an electronic-wallet vendor will yield consumer information such as credit card numbers. Many of the early electronic wallets were client-side wallets and required lengthy downloads, which continues to be one of their chief disadvantages. Server-side wallets, on the other hand, remain on a server and therefore require no download time or installation on a user's computer. A disadvantage of client-side wallets is that they are not portable. For example, a client-side wallet is not available when you make a purchase at a location other than the computer on which your wallet resides. Most Internet consumers are nomadic and enjoy the freedom of purchasing anywhere and any time.

For a wallet to be useful at many online sites, it should be able to populate the data fields in any merchants' forms at any site that the consumer visits. This accessibility means that the electronic wallet manufacturer and merchants from many sites must coordinate their efforts so that a wallet can "recognize" what consumer information goes into each of a given merchant's forms. This task can be daunting, but wallet manufacturers are making huge strides in achieving a universal wallet. However, that goal has not yet been achieved.

Exactly what information does an electronic wallet store? Minimally, electronic wallets store shipping and billing information, including a consumer's first and last names, street address, city, state, country, and ZIP or postal code. Most electronic wallets also can hold many credit card names and numbers, affording the consumer a choice of credit cards at the online checkout. Some electronic wallets also hold electronic cash from various suppliers, such as eCash. Finally, some wallets contain an encrypted digital certificate, which securely identifies the wallet's owner. Wallets that store digital certificates are particularly handy when you shop at a site that requests user authentication information, because the wallet can supply the certificate automatically.

Electronic wallets are particularly useful and save a lot of time. Here's why: When shoppers have filled their electronic shopping carts, they proceed to the checkout counter to confirm their choices. At the checkout counter, they are confronted with a one- or two-page form into which they must enter their name, address, credit card number, and other personal information. Figure 7-13 shows an example of one part of a multiscreen order form from 42nd St. Photo. A consumer must fill in all the information boxes to complete the checkout process. Nearly all the text boxes you see in the figure are required.

Please enter your details here. You are now on a Secure Server so you may send your details with confidence. Please wait while this page loads, and then enter all requested details.

Billing Details

First Name	*
Last Name	*
Company:	
Address	*
City:	*
State	
Zip / Postcode	*
Country	United States
Phone daytime	*
Email	*
Phone Evening	
Please enter a password	*

This password is useful for you to track your order, or use for your next order and save having to give us your details all over again. Enter any letters and number between 4 and 12 characters.

* = Required information If Billing address is different than shipping address, please contact your credit card company and inform them to note the account. Their toll free number is on the back of your credit card.

Figure 7-13   A typical electronic checkout counter form

Forms like the one in Figure 7-13 take some time to fill out. Repeatedly having to fill out long forms has cost the electronic industry millions of dollars, because significant numbers of people find the forms daunting and abandon their electronic shopping carts at the checkout counter. Surveys by **Forrester Research** Inc. and **Jupiter Communications** found that between 27% and 65% of shoppers selected the products they wanted to purchase but failed to complete the checkout process. **Brodia** discovered that Web consumers were less likely to abandon their shopping carts at the checkout counter when they used electronic wallets, which enter required information into checkout forms automatically. Other key findings were that consumers felt more secure about electronic wallets issued by major financial institutions. (Additionally, the Brodia study found that consumers appreciated the advantages of shopping on the Internet versus shopping at brick-and-mortar stores, but were concerned about the security of online shopping.) MasterCard recognizes the importance and convenience of electronic wallets. MasterCard offers its own electronic wallet, called the **MasterCard e-wallet**.

One of the first online merchants to recognize the need to simplify the end game of shopping—filling out the name, address, and shipping information—was Amazon.com. Its 1-Click^{sm} shopping speeds shoppers on their way by requiring them to fill out a form only once. Once 1-Click is activated for a consumer, he or she can click one button to fill in the shipping and credit card information automatically and complete the transaction.

The difference between similar single-click systems and a digital wallet is that the single-click systems each work in a particular store. Electronic wallets, in theory, work at many merchants' stores—every merchant that accepts the system. How do you know which merchants accept a particular wallet, or vice versa? Many wallets display a list of the electronic commerce sites that accept their wallet service. Next, you will learn about some of the electronic wallets available and what the World Wide Web Consortium (W3C) has done to standardize electronic wallets.

eWallet

EntryPoint, which merged with **Infogate,** produces **eWallet**, free wallet software that consumers download and install on their computer (it is a client-side wallet). It is not stored on a central server along with others' wallets. Like other wallet technologies, eWallet stores personal and payment information inside the electronic wallet. Interestingly, eWallet even provides a place in the electronic wallet for users to store their favorite photographs—just like a real wallet. When you want to purchase an item, you simply click the eWallet icon on your computer, enter your password, and drag the credit card of your choice from the eWallet to a checkout form. EWallet does the rest by completing the form based on the personal information you supplied when you installed the software. To protect your information, eWallet is encrypted and pass-word-protected. You can download eWallet by clicking the appropriate download icon link found on the eWallet home page (see Figure 7-14). Naturally, eWallet works with both Internet Explorer 4.0 and higher and Netscape Navigator 4.0 or higher.

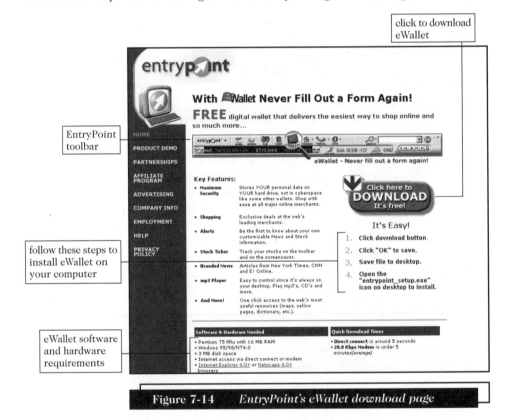

Figure 7-14 *EntryPoint's eWallet download page*

Microsoft Passport

Microsoft Passport Wallet comes preinstalled in Internet Explorer 4.0 and higher but not, of course, in Netscape Navigator. (Netscape Navigator users can install and use Microsoft Passport, though.) It functions in the same way as most other electronic wallets, by automatically completing order forms when you request it to do so. Passport is Microsoft's attempt to standardize wallets. All the personal data you enter into your Microsoft Passport, including your name, address, and credit card information, are encrypted and password-protected. Future versions of the wallet will be able to interact with electronic cash systems, Internet bank accounts, and other payment schemes. Currently, it works with the American Express charge card and credit cards from Diner's Club, Discover, MasterCard, and Visa. You fill out your personal information through the Content tab of the Internet Options command (in Version 4.0, Internet Options is in the View menu; in version 5.x, it is in the Tools menu). Clicking the Wallet button in the Personal information panel of the Content tab brings up the first of a few dialog boxes into which you type personal information. Clicking the Add button and following a few simple steps displays the panel shown in Figure 7-15, into which you enter credit card information. You complete the process by entering your address information in the next panel that appears, then typing your password in the last dialog box.

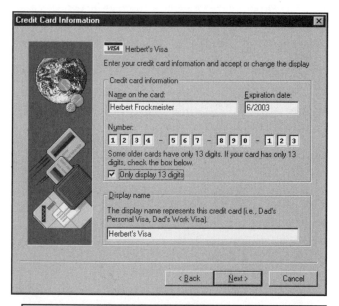

Figure 7-15 *Entering credit card information into a Microsoft Passport Wallet*

Passport consists of four integrated services: Passport single sign-in service (SSI), Passport Wallet Service, Kids Passport service, and public profiles. The sign-in service allows a user to sign in at a participating Web site using his or her username and password. The Passport Wallet service provides standard electronic wallet functions, such as secure storage and form completion of credit card and address information. When requested by a participating merchant, a consumer's secure information is released to the merchant so that the consumer does not need to enter data into a form. The Kids Passport service helps parents protect and control their children's online privacy, and the public profiles service allows consumers to create a public page of information about themselves.

Spending with your Microsoft Passport wallet is easy. First, you locate an item you want to purchase on the Internet at a site that is set up to interact with

Microsoft Passport. Then, when you go to the electronic checkout counter, the merchant's site software asks you whether you want to type your information directly or enter it from your wallet. Choosing the latter will then display the list of credit cards in your wallet. Pick one and type your password. Because you can have more than one "ship to" address, you also tell Microsoft Passport where you want the merchandise shipped. Passport and the merchant's site take over and you type nothing more. **Microsoft Passport merchant directory** shows merchants currently set up to accept Microsoft Passport. The list will undoubtedly grow quickly.

It is difficult to judge which of the several products available will win the electronic wallet competition. To survive, an electronic wallet's publisher will have to convince a critical mass of Internet merchants to interact with their particular wallet's software. Careful Internet merchants will wait to see which wallets consumers favor. The sheer marketing pressure that one or more wallet publisher can bring to bear on merchants may break this stalemate. The large corporations with lots of Internet influence, such as Microsoft, will not sit around and nervously wring their corporate hands.

The W3C Proposed Standard

The World Wide Web Consortium (W3C) has recently weighed in with its proposed standards for electronic wallets. These standards will impact every vendor's wallet offerings. The **W3C Electronic Commerce Interest Group** (ECIG) developed the **Common Markup for Web Micropayment Systems** public working draft. In that document, the ECIG states that it wants to stimulate comment on its proposed draft micropayment standard. ECIG asserts "there is no clear definition of a Web micropayment that encompasses all systems claiming to be micropayment systems." The proposed ECIG standard sets the guidelines for a system that provides an extensible and interoperable way to embed micropayment information in a Web page. An **extensible system** is one that people can easily enhance without voiding any earlier work on the system. The draft proposal goes on to identify the existing system micropayment types of online connections, stored-value systems such as Mondex (described in the next section), and semi-online systems (such as MilliCent or Micro Payments from IBM).

Merchants who want the widest Internet consumer audience must be willing to offer support for several different payments systems. To do so, the merchants' Web pages must embed, within each Web page, payment information that is specific to each payment system that different consumers have adopted (Microsoft Passport, eCash, and so on). This redundancy has motivated the W3C to develop a *common* Web page markup system that multiple payment systems all support. The W3C is moving quickly to propose a standard so that the merchants have a chance to examine and comment on the proposed standard before merchant systems become entrenched—each with their own distinct set of consumers. In other words, the W3C community would like agreement on a standard before it becomes too expensive for existing micropayment systems to change and those existing systems become *de facto* standards.

Currently, the ECIG draft standard proposes the following architecture: The client (the consumer's Web browser) initiates the micropayment activity by requesting information from the merchant server. The client browser consists of the browser itself, a module called Per Fee Link Handler (PFLH), and one or more electronic payment wallets. On the merchant side of the client/server duo is an HTTP server. The W3C document proposes new HTML tags to carry the embedded micropayment information. Figure 7-16 shows a list of the proposed micropayment tags. Soon, the W3C intends to consider the aggregate opinions and feedback from the Web community—the browsing public, Internet consumers, and merchant site producers—to form a final standard. Then it will be up to the individual electronic wallet and electronic cash vendors to develop and revise their software to conform to the new standard.

Field Name	Short Description	Format	Requirements
merchanturl	Identifies the merchant site.	URL	MUST be provided
merchantname	Specifies a merchant designation.	character string	MAY be provided
buyurl	Identifies what the client is buying.	relative URL	MUST be provided
textlink	Describes textually what the client is buying. The text source of the fee link.	character string	MUST be provided
imagelink	Describes graphically what the client is buying. The graphic source of the fee link. (textlink provides a textual equivalent of the image for accessibility.)	URL	MAY be provided
price	Specifies amount and currency.	character string	MUST be provided
duration	Indicates the time after purchase any URL can be retrieved with payment.	integer number	SHOULD be provided
longdesc	Describes in detail the content of the client's purchase.	character string	SHOULD be provided
requesturl	Identifies what the client is actually requesting.	realtive URL	MAY be provided
expiration	Indicates a date until which the offer from the merchant is valid.	character string: YYYY-MM-DDThh:mm:ssTZD	MAY be provided
specific field	Provides information unique to each payment system.	URL and character string	MAY be provided

Figure 7-16   W3C proposed micropayment HTML tags

265

The ECML Standard

A consortium of several high-tech companies and credit card companies proposed another standards initiative to replace the competing electronic wallet standards with a single standard. The consortium of America Online, Brodia, Compaq, CyberCash, Dell, IBM, Microsoft, Visa U.S.A., and MasterCard has agreed on a technology called **ECML**, or electronic commerce modeling language. It is a break-through because users enter credit card and address information once into their ECML-capable electronic wallet. (Any existing wallet can be redesigned to follow the ECML standard, though none currently do because ECML is still in the proposal phase.) Users control access to the ECML electronic wallet, and the wallet will be accepted at *all* commerce sites if the consortium is successful in pushing the new technology to widespread acceptance. So far, several prominent electronic wallet makers have accepted and implemented the standard. They include Brodia, CyberCash, IBM, Microsoft, and Trintech.

The ECML standard will expedite online processing for customers by simplifying the form-filling procedure. Currently, the field names for various forms vary from one online merchant to another. For example, one merchant may refer to the telephone number field as "phone" whereas another merchant may call it "PhoneNumber." The ECML format will work by providing uniform field names upon which all merchants can build their input forms. An ECML-compliant Web site would allow a customer to enter billing, shipping, and payment information once. If the customer visits another ECML-compliant site, the same customer information is already filled out when the customer proceeds to the checkout counter. Figure 7-17 displays some of the ECML-standard fields and their characteristics.

266

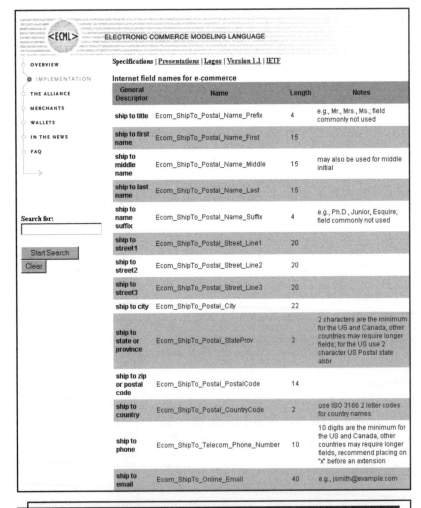

Figure 7-17   *Some ECML-standard fields*

Falling short of some expectations, the ECML standard does not provide standardized names for fields such as age, shopping preferences, or birthday—information that merchants usually collect to provide customers with a more personalized shopping experience. The ECML standard probably will include more advanced features in the future—features such as purchase receipts and package tracking. Clearly, the ECML committee's goal is to provide a leading, easy-to-implement system to spawn ECML-compliant Web sites quickly.

Experts agree that ECML is the best news in electronic commerce recently, and represents a real turning point for online business. Keep a sharp eye out for developments in ECML. Meanwhile, click the **ECML** links in the Online Companion for the latest news in this evolving field.

STORED-VALUE CARDS

How many plastic cards do you carry in your wallet: one, two, five? For most of us, the number is nearly a dozen when you count credit cards, debit cards, charge cards, driver's license, health insurance card, employee or student identification card, and others. Now consider what it would be like to reduce all those cards to a single plastic card carrying all the information of the dozen or so cards you carried before. Would that lighten your load and make identification and purchasing easier for you? Of course it would. That convenience has been pushing several vendors around the world to develop smart cards.

A **stored-value card** can be an elaborate smart card or a simple ("dumb") plastic card with a magnetic strip that records the currency balance. One of the most common stored-value cards, and one you are likely to have used, is a prepaid phone, copy, subway, or bus card. You can recharge most magnetic stripe cards by inserting them into special machines, inserting currency into the machine, and withdrawing the card, whose stripe now reflects an increased cash value. Unlike its smart-card cousins, magnetic stripe cards are passive. That is, they cannot send information such as a customer's digital certificate, nor can they receive and automatically increment or decrement the value of cash stored on the card. A smart card is better suited for Internet payment transactions because it has limited processing capability.

What Is a Smart Card?

A **smart card** is a plastic card with an embedded microchip containing information about you—a physical, electronic wallet. Credit, debit, and charge cards currently store *limited* information about you on a magnetic stripe. And, unlike a smart card, a credit card does not contain cash—it only contains a number of an account that can be charged. A smart card can store over 100 times more information than a magnetic stripe plastic card. A smart card contains private user information such as financial facts, private encryption keys, account information, credit card numbers, health insurance information, and so on.

Smart cards are better protected from misuse than, for example, conventional credit cards, because the information is encrypted. For example, conventional credit cards clearly show your account number on the face of the card. The card number along with a forged signature is all that a thief needs to purchase items and charge them against your card. With a smart card, credit theft is practically impossible because a key to unlock the encrypted information is required, and there is no external number that a thief can identify and no physical signature that a thief can forge. In addition, smart cards provide the advantages of portability and convenience.

Smart cards have been around for over a decade. Popular in Europe, Australia, and Japan, smart cards so far have not been as successful in the United States. In Europe and Japan, smart cards are being used to pay for pay phones and television. The cards are very popular in Australia, too, where practically every retail counter and restaurant cash register has a smart card reader on it.

Smart cards are beginning to appear in the United States in rapidly increasing numbers. In San Francisco, the Bay Area **Metropolitan Transportation Commission** voted to award a multimillion dollar contract to a Motorola-led consortium to build the first integrated ticketing system for public transportation in the United States—a completely smart card-based system. A transportation smart card would allow commuters to ride most modes of public transit available in the city, including trains, buses, cabs, and ferries, by simply waving a single card near a reader device in transit vehicles or in stations. People using the **TransLink** system could "recharge" their smart cards at several retail outlets or directly from their bank accounts. Visa recently introduced its smart card, the **Smart Visa card**, which will likely be very popular. The American Express smart card, called **Blue** is sure to be a huge sucess too.

In the United States, the **Smart Card Forum** is promoting the benefits of smart cards. The organization promotes the widespread acceptance of multiple-application smart card technology. It is composed of leaders from 195 corporations, including companies in banking, financial services, computer technology, health care, telecommunications, and a number of government agencies. The forum focuses on information exchange and member interaction. Every member of the Smart Card Forum recognizes that smart cards will succeed in the United States only if a critical mass of smart cards supports applications—both physical and Internet-based—of interest to consumers. The forum believes that the key to gaining widespread acceptance of smart cards lies in achieving compatibility between smart cards, card-reader devices, and applications. It is likely that the Smart Card Forum will have a significant presence and succeed in increasing consumers' awareness of smart cards and their advantages.

Mondex Smart Card

Mondex is a smart card that holds and dispenses electronic cash. It is a product of MasterCard International. As it gains acceptance on the Internet and in the general marketplace, the Mondex smart card will allow other applications to reside on its microchip. The Mondex card was invented in 1990. (Click **Mondex history** to review a year-by-year history of Mondex.) Mondex's Hong Kong pilot program took place in 1996. By the spring of 1997, more than 45,000 customers in Hong Kong carried Mondex cards, and approximately 400 Hong Kong merchants supported the system. Mondex reported that most people used Mondex for purchases under $100 (over 65%). Very few people purchased goods over $1000 with the Mondex cards. It is the largest pilot program of an electronic cash product to date. Though Mondex did not prove to be a huge success in that pilot study, it persists today and may yet become the smart card of choice.

Mondex faces several challenges. It requires special equipment; merchants who accept Mondex must have a specific card reader at their checkout counter. A

Mondex card must briefly be in physical contact with a special card reading and writing device during a merchant or "recharge" transaction. Internet users can transfer cash over the Internet using Mondex, but they must attach a Mondex reader to their PC in order to use the card. These requirements have proven to be barriers to the widespread use and success of Mondex.

Containing a microcomputer chip, Mondex cards can accept electronic cash directly from a user's bank account. Cardholders can spend their electronic cash with any merchant who has a Mondex card reader. Two cardholders can even transfer cash between their cards over a telephone line. That is an advantage of Mondex—a single card will work both in the online world of the Internet and the offline world of ordinary merchant stores while simultaneously being impervious to the theft threats of credit cards.

Another advantage of Mondex is that the cardholder always has the correct change for vending machines of various types. Coca Cola has reported that as much as 25% of its vending machine sales are probably lost because consumers do not have change. Mondex electronic cash supports micropayments as small as 3 cents.

Mondex has some disadvantages, too. The card carries real cash in electronic form, and the risk of theft of the card may deter users from loading it with very much money. Mondex can't compare to credit cards in the area of deferred payment—you can defer paying your charge or credit card bill for almost a month without incurring any interest charges. Mondex cards dispense their cash immediately.

A Mondex transaction has several steps (most of which are transparent to users). These steps ensure that the transferred cash safely reaches the correct destination. The following are the steps in using a Mondex card to transfer electronic cash from buyer to seller:

1. The card user inserts the Mondex card into a reader. The merchant and the card user are both validated to ensure that both the user and the merchant are still authorized to make transactions.
2. The merchant's terminal requests payment while simultaneously transmitting the merchant's digital signature.
3. The customer's card checks the merchant's digital signature. If the signature is valid, then the transaction amount is deducted from the cardholder's card.
4. The merchant's terminal checks the customer's just-sent customer digital signature for authenticity. If the cardholder's signature is validated, the merchant's terminal acknowledges the cardholder's signature by again sending back to the cardholder's card the merchant's signature.
5. Once the electronic cash is deducted from the cardholder's card, the same amount is transferred into the merchant's electronic cash account. Waiting until after the transaction amount is deducted from the cardholder's card ensures that electronic cash is neither created nor lost. That is, serializing the deduction and crediting events before signifying a complete transaction eliminates the creation or loss of cash if the system malfunctions in the middle of the process.

Figure 7-18 shows the larger process, beginning with the cardholder filling his or her card with electronic cash through the purchase transaction, and ending with the merchant exchanging accumulated electronic cash for credit in his or her bank account.

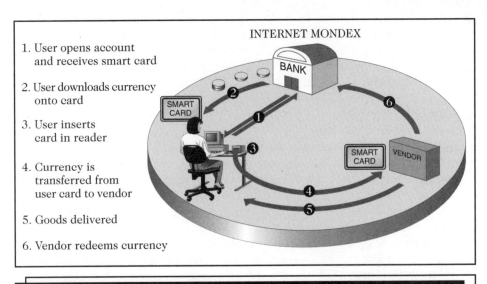

INTERNET MONDEX

1. User opens account and receives smart card

2. User downloads currency onto card

3. User inserts card in reader

4. Currency is transferred from user card to vendor

5. Goods delivered

6. Vendor redeems currency

BANK

SMART CARD

SMART CARD   VENDOR

Figure 7-18   *Mondex smart card processing*

Currently only a few cities in the United States (see the **Mondex USA** link on the Online Companion) accept Mondex. Worldwide, Mondex is available in several countries on the North and South American continents, Europe, Africa, the Middle East, Asia, Australia, and New Zealand. The success of Mondex is yet to be determined— much of the system's success depends on the W3C guidelines and whether Mondex can adapt to those specifications if and when they are approved.

CREDIT AND CHARGE CARDS

Credit cards are by far the most popular form of online payments for consumers. For starters, most consumers already have a credit card, or are at least familiar with how they work. Credit cards are widely accepted by merchants around the world and provide assurances for both the consumer and the merchant. A consumer is protected by an automatic 30-day period in which he or she can dispute an online credit card purchase. A merchant has a high degree of confidence that a credit card can be safely accepted from an unseen purchaser. Paying for online purchases with a credit card is just as easy as in a brick-and-mortar store. Merchants who already accept credit cards in an offline store can accept them immediately for online payment because they already have a merchant credit card account. Online purchases require an extra degree of security not required in offline purchases, because a card holder is not present and cannot be identified as easily as he or she can when standing at the cash register. Online credit card services must somehow authenticate the purchaser as well as protect sensitive information as it is transmitted on the Internet.

A **credit card**, such as a Visa or a MasterCard, has a preset spending limit based on the user's credit limit; a user can pay off the entire credit card balance or pay a minimum amount each billing period. Credit card issuers charge interest on any unpaid balance. A **charge card**, such as one from American Express, carries no preset spending limit, and the entire amount charged to the card is due at the end of the billing period. Charge cards do not involve lines of credit and do not accumulate interest charges. Because the distinction between credit cards and charge cards is unimportant in the discussion of processing credit and charge cards for electronic commerce, the collective term *payment card* will refer to both types throughout the remainder of this chapter.

Advantages and Disadvantages of Payment Cards

Payment cards have several features that make them an attractive and popular choice both with consumers and merchants in online and offline transactions. For merchants, payment cards provide fraud protection. When a merchant accepts payment cards for online payment—called **card not present** because the merchant's location and the purchaser's location are different—he or she can authenticate and authorize purchases using a payment card processing network. For consumers, payment cards are advantageous because the Consumer Credit Protection Act limits the cardholder's liability to $50 if the card is used fraudulently. Once the cardholder notifies the card's issuer of the card theft, the cardholder's liability ends. Frequently, the payment card's issuer waives the $50 consumer payment when a stolen card is used to purchase goods.

Perhaps the biggest advantage of using payment cards is their worldwide acceptance. You can pay for goods with payment cards anywhere in the world, and the currency conversion, if needed, is handled by the card issuer—the consumer doesn't have to deal with it. For online transactions, payment cards are particularly advantageous. When a consumer reaches the electronic checkout, he or she enters the payment card's number and his or her shipping and billing information in the appropriate fields to complete the transaction. The consumer does not need any special hardware or software to complete the transaction.

Payment cards have very few disadvantages, but they do have one when compared to cash. Payment card service companies charge merchants per-transaction fees and monthly processing fees. These fees can add up, but online and offline merchants view them as a cost of doing business. Any merchant who does not accept payment cards for purchases is probably losing significant sales because of it. The consumer pays no direct fees for using payment cards, but the prices of goods and services are slightly higher than they would be in an environment free of payment cards altogether.

Payment cards provide built-in security for merchants because merchants have a higher assurance that they will be paid through companies that issue payment cards than through the sometimes-slow direct invoicing process. To process payment card orders, a merchant must first set up a merchant account. Using payment cards avoids the additional expense and trouble of paper, which is part of invoicing systems (unless, of course, you use a business-to-business electronic system of invoicing and payment). The elaborate series of actions associated with using a payment card are often transparent to the consumer. Several groups and individuals are involved: the merchant, the merchant's bank, the customer, the customer's bank, and the company that issued the customer's payment card. All of these entities must work together in order for customers' charges to be credited to merchants' accounts (and vice versa when a customer receives a payment card credit for returned goods).

Payment Acceptance and Processing

Most people are familiar with the use of payment cards: When you purchase one or more items, the clerk runs your card through the online payment card terminal and your card account is charged immediately. The process is slightly different on the Internet, though the purchase and charge processes follow the same rules.

Unless you have been in the mail order or Internet business before, you may not realize that laws prohibit merchants from accepting payment card credits to their account until the products are shipped. In a brick-and-mortar store, you walk out of the store with the purchase in your possession, so charging and shipment occur nearly simultaneously. Online stores cannot charge your payment card until the day that they pack and ship your merchandise to you.

Payment card transactions follow these general steps once the merchant receives a consumer's payment card information, which is sent via an SSL-protected Web page:

1. The merchant must authenticate the payment card to ensure it is both valid and not stolen.
2. The merchant can check with the consumer's payment card issuer to ensure that funds are available and put a hold on the funds needed to satisfy the current charge.
3. Often a few days following the consumer's request for purchase, settlement occurs, which means that funds travel through the banking system into the merchant's account *after* the purchase has been shipped.

Of course, if the Internet consumer is downloading soft goods—digital files such as programs or data files—the merchant can submit the consumer's payment card immediately. You must also consider more complicated payment card transactions, such as partial order handling (wherein some items are available and some are back-ordered). Merchants have to be prepared to handle returns and issue RMA numbers (return materials authorization) for goods that are defective or incorrect.

Open and Closed Loop Systems

For some payment card systems, banks and other financial institutions serve as a broker between card users and the merchants accepting the cards. These types of arrangements are called **closed loop systems** because no other institution (local bank, national bank, or other clearinghouse) is involved in the transaction. American Express and Discover Card are examples of closed loop systems because there is exactly one franchise for each of those systems—not many franchisees.

Open loop systems can—and usually do—involve three or more parties. Suppose an Internet shopper uses his Visa card issued by the First Bank of Tucumcari to purchase an item from Widget Web Wonders, whose merchant account is at the Hackensack Bank and Commerce. Besides the two banks—the customer's and the merchant's—a third party, called the acquiring bank (see the next section), is involved in an open loop system. The acquiring bank passes authorization requests from Widget Web Brokers to the First Bank of Tucumcari (in this case) to obtain authorization for the credit purchase from the customer's bank. A response is sent back to the acquiring bank and on to the merchant. Similarly, the acquiring bank is responsible for contacting the merchant's bank and the many customers' banks to process sales drafts. Whenever a third party processes a transaction, the system is called an **open loop system**. Systems using Visa or

MasterCard are the most visible examples of open loop systems because many banks issue both cards. Unlike American Express or Discover, neither Visa nor MasterCard issue cards directly to consumers. Member banks are responsible for handling all the details of establishing customer credit limits.

Setting up a Merchant Account

A **merchant bank** or **acquiring bank** is a bank that does business with merchants (both Internet and non-Internet) who want to accept payment cards. In other words, to process payment cards for Internet transactions, an online merchant must set up a merchant account. Merchants receive a numbered account into which they deposit the accumulated card sales totals. When the merchant's bank acquires the sales slips, it credits their value to the merchant's account. A business must provide a potential merchant bank certain business information before the bank will provide an account through which the business can process payment card transactions. Typically, a new business must supply a business plan, details about existing bank accounts, and a credit history (business and personal). The merchant bank wants to have confidence that the merchant has a good prospect of staying in business and wants to minimize its risk by working with a well-focused store with a bright future. An online business that appears disorganized is less attractive to a merchant bank than a well-organized online business.

Several third-party Internet and Web-based services are available to handle all the details of processing payment card transactions. The next section discusses payment card processing options for Internet stores.

Processing Payment Cards Online

Software packaged with your electronic commerce software can handle payment card processing automatically, or you can contract with a third party to handle all your payment card processing. Many third-party, payment-handling organizations can also perform fulfillment operations in which they pick, pack, and ship products to your customers. If you choose that service, you reduce your number of employees and virtually eliminate your shipping department. Most of your work will be to maintain your store's Web presence and ensure that your suppliers provide sufficient inventory to meet customer demand.

Several companies provide payment-processing services. **Internetsecure**, for example, allows merchants to concentrate on business while it provides secure payment card payment services. Internetsecure supports payments with Visa and MasterCard for Canadian and United States accounts. The company provides risk management and fraud detection, and handles transactions from online merchants using existing bank-approved payment card processing infrastructure, secure links, and firewall technology to ensure complete security. Internetsecure notifies the merchant of all approved orders and supplies authorization codes to customers who purchase soft goods that are downloaded upon payment card approval. Internetsecure is responsible for ensuring that all transactions that it processes for payment card companies are credited to the merchant's account.

CyberCash provides merchant payment card processing services with the **ICVERIFY**, **PCAuthorize**, and **WebAuthorize** programs. ICVERIFY is intended for Windows systems. PCAuthorize is intended for smaller commerce service providers, and WebAuthorize is for large, enterprise-class merchant sites. PCAuthorize (which is representative of the two systems) automatically accepts merchant forms containing consumers' payment card information and captures the data and deposits the amount in the merchant's payment card account. Once the merchant receives payment from a customer through a Web page form, the merchant passes the Web page with customer payment card information on to the processing bank. PCAuthorize and WebAuthorize both connect directly to the bank network either by dial-up lines or private, leased telephone lines. The bank network receives the customer information, including payment card number, and performs the credit authorization with the issuing bank. Then the issuing bank deposits the money in the merchant's bank account. The merchant's Web site receives confirmation of the acceptance of the consumer transaction. After receiving notification of acceptance or rejection of the transaction, the merchant Web site confirms the sale to the customer over the Internet. In addition, the merchant site sends an e-mail confirmation of the sale to the consumer with details about the purchase price and shipping information. Figure 7-19 shows a graphic representation of the process.

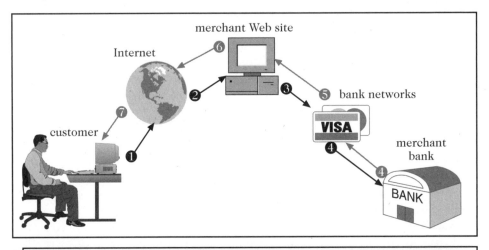

Figure 7-19   *Processing a payment card order*

Other payment card processing companies include **iAuthorizer** and **Authorize.Net**. IAuthorizer is a complete payment service that links a Web site directly to the payment portal. It provides order forms, CGI scripts, shopping cart features, and transaction processing. Transactions are processed on a secure host at the iAuthorizer center, and iAuthorizer merchants access their reporting and administration information through a secure system that requires a merchant to type a username and password for access. Both merchants and the merchants' customers receive notification of transaction approvals by e-mail. Authorize.Net is an online, real-time payment card processing service that allows merchants to link their sites to the Authorize.Net system by simply inserting a small block of HTML code into their transaction page. With Authorize.Net, a customer's order is encrypted and transferred to the Authorize.Net server. The server, in turn, relays the transaction to a bank network through a private, leased line. Merchants must have an Authorize.Net account to use the service. Check the Online Companion links for more details about these services.

Secure Electronic Transaction (SET) Protocol

Secure electronic transaction, or **SET**, is a secure protocol jointly designed by MasterCard and Visa with the backing of Microsoft, Netscape, IBM, GTE, SAIC, and other companies. The purpose of SET is to provide security for card payments as they traverse the Internet between merchant sites and processing banks. Though the Secure Socket Layers (SSL) protocol transmits payment data and other sensitive information securely between merchants and consumers, SSL does not verify that the consumer is the payment holder who owns the payment card. (The GartnerGroup published a report comparing SET and SSL. Click **GartnerGroup report on SET versus SSL** to download its PDF-format report.) While Visa and MasterCard have publicly stated that the goal of proposing the SET protocol is to establish a single method for consumers and merchants to use to conduct payment card transactions on the Internet, acceptance of the standard has been slow. **SETCo** is the consortium that manages and promotes the SET standard. Figure 7-20 shows SETCo's home page. The left panel of SETCo's home page contains links you can click to access frequently asked questions, the SET standard, and **SET downloadable documents**.

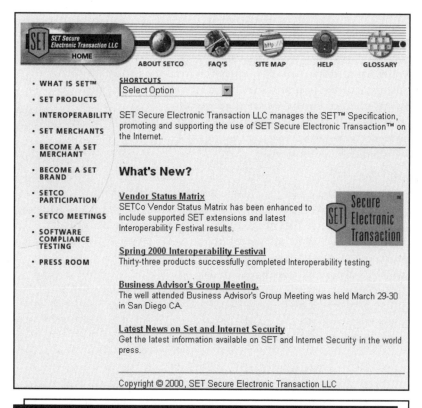

Figure 7-20 *SETCo's home page*

The SET specification uses public key cryptography (see Chapter 6) and digital certificates for validating both consumers and merchants. Specifically, the SET protocol provides confidentiality (secrecy), data integrity, user and merchant authentication, and consumer nonrepudiation. (Chapter 6 describes these attributes in detail.) **The SET Standard** describes the SET protocol in great detail. You can download either PDF (Portable Document Format) or Word formats of the specification. **SET goals** contains a recap of the SET protocol goals, and **VeriFone and SET** describes SET's security features thoroughly. Briefly, a SET-protected payment transaction works like this:

1. A shopper makes a purchase from a merchant who supports the SET specification. Using the browser's installed electronic wallet, a user transmits encrypted financial information from a wallet along with his or her digital certificate.
2. The merchant's Web server transfers the SET-encoded transaction to a payment card–processing center, which decrypts the transaction and processes it. At the same time, a certification authority such as VeriSign certifies the digital certificate as belonging to the sender.
3. The payment card–processing center routes the transaction to the financial institution that issued the consumer's payment card for approval.
4. Finally, the merchant receives notification from the consumer's bank that the transaction is approved. Then, the consumer's payment card account is charged for the transaction amount.
5. The merchant ships the merchandise and adds the transaction amount to the batch of payment card transactions that are later transmitted to the merchant's bank for deposit.

SET has received a lukewarm reception in the United States and, so far, has not attracted a large number of merchants and consumers. According to Art Kranzley, senior vice president of electronic commerce at MasterCard, 80% of SET activities are in Asian and European nations. Part of the problem with the acceptance of SET is that apparently it is not as easy to implement nor as inexpensive as most banks and merchants had expected. The typical reaction of many banks to SET is that of the British bank Barclays, whose information technology director, Alex Stevenson, says that SET is "rather clumsy, not tried and tested, and we simply don't need it." The future might prove brighter for SET, though. After a few years of testing and trials, SET's supporters believe it is ready for widespread deployment. Whether it achieves its goal will be determined in the next few years.

Summary

Online stores can accept a variety of forms of payment. Electronic cash, one form of online payment, has been slow to catch on in the United States. More than one organization has faltered in the last few years in introducing electronic cash to the online world. CyberCash is among the electronic cash survivors, and more companies will enter the electronic cash arena. Electronic cash is especially useful for making micropayments—Internet payments of under $1—because the cost of processing payment cards for transactions of under $10 obliterates any profit on small items. Electronic cash shares several benefits with real cash. Electronic cash is portable, easily divisible into small units, and usable for international transactions. Electronic cash can be stored online or offline. A third party, such as a bank, stores online electronic cash. The consumer holds offline cash in specially designed wallets.

Electronic wallets provide convenience to online shoppers because they hold payment card information, electronic cash, and personal consumer identification. Electronic wallets eliminate the need for consumers to reenter payment card and shipping information at a site's electronic checkout counter. Instead, the electronic wallet automatically fills in form information at sites that recognize the particular wallet software's technology. One persistent problem with electronic wallets is the lack of an internationally accepted standard. Several electronic wallets are available. Merchant sites currently have to decide whether to support several wallets or simply to support a single wallet. Soon, the W3C and a newly formed ECML standards group hope to create a single standard to which all wallet organizations will adhere. With a single wallet standard, merchants will be more willing to install electronic wallet–friendly software on their commerce sites.

Smart cards are physical, plastic devices that contain an embedded microcomputer chip that stores financial information about the cardholder. Smart cards are intended to replace the collection of plastic cards people now carry, including payment cards, drivers' licenses, and insurance cards. Trials of smart cards in a few cities have proved, so far, disappointing. Recent indicators point to a sharp rise in smart card popularity in the United States as Visa and American Express throw their weight behind their own brands of smart cards. Mondex, the most visible of the smart card companies, hopes that its card will be used around the world. Unlike electronic cash or payment cards, smart cards require extra hardware—card readers—in order to work. To use a smart card on your PC, you need a Mondex card reader to submit electronic cash from your smart card.

Credit and charge (payment) cards are the most popular forms of payment on the Internet. They are ubiquitous, convenient, and easy to use. Payment cards do not require any special consumer hardware to operate—you simply type in your payment card number in an online form. A hierarchy of banks process payment card payments. The merchant is credited with the transaction amount only when the consumer items are shipped, and not before.

MasterCard and Visa have developed an Internet protocol called Secure Electronic Transaction (SET) for the electronic commerce industry to facilitate secure payment card transactions over the Internet. Unlike the SSL protocol, SET provides cardholder authentication, ensuring that the card user and the cardholder named on the face of the card are the same person. SET is gaining popularity and is poised to become the payment card security protocol of choice soon.

Key Terms

Acquiring Bank
Anonymous Electronic Cash
Card Not Present
Charge Card
Client-Side Electronic Wallet
Closed Loop System
Credit Card
Double Spending
E-Card
Electronic Cash
Electronic Wallet
Extensible System

Float
Interoperable Software
Merchant Bank
Micropayments
Open Loop System
Personal Identification Number (PIN)
Scrip
Secure Electronic Transaction (SET)
Server-Side Electronic Wallet
Smart Card
Stored-Value Card
Vendor-Specific Scrip

Review Questions

1. Explain in three or four sentences the advantages and disadvantages of electronic cash.
2. Discuss why anyone with a credit card would want to use an electronic cash system on the Internet. That is, what niche might electronic cash fill? Are there problems with electronic cash and international sales? Write at least 100 words.
3. Compare a scrip-based electronic cash system to an electronic cash system that does not use scrip. What are the advantages and disadvantages of each? List one example of each by name. Write no more than 15 sentences.
4. Explain in three or four sentences what software wallets are and why they are useful. Is there a standard for wallet software or one being proposed? Explain why an electronic wallet standard is or is not useful.
5. Explain the advantages and disadvantages of a smart card such as Mondex.

Exercises

1. Matt Remes has formed a small business and has just completed building an electronic commerce Web site that sells subscriptions to special interest newsletters. The titles range from *Apple Growers Digest and Newsletter* to *Wilderness Backpacking Newsletter*. Many organizations and individuals produce the newsletters, and Matt's role is to raise the visibility of these sometimes-obscure publications produced in out-of-the-way places. All the newsletters are published and available biweekly or monthly. Unlike traditional subscription services, Matt's business has an agreement from all newsletter publishers that he can sell subscriptions for single issues or subscriptions for periods of up to three years. He does not want to allow subscribers to use their payment cards to purchase a subscription that is less than two years in duration. But he finds that nearly 60% of the first-time customers on his site prefers to sample issues before committing to a subscription of a year or more. Discuss this case and present possible solutions to the problem. What existing systems could you suggest that Matt pursue to provide his subscribers with a simple system that does not depend on payment cards? Can you list other Internet sites that offer similar services?

Locate a few similar sites and print their home pages. Use **HotBot**, **Northern Light**, **Lycos**, **AltaVista**, **Google**, or another good search engine to locate a dozen newsletters that are published online. Look for mailing lists (called listprocs or **listservs**) that either sell newsletter subscriptions or that are newsletter publishers in any field. **Publicly Accessible Mailing Lists (PAML)** provides comprehensive lists of listprocs and listservs by category and name.

2. Bonnie Carson has owned and managed her gift and card shop in the Central Shopping Mall for three years. Business has been good, but she'd like to expand her reach. So, a year ago, she hired a Web designer and built a Web site hosted by a national Internet Service Provider. Part of the monthly ISP fee for her merchant site includes the software needed to process credit card purchases. She has obtained a merchant account with a national credit card processing company. Bonnie's Web-based business is beginning to pick up. She wants to provide more payment options to her customers. Recently Bonnie heard about a new type of Web scrip or Web currency called Flooz. Slang for "cash," Flooz can be spent at a growing number of stores on the Internet. Users purchase Flooz online at **Flooz.com**, much like they would travelers' checks. Then, they can shop with their Flooz scrip at selected merchants. Bonnie has hired you to investigate this new cyber cash and report back to her. She wants you to write a 300-word report containing the following information: the name of the founder(s), a brief description of how a customer purchases Flooz, a list of a dozen online stores that accept Flooz, and a brief discussion of the procedure for spending Flooz online. Explain how the Flooz company, itself, makes money. In other words, what's in it for Flooz.com? Indicate the exchange rate for Flooz in terms of dollars. Discuss the advantages and disadvantages of using Flooz compared to using a credit card for online purchases. In other words, is spending Flooz particularly less secure than using a credit

card? What about international trade? Can you figure out and report on the name of the Web server software that Flooz.com uses? Be sure to include URLs for any especially important sites —and certainly include one for the Flooz site itself. Use the links in the Online Companion under Exercise 2 to help you in your research.

3. Evan Moskowitz and you have formed an Internet company called Teach-U-Comp to market and sell computer courses online. The first courses you will offer online are all on computer programming languages, including Visual Basic, Java, and C++. Students can sign up for as many courses as they would like, and each course takes four weeks to complete. Each course costs $55, and students receive continuing education units (CEUs) based on the duration of the course and its difficulty. Evan is tied up with the details of creating the online content and installing the course delivery software, and he has asked you to investigate and report back to him about the feasibility of implementing electronic wallet payment systems in addition to the existing credit card payment system. Students could download an electronic wallet that Teach-U-Comp supports, install it on their own computer, and then use the wallet to pay for their courses as well as to shop at other electronic commerce sites that support the selected electronic wallet software. Your job is to investigate electronic wallet software and write a 400-word report about what you found. Begin by looking at **Microsoft Passport**. If you have Internet Explorer installed, you can enter information into the wallet, or you can simply describe its features briefly. Second, look at **Gator**. Unless restrictions prohibit you from doing this, download the Gator software and install it. Describe in another paragraph the process of downloading and installing the wallet. Finally, review the current status of the electronic commerce modeling language (**ECML**). Be sure to include URLs of any significant sites you explore.

For Further Study and Research

Andrews, W. 1999. "The Digital Wallet: A concept revolutionizing e-commerce," *Internet World*, October 15, 35–45.

Berst, J. 1999. "Berst-Kept Secret: How PowerWallet Could Change Ecommerce," *ZDNET*, July 15. Available online at: (http://www.zdnet.com/anchordesk/story/story_3614.html).

Berst, J. 1999. "Ecommerce Breakthrough. Finally, an Easy Way to Pay Online," *PC Week*, 16(24), June 14.

Brands, S. 1994. "A Proposal for an Internet Cash System," *CWI*.

Bryant, A. 2000. "Plastic Is Getting Smarter," *Newsweek*, CXXXVI(16), October 16, 80.

Chaum, D. 1987. "Security Without Identification: Transaction Systems to Make Big Brother Obsolete," *Communications of the ACM*, 28(10), October.

Chaum, D. 1992. "Achieving Electronic Privacy," *Scientific American*, 267(2), August, 96–101.

Chaum, D. 1996. "Digital Signatures and Smart Cards," *Proceedings of the 3rd International Smart Card Conference*, Amsterdam, March.

Cipparone, M. 1996. "DigiCash Convertibility: A Look into the Future," *Journal of Internet Banking and Commerce*, 1(1), January.

Computerworld. 2000. "Web Ad Spending Up," 34(21), May 22, 65.

DeCarmo, L. 2000. "Security Protocols and Performance," *Dr. Dobb's Journal*, 25(11), November, 40–48.

Dodge, J. 1999. "Bill-Paying Via the Internet Offers Wins for All (Except the Postman)," *The Wall Street Journal Interactive Edition*, March 16.

Essick, K., T. Busse, and R. Guth. 1998, "Ready, SET, Wait," *Info World*, 20(21), May 25, 1.

eWeek. 2000. "Dot-coms feel the pinch," July 24, 55.

Fry, J. 1997. "Survey Finds Rise in Net Use; On-Line Buying Stays Stalled," *The Wall Street Journal Interactive Edition*, March 12.

Gallbraith, J. 1995. *Money: Whence it Came, Where it Went*. London: Penguin Books.

Graven, M. 2000. "Statements and Payments," *PC Magazine*, 19(13), July, 155–156.

Haskins, W. 2000. "Cashing In on E-Shopping," *PC Magazine*, 19(19), November 7, 85.

Janal, D. 1998. *Risky Business: Protect Your Company from Being Stalked, Conned or Blackmailed on the Web*. New York: John Wiley & Sons.

Jerome, M. and W. Taylor. 2000. "Pay Up," *PC Computing*, April, 37. Available online at: (http://www.zdnet.com/smartbusinessmag/stories/all/0,6605,2453035,00.html).

Johnson, A. 2000. "Looking for Big Profits In Small Purchases," *Computerworld*, 34(20), May 15, 78.

Kerstetter, J. 1998. "E-wallet Eliminates Downloads," *PC Week*, December 21.

Levy, S. 1994. "E-Money (That's What I Want)," *Wired*, 2(12), December, 174.

Merkow, M., J. Breithaupt, and K. Wheeler. 1998. *Building SET Applications for Secure Transactions*. New York: John Wiley & Sons.

Muller, J. 1997. "Selected U.S. Legal Issues in Issuance of Electronic Money," *Journal of Internet Banking and Commerce*," 2(2), March.

Nerurkar, U. 2000. "Security Analysis & Design," *Dr. Dobb's Journal*, 25(11), November, 50–56.

Rist, O. 1998. "Building E-Commerce," *Network Computing Online*, December 15.

Rubin, A. 2000. "Kerberos Versus the Leighton-Micali Protocol," *Dr. Dobb's Journal*, 25(11), November, 21–26.

"Show Me the E-Money," *Wired*, June 3, 1999.

Sprenger, P. 1999. "SF Moving to Smart Card Transit," *Wired News*, May 27.

Stallings, W. 2000. "The SET Standard & E-Commerce," *Dr. Dobb's Journal*, 25(11), November, 30–36.

Tracey, B. 1998. "The Color of Money," *The Wall Street Journal Interactive Edition*, November 16.

Ulfelder, S. 2000. "Timing Is Everything To Industry Veteran," *Computerworld*, 34(21), May 22, 85.

Weber, T. 1998. "Internet Titans Are Drawn by Lure of Electronic Wallet," *The Wall Street Journal Interactive Edition*, December 18.

Willmott, D. 1999. "The Internet Economy Will Take Over," *PC Magazine*, 18(12), June 22, 132.

Wingfield, N. 1999. "Single Standard Is Set in Online Wallet," *The Wall Street Journal*, June 14, B10.

STRATEGIES FOR MARKETING, SALES, AND PROMOTION

INTRODUCTION

The stock brokerage business was a staid affair for many years. Well-heeled customers conferred with their personal stockbrokers in well-appointed offices to decide which securities to buy and sell. Brokers charged commissions, often hundreds of dollars per trade, that covered the costs of providing account management and research performed by teams of highly paid financial analysts.

Beginning in the 1970s, firms such as **Charles Schwab** and **Quick & Reilly** began offering brokerage services for investors who preferred to do their own research and had no desire to visit their brokers' plush offices. Over the next 20 years, these discount firms took a significant part of the market from the more traditional brokerage firms. Many of the traditional firms merged or disappeared, but the industry leader, **Merrill Lynch**, continued to grow by consolidating its position and aggressively pursuing the most lucrative customers.

As commercial activity moved to the Internet in the mid-1990s, new firms, such as **Ameritrade** and **E*Trade**, began offering brokerage services on the Web at commission rates that were very low—even below discount broker rates. E*Trade used an aggressive "Boot Your Broker" theme in its ads. Many of the discount brokers offered Web-based trading services to compete with the new entrants. Merrill Lynch, however, remained steadfast in its refusal to compete with the discounters and online brokers. Throughout 1997 and 1998, Merrill Lynch executives were regularly quoted in the news, pointing out the dangers that investors could encounter when dealing with discount and online brokers. They vowed that Merrill Lynch would never compete in that business. Having been a successful firm for 85 years and with over $1 trillion in customer assets, Merrill Lynch believed that it could withstand the challenges of these upstarts. However, in 1998, the market value of Charles Schwab surpassed that of Merrill Lynch. It appeared that Merrill Lynch was losing the connection it had successfully maintained with its customers and was no longer providing what they wanted from a brokerage firm.

In June of 1999, Merrill Lynch announced that it would begin offering Internet trading with a pricing structure that would be competitive with those of the discount firms. It had fought the inevitable long enough and realized it had no choice but to modify its business model. The 14,800 Merrill Lynch stockbrokers, who had been earning an average of $300,000 per year in commissions, were deeply concerned about the new business model. Top executives at Merrill Lynch admitted that broker compensation would drop in the near term, but stated that they would offer alternate compensation and stock options to the brokers most affected by the changes. When the Internet changes an industry, even a firm with Merrill Lynch's history and resources must adjust to the new customer expectations, no matter how painful the adjustment.

LEARNING OBJECTIVES

In this chapter, you will learn about:
- Establishing an effective business presence on the Web
- Promoting your Web site
- Meeting the needs of Web site visitors
- Creating trust and building loyalty in Web site visitors
- Testing usability in Web site design
- Identifying and reaching customers on the Web
- Choosing successful marketing approaches for the Web
- Understanding the elements of branding
- Considering branding strategies and costs
- Choosing a business model for selling on the Web

CREATING AN EFFECTIVE WEB PRESENCE

Businesses have always created a presence in the physical world by building stores, factories, warehouses, and office buildings. An organization's **presence** is the public image it conveys to its stakeholders. The **stakeholders** of a firm include its customers, suppliers, employees, stockholders, neighbors, and the general public. Companies tend not to worry much about the image they project until they grow to a significant size—until then, they are too focused on survival to spare the effort. On the Web, presence can be much more important. The only contact that customers and other stakeholders have with a firm on the Web might be through its presence there. Creating an effective Web presence can be critical even for the smallest and newest firm operating on the Web.

Identifying Web Presence Goals

When a business creates a physical space in which to conduct its activities, its managers focus on very specific objectives. Few of these objectives are image-driven. They must find a location that will be convenient for customers to find, with sufficient floor space and features to allow the selling activity to occur, and they must balance the need for room to store inventory and provide employee working space with the costs of obtaining that space. The presence of a physical business location results from satisfying these many other objectives and is rarely a main goal of designing the space.

On the Web, businesses and other organizations have the luxury of intentionally creating a space that creates a distinctive presence. A firm's physical location must satisfy so many other business needs that it often fails to convey a recognizable presence. A Web site can perform many image-creation and image-enhancing tasks very effectively—it can serve as a sales brochure, a product showroom, a financial report, an employment ad, or a customer contact point. Each entity that establishes a Web presence should decide which tasks the Web site must accomplish and which tasks are the most important to include.

Different firms, even those in the same industry, might establish different Web presence goals. For example, **Coca Cola** and **Pepsi** are two companies that have established very strong brand images and are in the same business, but have developed very different Web presences. These two companies change their Web pages frequently, but the Coca Cola page usually includes a trusted corporate image such the Coke bottle. Alternatively, the Pepsi page is usually filled with hyperlinks to a variety of activities and product-related promotions.

These Web presences convey the images each company wishes to project. Each presence is consistent with other elements of the marketing efforts of these companies: Coca Cola's traditional position as a trusted classic, and Pepsi's position as the upstart product favored by a younger generation.

Achieving Web Presence Goals

An effective site is one that creates an attractive presence that meets the objectives of the business or other organization. These objectives include:

- Attracting visitors to the Web site
- Making the site interesting enough that visitors stay and explore

Strategies for Marketing, Sales, and Promotion

- Convincing visitors to follow the site's links to obtain information
- Creating an impression consistent with the organization's desired image
- Building a trusting relationship with visitors
- Reinforcing positive images that the visitor might already have about the organization
- Encouraging visitors to return to the site

The **Toyota** site that appears in Figure 8-1 is a good example of an effective Web presence. The site provides a product showroom feature, links to detailed information about each product line, links to dealers, and links to information about the company and the ancillary services it offers, such as financing and insurance. The page offers a help link and information about how to contact the company; it also has a site search feature. A good example of how Toyota has created a presence with this page that is consistent with its corporate philosophy is the statement that appears on the page: "…we've built a Web site that illustrates why Toyota's Cars and Trucks are ideal for your life…" To the extent that the Web site fulfills that promise, it is an effective extension of the corporate presence that Toyota wants to convey through the Internet to customers and potential customers.

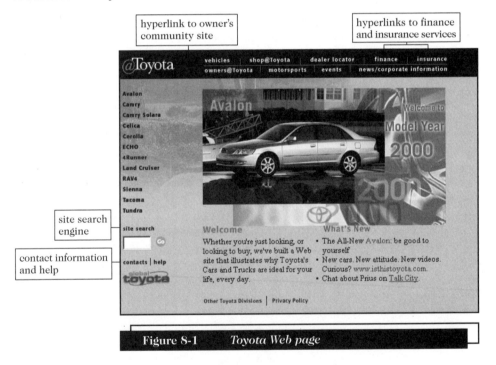

Figure 8-1   *Toyota Web page*

In contrast, **Quaker Oats** has created Web sites that do not offer any real sense of corporate presence, although they do provide a good selection of information about the firm. Figure 8-2 shows the Quaker Oats Web site as it appeared until 1999.

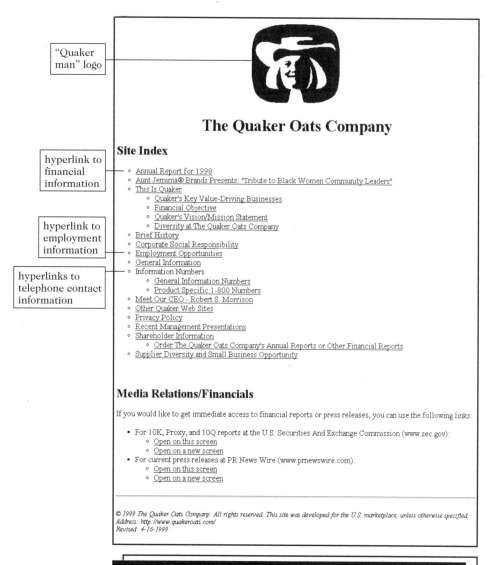

The labels pointing to the figure read:

"Quaker man" logo

hyperlink to financial information

hyperlink to employment information

hyperlinks to telephone contact information

The Quaker Oats Company

Site Index

- Annual Report for 1998
- Aunt Jemima® Brands Presents: "Tribute to Black Women Community Leaders"
- This Is Quaker
  - Quaker's Key Value-Driving Businesses
  - Financial Objective
  - Quaker's Vision/Mission Statement
  - Diversity at The Quaker Oats Company
- Brief History
- Corporate Social Responsibility
- Employment Opportunities
- General Information
- Information Numbers
  - General Information Numbers
  - Product Specific 1-800 Numbers
- Meet Our CEO - Robert S. Morrison
- Other Quaker Web Sites
- Privacy Policy
- Recent Management Presentations
- Shareholder Information
  - Order The Quaker Oats Company's Annual Reports or Other Financial Reports
- Supplier Diversity and Small Business Opportunity

Media Relations/Financials

If you would like to get immediate access to financial reports or press releases, you can use the following links:

- For 10K, Proxy, and 10Q reports at the U.S. Securities And Exchange Commission (www.sec.gov):
  - Open on this screen
  - Open on a new screen
- For current press releases at PR News Wire (www.prnewswire.com):
  - Open on this screen
  - Open on a new screen

Figure 8-2   *Quaker Oats Web page*

The site is a straightforward presentation of links to information about the firm. Note that the Quaker Oats Web page includes 24 links to financial information, employment opportunities, current press releases, and other information about the company. It includes links to contact information for the firm. It even has the corporate logo. Although this Quaker Oats site provides access to much of the same information as the Toyota site, it offers the visitor a completely different experience and impression. In 1999, Quaker changed its Web page to include pictures of some of its products and to improve its general appearance and user-friendliness. The overall impression of this new site is much more lively and fun than the original site. The updated Quaker Oats home page appears in Figure 8-3.

"Quaker man" logo

hyperlinks to telephone contact information

hyerlink to financial information

hyperlink to employment information

Figure 8-3   *Quaker Oats updated home page*

Although the new site is more colorful and more interesting, the basic information offered is essentially the same as that offered by the previous version. The new site offers this information more efficiently, using seven links instead of 24. In both the old and the new sites, Quaker Oats uses versions of its "Quaker man" logo to convey a highly recognizable image that is strongly associated with its company. Even though the new site is the result of a major overhaul, Quaker decided to include this image because it conveys the Quaker brand so well.

The Toyota and Quaker Oats examples illustrate that the Web can integrate an opportunity for enhancing the image of a business with the provision of information. For some organizations, this integrated image-enhancement capability is a key goal of their Web presence efforts. Not-for-profit organizations are an excellent example of this. They can use their Web sites as a central resource for integrated communications with their varied and often geographically dispersed constituencies.

A key goal for many not-for-profit organizations is information dissemination. The Web allows them to integrate information dissemination with fund-raising. Web pages also provide a two-way contact channel with persons engaged in the organization's work who are not working directly for the organization—many not-for-profits rely on volunteers and coordination with other organizations to accomplish their goals. This combination of information dissemination and a two-way contact channel is a key element in any successful electronic commerce Web site. Interestingly, not-for-profit organizations are far ahead of most businesses in accomplishing this combination of elements in their Web presences. Figure 8-4 shows the home page of the **American Civil Liberties Union** (ACLU), which is devoted to the advocacy of individual rights in the United States.

American Civil Liberties Union
FREEDOM NETWORK

▼ Features

ACT NOW!

JOIN THE ACLU

JOIN OUR ALERT LIST

NEWS

IN CONGRESS

IN THE COURTS

IN THE STATES

THE LIBRARY

THE STORE

STUDENTS

VOICES OF LIBERTY

ABOUT THE ACLU

Arrest the Racism

Clinton Orders Federal Agencies to Investigate Racial Profiling

High Court Victory!

Justices Strike Down Chicago Loitering Law

Free Speech in Peril

Your Voice Can Make *The* Difference!

▶ More Features

▼ The Issues

☐ Criminal Justice
☐ Cyber-Liberties
☐ Death Penalty
☐ Drug Policy
☐ Free Speech
☐ HIV/AIDS
☐ Immigrants Rights
☐ Lesbian & Gay Rights
☐ National Security
☐ Police Practices
☐ Prisons
☐ Privacy
☐ Racial Equality
☐ Religious Liberty
☐ Reproductive Rights
☐ Students Rights
☐ Voting Rights
☐ Women's Rights
☐ Workplace Rights

INDEX   SEARCH   FEEDBACK   JOIN

Figure 8-4   *American Civil Liberties Union Web page*

This page allows interested visitors to learn more about the ACLU and to join the organization if their interest is piqued by what they see. The "Feedback" link at the bottom of the page leads to a form that visitors can use to report a civil liberties violation, obtain assistance with legal research, ask questions about ACLU membership, or request permission to reprint ACLU publications.

Not-for-profit organizations can use the Web to stay in touch with existing stakeholders and identify new opportunities for serving them. Organizations as diverse as the **American Red Cross**, the **California Voter Foundation**, the **Public Broadcasting System**, and the **Union of Concerned Scientists** have created effective Web presences. Organizations such as **Amnesty International** and the **United Nations** also use the Web to create international communities of interested persons.

Political parties want to offer information about party positions on issues, recruit members, keep existing members informed, and provide communication links to visitors who have questions about the party. All the major U. S. political parties have Web sites, including the **Democratic**, **Green**, **Libertarian**, **Reform**, **Republican**, and **Socialist** parties. In addition, political organizations that are not affiliated with a specific party, such as the **Non-Partisan Center for Responsive Politics**, also accomplish similar goals with their Web presences.

Museums are finding that the Web offers a good way to introduce interested visitors to their collections and create a presence that is congruent with their image. The **Museum of Modern Art** (MoMA) in New York has created a page, which appears in Figure 8-5, that features a clean and functional design—two tenets of the modern art movement to which it is devoted.

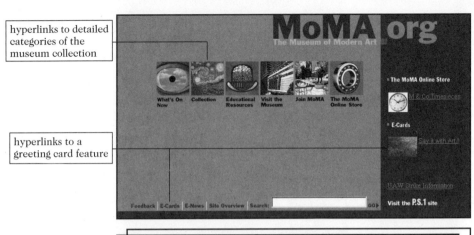

hyperlinks to detailed categories of the museum collection

hyperlinks to a greeting card feature

Figure 8-5　　*Museum of Modern Art Web page*

The MoMA page includes a directory of the entire museum collection, and, on the right side of the page, links to items offered by the museum's online store. The page includes two links to E-Cards that let visitors send electronic cards with MoMA artwork images. The page succeeds at engaging the browsing visitor on several levels, yet provides clear paths to museum resources for visitors who know what they are seeking when they access the Web site.

How the Web Is Different

When firms first started creating Web sites in the mid-1990s, they often built simple sites that conveyed basic information about their businesses. Few firms conducted any market research to see what kinds of things potential visitors might want to obtain from these Web sites and even fewer considered what business infrastructure adjustments would be needed to service the site. For example, few firms had e-mail address links on their sites. Those firms that did include an e-mail link often under-staffed the department responsible for answering visitors' e-mail messages. Thus, many site visitors sent e-mail messages that were never answered.

This failure to understand how the Web is different from other presence-building media is one reason that so many businesses fail to achieve their Web objectives. To learn more about this issue, see Jakob Nielsen's **Failure of Corporate Websites** page in the Online Companion.

Most Web sites that are designed to create an organization's presence in the Web medium include links to a fairly standard information set. The site should give the visitor easy access to a history, a statement of objectives or mission statement, information about products or services, financial information, and a way to communicate with the organization. Sites achieve varying levels of success based largely on how they offer this information. Presentation is important, but so is realizing that the Web is an interactive medium.

A number of Web designers and consultants have taken firms to task for their uninspired use of the Web's interactive nature. Some of these criticisms appear in the print media, but many appear in online newsletters or e-zines. An **e-zine** is an electronic magazine published on the Web. For example, Christopher Locke publishes **Entropy Gradient Reversals**, an e-zine that regularly discusses issues surrounding site design. He argues that large corporations should encourage their employees to engage in unrestricted online dialog with the firm's customers, suppliers, and other stakeholders. He believes that this dialog would be in the spirit of the Internet and would help companies create more honest and real personalities as part of their Web presences.

David Weinberger presents similar arguments in his online **Journal of the Hyperlinked Organization**. With two other consultants, Locke and Weinberger have published the **Cluetrain Manifesto** Web site, which presents 95 theses directed at major businesses and other organizations that use the Web. These theses include some short direct statements, such as "Markets are conversations" and "Hyperlinks subvert hierarchy," and other more lengthy bits of advice for Web presence creators. The authors of the Web site have also published a book, also titled *The Cluetrain Manifesto*, which includes additional material that does not appear on the Web site. The Cluetrain Manifesto has generated considerable discussion and has been electronically "signed" by hundreds of visitors to the site, many of whom have added comments.

The main point that these consultants make is that large firms must acknowledge and use the Web's capability for two-way, meaningful communication with their customers. They further argue that use of this communication process is not optional; companies that fail to communicate effectively through this channel will lose customers to competitors that do.

Meeting the Needs of Web Site Visitors

Businesses that are successful on the Web realize that every visitor to their Web site is a potential customer. Thus, an important concern for businesses crafting a Web presence is the variation in important visitor characteristics. People who visit a Web site seldom arrive by accident; they are there for a reason. Unfortunately for the Web designer trying to make a site that will be useful for everyone, those visitors arrive for many different reasons, including:

- Learning about products or services that the company offers
- Buying the products or services that the company offers
- Obtaining information about warranties or service and repair policies for products they have purchased
- Obtaining general information about the company or organization
- Obtaining financial information for making an investment or credit-granting decision
- Identifying the people who manage the company or organization
- Obtaining contact information for a person or department in the organization

Creating a Web site that meets the needs of visitors with such a wide range of motivations can be challenging. Not only do Web site visitors arrive with different needs that they hope to meet, they arrive with different experience and expectation levels. In addition to the problems posed by the diversity of visitor characteristics, technology issues can also arise. These Web site visitors will be connected to the Internet through a variety of communication channels that provide different band-widths and data transmission speeds. They also will be using several different Web browsers. Even those who are using the same browser can have a variety of configurations. The wide array of browser add-in and plug-in software adds yet another dimension to visitor variability. Considering and addressing the implications of these many visitor characteristic variations when building a Web site can help convert those visitors into customers.

One of the best ways to accommodate a broad range of visitor needs is to build flexibility into the Web site's interface. Many sites offer separate versions with and without frames and give visitors the option of choosing either one. Some sites offer a text-only version. As researchers at the **Trace Center** note, this can be an especially important feature for visually impaired visitors who use special browser software, such as the **IBM Home Page Reader**, to access Web site content. The **W3C Web Accessibility Initiative** site includes a number of useful links to information regarding these issues.

If the site design uses graphics, the site can give the visitor the option to select smaller versions of the images so that the page will load on a low-bandwidth connection in a reasonable amount of time. If the site includes streaming audio or video clips, it can give the visitor the option to specify a connection type so that the streaming media adjusts itself to the bandwidth for that connection.

A good site design lets visitors choose among information attributes, such as level of detail, forms of aggregation, viewing format, and downloading format. Many electronic commerce Web sites give visitors a selectable level of detail by presenting product information by product line. The site presents one page for each line of products. A product line page contains pictures of each item in that product line accompanied by a brief description. By using hyperlinked graphics for the product pictures, the site offers visitors the option of clicking the product picture, which opens a page of detailed specifications for that product.

Web sites can also offer visitors multiple information formats by including links to files in those formats. For example, the page offering financial information could include links to an HTML file, an Adobe PDF file, and an Excel spreadsheet file. Each of these files would contain the same financial information; however, the multiple file types allow visitors to choose the format that best suits their immediate needs. Visitors looking for a specific financial fact might choose the HTML file so that the information would appear in their Web browsers. Other visitors who want a copy of the entire annual report as it was printed would select the PDF file and either view it in their browsers or download and print the file. Visitors who want to conduct analyses on the financial data would download the spreadsheet file and perform calculations using the data in their own spreadsheet software.

290

To be successful in conveying an integrated image and offering information to potential customers, businesses should try to meet the following goals when constructing their Web sites:

- Convey an integrated image of the organization.
- Offer easily accessible facts about the organization.
- Allow visitors to experience the site in different ways and at different levels.
- Provide visitors with a meaningful, two-way (interactive) communication link with the organization.
- Sustain visitor attention and encourage return visits.
- Offer easily accessible information about products and services, and how to use them.

Trust and Loyalty

When companies first started selling on the Web, many of them believed that their customers would use the abundance of information to find the best prices and disregard other aspects of the buying experience. For some products this may be true; however, most products include an element of service. When customers buy a product, they are also buying that service element. A seller can create value in a relationship with a customer by nurturing customers' trust and developing it into loyalty.

Even when products are commodity items, the service element can be a powerful differentiating factor for which customers will pay extra. These services include such things as delivery, order handling, help with selecting a product, and after-sale support. Since many of these services are things that a potential customer cannot evaluate before purchasing a product, the customer must trust the seller to provide an acceptable level of service.

When a customer has an experience with a seller that provides good service, that customer begins to trust the seller. When a customer has multiple good experiences with a seller, that customer feels loyal to the seller. Thus, the repetition of satisfactory service can build customer loyalty that will prevent a customer from seeking out alternative sellers that offer lower prices.

A company that has intentionally built customer loyalty is the **Vanguard Group**. Vanguard manages a variety of mutual funds. It has spent more than $100 million on developing its Web site. Customers can use this site to obtain account information, manage their current investments, and make further investments in Vanguard mutual funds. Unlike many mutual fund management companies, however, Vanguard does not use the site to tout its products—its Web site is a tool that Vanguard has used to build customer loyalty. The stated goal of Vanguard's Web presence is to educate its customers and provide them with a high level of service. Many times, information on the site will discourage customers from buying mutual fund shares that are inappropriate for their investment goals. Vanguard would rather help its customers make good investment decisions than pick up a quick profit by selling customers on particular investments. The company believes that loyal, trusting customers are its most important asset.

Many companies that have begun doing business on the Web have spent large amounts of money to obtain customers. If they do not provide levels of customer service that lead customers to develop trust in and loyalty to the firm, the companies are unlikely to recover the money they spent to attract the customers in the first place, much less earn a profit.

Customer service is a problem for many corporate sites. Recent research indicates that customers rate most retail electronic commerce sites to be average or low in customer service. A primary weak spot for many sites is the lack of integration between the companies' call centers and their Web sites. As a result, when a customer calls with a complaint or problem with a Web purchase, the customer service representative often does not have information about Web transactions and is unable to resolve the caller's problem.

Usability Testing

Research indicates that few businesses accomplish all of the goals for a Web site in their current Web presences. Even sites that succeed in achieving most of these goals often fail to provide sufficient interactive contact opportunities for site visitors. Most firms' Web sites give the general impression that the firm is too important and its employees are too busy to respond to inquiries. This is no way to encourage visitors to become customers! In general, firms are only now starting to perform **usability testing** on their Web sites. As the practice of usability testing becomes more common, more Web sites will meet the goals outlined previously in this chapter.

Companies that have done usability tests, such as **Eastman Kodak**, **T. Rowe Price**, and **Maytag**, have found that they can learn a great deal about meeting visitor needs by conducting focus groups and watching how different customers navigate through a series of Web site test designs. Analysts at **Cambridge Technology Partners** and **Forrester Research** agree that the cost of usability testing is so low compared to the total cost of a Web site design or overhaul that it should almost always be included in such projects. Two pioneers of usability testing are Ben Shneiderman and Jakob Nielsen. Dr. Shneiderman founded the **University of Maryland Human-Computer Interaction Lab** and has published a number of books on interface design. Dr. Nielsen's **Alertbox** Web site includes much information about how to conduct usability testing and use its results to improve Web site design and operation.

T. Rowe Price, a leading mutual fund, asset management, and securities brokerage firm, built its first Web site in 1996. Within two years, the site contained so much information that visitors had a difficult time getting to the data they were seeking. For example, users had to click through five or six pages to get to investor reports. The company rolled out a redesigned Web site in late 1998. One of the design team's goals was to ensure that customers would never need to click through more than two pages to get the information they were seeking. The new design was focused on function rather than visual appeal and included a greater number of links on the home page. Kodak also redesigned its site in response to user feedback. Kodak's redesign resulted in a Web site structure similar to that of the T. Rowe Price design, with a large number of links. Kodak has continued to improve its Web page; a recent version of the page appears in Figure 8-6.

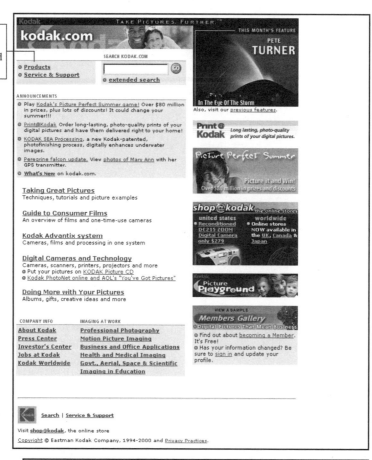

hyperlinks that are important to most site visitors are placed near top of page

Figure 8-6 *Kodak's redesigned home page*

Both Kodak and T. Rowe Price sell products that have a large intangible element. The benefits of high-quality photographs and sound investment plans are complex and can be hard for consumers to evaluate. Thus, both the Kodak and T. Rowe Price Web sites offer visitors a large number of links on the home pages so that visitors can explore and identify benefits in a different way.

In contrast to the experience of Kodak and T. Rowe Price, Maytag's usability research found that customers wanted fewer links on the home page. The tangible nature of Maytag's main product line is different from T. Rowe Price's intangible financial products and the intangible benefits of Kodak's cameras and photo finishing products. Thus, the new Maytag Web site design includes one or two graphic hyperlinks to featured products, a drop-down list of links to information about other products, and a few other links at the top of the home page, as shown in Figure 8-7.

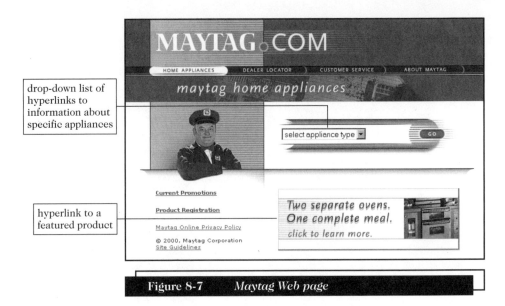

drop-down list of hyperlinks to information about specific appliances

hyperlink to a featured product

Figure 8-7   *Maytag Web page*

A common theme in these site redesigns is the attention paid to creating a site that meets the needs of potential site visitors. A 1999 *PC Week* article analyzed these redesign projects and identified important usability hints that Web designers should consider when they create any type of Web site, including the following:

- Design the site around how visitors will navigate the links, not around the company's organizational structure.
- Allow visitors to access information quickly.
- Avoid using inflated marketing statements in product or service descriptions.
- Avoid using business jargon and terms that visitors might not understand.
- Build the site to work for visitors who are using the oldest browser software on the oldest computer connected through the lowest bandwidth connection—even if this means creating multiple versions of Web pages.
- Be consistent in use of design features and colors.
- Make sure that navigation controls are clearly labeled or otherwise recognizable.
- Test text visibility on smaller monitors.
- Check to make sure that color combinations do not impair viewing clarity for color-blind visitors.

After you have designed the site with the preceding in mind, you should conduct usability tests by having visitors test several versions of the design to find information.

Web marketing consultant Kristin Zhivago has a number of recommendations for Web sites that are designed for electronic commerce. She encourages Web designers to create sites focused on the customer's buying process rather than the company's perspective and organization. For example, she suggests that companies examine how much information their Web site provides and how useful that information is for customers. If the site does not provide substantial "content for your click" to visitors, they will not become customers.

Using these guidelines when you create your site can help make visitors' Web experiences more efficient, effective, and memorable—in a positive way. Usability is an important element of creating an effective Web presence. For an interesting look at Web design issues, you can visit the **Webby Awards** site. The Webby Awards are given to sites that "exemplify the kinds of sites that Internet users should visit every day for information and entertainment" as judged by a panel of Web designers, journalists, and industry leaders.

IDENTIFYING AND REACHING CUSTOMERS

An important element of corporate Web presence is connecting with site visitors who are customers or potential customers. In this section, you will learn how a Web site can help firms identify and reach out to customers.

The Nature of Communication on the Web

Most businesses are familiar with two general ways of identifying and reaching customers: personal contact and mass media. In the **personal contact** model, the firm's employees individually search for, qualify, and contact potential customers. This personal contact approach to identifying and reaching customers is sometimes called **prospecting**. In the **mass media** approach, firms prepare advertising and promotional materials about the firm and its products or services. They then deliver these messages to potential customers by broadcasting them on television or radio, printing them in newspapers or magazines, posting them on highway billboards, or mailing them.

Some experts distinguish between broadcast media and addressable media. **Addressable media** are advertising efforts directed to a known addressee and include direct mail, telephone calls, and e-mail. Since few users of addressable media actually use address information in their advertising strategies, in this book we consider addressable media to be mass media. Many businesses use a combination of mass media and personal contact to identify and reach customers. For example, Prudential uses mass media to create and maintain the public's general awareness of its insurance products and reputation, while its salespersons use prospecting techniques to identify potential customers. Once an individual becomes a customer, Prudential maintains contact through a combination of personal contact and mailings.

The Internet is not a mass medium, even though a large number of people now use it and many companies seem to view their Web sites as billboards or broadcasts. Nor is the Internet a personal contact tool, although it can provide individuals the convenience of making personal contacts through e-mail and newsgroups. Jeff Bezos, founder of Amazon.com, has described the Web as the ideal tool for reaching what he calls "the hard middle"—markets that are too small to justify a mass media campaign, yet too large to cover using personal contact. Figure 8-8 illustrates the position of the Web as a customer-contact medium, between the large markets addressed by mass media and the highly focused markets addressed by personal contact selling and promotion techniques.

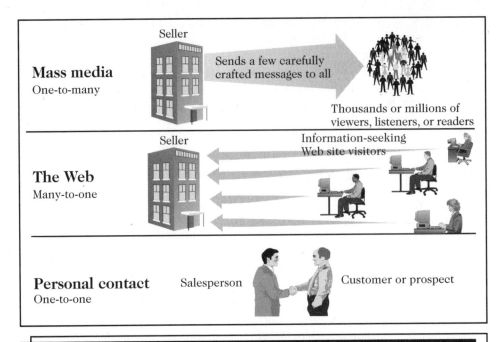

Mass media One-to-many	Seller	Sends a few carefully crafted messages to all	Thousands or millions of viewers, listeners, or readers
The Web Many-to-one	Seller		Information-seeking Web site visitors
Personal contact One-to-one		Salesperson	Customer or prospect

Figure 8-8 *Mass media, personal contact, and the Web*

To better understand the differences shown in Figure 8-8, we will work through a scenario in which you have heard about a new book but would like to learn more about it before buying it. Consider how your information acquisition process would vary, depending on the medium you used to gather the information:

- *Mass media.* You might have been exposed to general promotional messages from book publishers that have created impressions about quality associated with particular book brands. If your existing knowledge includes a brand identity for the book's publisher, these messages may influence your perceptions of the book. You may have been exposed to an ad for the title on television or radio, or in print. You may have heard the book's author interviewed on a radio program or might have read a review of the book in a publication such as *The New York Times Book Review* or *Booklist* magazine. Notice that most of these process elements involve you as a passive recipient of information. This communication channel is labeled "Mass media" and appears at the top of Figure 8-8. Communication in this model flows from one advertiser to many potential buyers and thus is called a **one-to-many communication model**. The defining characteristic of the mass media promotion process is that the seller is active and the buyer is passive.
- *Personal contact.* Small dollar-value items are not frequently sold through this medium because the costs of devoting a salesperson's efforts to a small sale are prohibitive. However, in the case of books, local bookshop owners and employees often devote considerable time and resources to developing a close relationship with their customers.

Although each individual book sale is a small-value transaction, people who frequent local bookshops tend to buy large numbers of books over time. Thus, the bookseller's investment in developing personal contacts is often rewarded. In the scenario, you may visit your local bookshop and strike up a conversation with a knowledgeable bookseller. In the personal contact model, this would most likely be a bookseller with whom you have already established a relationship. The bookseller would offer an opinion on the book based on having read that book, books by the same author, or reviews of the book. This opinion would be expressed as part of a two-way conversational interchange. This interchange usually includes a number of conversational elements, such as discussions about the weather, local sports, or politics, that are not directly related to the transaction you are considering. These other interchanges are part of the trust-building and trust-maintaining activities that you undertake to maintain the relationship element of the personal contact model. The underlying **one-to-one communication model** appears at the bottom of Figure 8-8 and is labeled "Personal contact." The defining characteristic of information gathering in the personal contact model is the wide-ranging interchange that occurs within the framework of an existing trust relationship. Both the buyer and the seller (or the seller's representative) actively participate in this exchange of information.

- *The Web*. To obtain information about a book on the Web, you could search for Web site references to the book, the author, or the subject of the book. You would likely identify a number of Web sites that offered such information. These sites might include those of the book's publisher, firms that sell books on the Web, independent book reviews, or discussion groups focused on the book's author or genre. **The New York Times Book Review** and **Booklist** magazine, both staples of mass media book promotion, now have online Web editions. Book review sites that did not originate in a print edition, such as the **BookBrowser**, have also appeared on the Web. Most Web-based book-sellers maintain searchable space on their sites for readers to post reviews and comments about specific titles. If the author of the book is famous, there might even be independent Web fan sites devoted to him or her. If the book is about a notable person, incident, or time period, you might find Web sites devoted to those notable topics that include reviews of books related to the topic. You could examine any number of these resources to any extent you desired. You might encounter some advertising material created by the publisher while searching the Web. However, if you choose not to view the publisher's ads, you will find it as easy to click the Back button on your Web browser as it is to surf television channels with your remote control.

The Web affords you many communication channels. Figure 8-8 shows only one of the communication models that can occur when using the Web to search for product information. The model labeled "The Web" in Figure 8-8 is the **many-to-one communication model**. The Web gives you the flexibility to use a one-to-one model (as in the personal contact model) in which you communicate via the Web with an individual working for the seller, or even to engage in **many-to-many communications** with other potential buyers. The defining characteristic of a product information search on the Web is that the buyer actively participates in the search and controls the length, depth, and scope of the search.

Measuring the Effectiveness of Web Site Advertising

As more companies rely on their Web sites to make a favorable impression on potential customers, the issue of measuring Web site effectiveness has become important. Mass media efforts are measured by estimates of audience size, circulation, or number of addressees. When a company purchases mass media advertising, it pays a dollar amount for each thousand persons in the estimated audience. This pricing metric is called **cost per thousand** and is often abbreviated **CPM**.

On the other hand, measuring Web audiences is more complicated because of the Web's interactivity and because the value of a visitor to an advertiser depends on how much information the site gathers from the visitor (for example, name, address, e-mail address, telephone number, and other demographic data). Since each visitor voluntarily provides or refuses to provide these bits of information, all visitors are not of equal value. Internet advertisers have developed some Web-specific metrics, described in this section, for site activity, but these are not generally accepted and are currently the subject of debate.

A **visit** occurs when a visitor requests a page from the Web site. Further page loads from the same site are counted as part of the visit for a specified period of time. This period of time is chosen by the administrators of the site and depends on the type of site. A site that features stock quotes might use a short time period, because visitors may load the page to check the price of one stock and reload the page 15 minutes later to check another stock's price. A museum site would expect a visitor to load multiple pages over a longer time period during a visit and would use a longer visit time window. The first time that a particular visitor loads a Web site page is called a **trial visit**; subsequent page loads are called **repeat visits**. Each page loaded by a visitor counts as a **page view**. If the page contains an ad, the page load is called an **ad view**. Some Web pages have banner ads that continue to load and reload as long as the page is open in the visitor's Web browser. Each time the banner ad loads is called an **impression**, and if the visitor clicks the banner ad to open the advertiser's page, that action is called a **click**, or a **click-through**. Banner ads are often sold on a CPM basis where the "thousand" is 1000 impressions. Rates vary greatly and depend on how much demographic information the Web site obtains about its visitors, but most are within the range of $1 to $100 CPM.

One of the most difficult things for companies to do as they move onto the Web is to determine the costs and benefits of advertising on the Web. Many companies are experimenting with new metrics they have created that consider the number of desired outcomes that their advertising yields. For example, instead of comparing the number of click-throughs that companies obtain per dollar of advertising, they measure the number of new visitors to their site that buy for the first time after arriving at the site via a click-through. They can then calculate the advertising cost of acquiring one customer on the Web and compare that to how much it costs them to acquire one customer through traditional channels.

New Marketing Approaches for the Web

The Web is an intermediate step between mass media and personal contact, but it is a very broad step. Using the Web to communicate with potential customers offers many of the advantages of personal contact selling and many of the cost savings of mass media. Figure 8-9 shows how these three information dissemination models compare on another important dimension: trust.

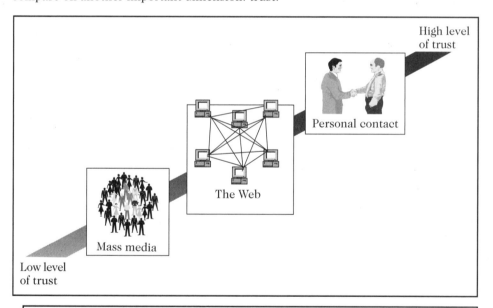

Figure 8-9   *Information acquisition approaches: levels of trust*

After years of being barraged by television and radio commercials, most people have developed a resistance to the messages conveyed in the mass media. The impact on an audience of the shouted expression "New and improved!" has diminished to almost nothing in most cases. The overuse of superlatives has caused most people to distrust or ignore the messages contained in mass media. Television remote controls have a mute button and make channel surfing easy, for a reason.

Attempts to re-create mass media advertising on the Web are doomed to fail for the same reasons—many people will ignore or resist messages that lack content of any specific personal interest to them.

Companies can use the Web to capture some of the benefits of personal contact, yet avoid some of the costs inherent in that approach. Most experts agree that it is better to scale up the trust-based model of personal contact selling to the Web than to scale down the mass marketing approach. In a 1996 marketing report, the **GartnerGroup** concluded that customer-centered marketing strategies would be an excellent fit for the Internet marketplace that was then emerging. The GartnerGroup also noted that rising consumer expectations and reduced product differentiation had led to increased competition and a splintering of mass markets. Both of these results were reducing the effectiveness of mass media advertising.

Advertisers' response to this decrease in effectiveness was to identify specific portions of their markets and target them with specific advertising messages. This practice, called **market segmentation**, divides the pool of potential customers into **segments**. Segments are usually defined in terms of demographic characteristics such as age, gender, marital status, income level, and geographic location. Thus, for example, unmarried women between the ages of 19 and 25 might be one market segment.

In the early 1990s, firms began identifying smaller and smaller market segments for specific advertising and promotion efforts. This practice of targeting very small market segments is called **micromarketing**. However, the economies of scale that saved so much cost in traditional mass media advertising campaigns were not available when those methods were used to target very small market segments. This hampered the success of micromarketing strategies. Even though micromarketing was an improvement over mass media advertising, it still used the same basic approach and suffered from the weaknesses of that model.

Technology-Enabled Relationship Management

The nature of the Web, with its two-way communication features and traceable connection technology, allows firms to gather much more information about customer behavior and preferences than they can using micromarketing approaches. Now, companies can measure a large number of things that are happening as customers and potential customers gather information and make purchase decisions. The idea of technology-enabled relationship management has become possible when promoting and selling via the Web. **Technology-enabled relationship management** occurs when a firm obtains detailed information about a customer's behavior, preferences, needs, and buying patterns *and* uses that information to set prices, negotiate terms, tailor promotions, add product features, and otherwise customize its entire relationship with that customer. Although companies can use technology-enabled relationship management concepts to help manage relationships with vendors, employees, and other stakeholders, most companies currently use these concepts to manage customer relationships. Thus, technology-enabled relationship management is often called **customer relationship management (CRM)** or **electronic customer relationship management (eCRM)**. Figure 8-10 compares technology-enabled relationship management to traditional seller-customer interactions on seven dimensions.

Dimensions	Technology-enabled relationship management	Traditional relationships with customers
Advertising	Provide information in response to specific customer inquiries	"Push and sell" a uniform message to all customers
Targeting	Identifying and responding to specific customer behaviors and preferences	Market segmentation
Promotions and discounts offered	Individually tailored to customer	Same for all customers
Distribution channels	Direct or through intermediaries; customer's choice	Through intermediaries chosen by the seller
Pricing of products or services	Negotiated with each customer	Set by the seller for all customers
New product features	Created in response to customer demands	Determined by the seller based on research and development
Measurements used to manage the customer relationship	Customer retention; total value of the individual customer relationship	Market share; profit

Figure 8-10   *Technology-enabled relationship management and traditional relationships with customers*

Business researchers Jeffrey Rayport and John Sviolka observed that firms today do business in both a physical world and in a virtual, information world. Rayport and Sviolka distinguish between commerce in the physical world, or marketplace, and commerce in the information world, which they term the **marketspace**. In the information world's marketspace, digital products and services are delivered through electronic communication channels, such as the Internet.

In Chapter 1, you learned that the value chain model described the primary and support activities that firms use to create value. This value chain model is valid for activities in the physical world and in the information world. However, value creation requires different processes in the marketspace. By understanding that value creation in the marketspace is different, firms can identify value opportunities effectively in both the physical and information worlds.

For years, businesses have viewed information as a part of the value chain's supporting activities, but they have not considered how information itself might be a source of value. In the marketspace, firms can use information to create new value for customers. For example, **CDnow**, an online seller of music CDs, provides a number of valuable customer services that are derived purely from activities in the information world. CDnow will, if the customer so chooses, store an order history, provide recommendations based on previous purchases, and show current information on performing artists in which the customer is interested.

Teenagers and young adults—people who often want more CDs than they can afford to buy right away—are important CDnow customers. Thus, CDnow's site offers a "Wish List" feature. Customers can store the names of CDs that they want to purchase later. Many CDnow customers order CDs as gifts for others, so the Web site provides a way to store gift recipient names and addresses to make repeat shopping easier. The CDnow site also has a "Gift Registry" in which customers can list CDs they would like to receive as gifts. Gift-givers can then obtain gift CD preferences for persons in the registry by entering the recipient's e-mail address.

When visitors register with the CDnow site, they are given a my CDnow page that is automatically configured to provide useful links. CDnow determines the links to include on the page by examining the visitor's history of purchases, Wish List and Gift Registry selections, and music interests that the visitor can expressly choose by following the link to Preferences and making selections. Figure 8-11 shows a my CDnow page that includes these marketspace features.

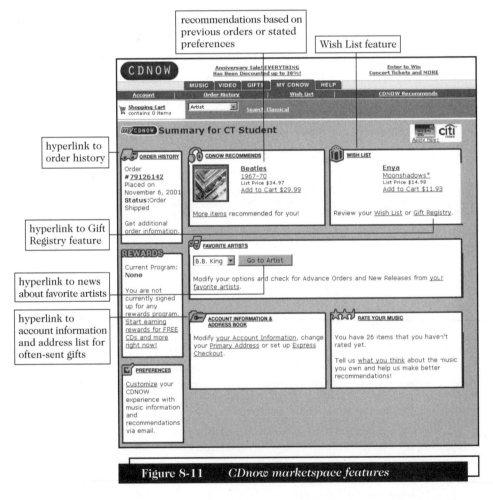

Figure 8-11   CDnow marketspace features

All the features that CDnow has built into its site add value for its customers. None of the features mentioned here occurs in the physical marketplace; they all occur in the virtual information world of the marketspace. CDnow can provide these features only to customers who are willing to interact with the site or allow it to track their behaviors. Note that all of these marketspace site features are optional and must be actively selected by the individual customer. CDnow is implementing technology-enabled relationship management on five (advertising, targeting, promotions, distribution, and measurement) of the seven dimensions listed in Figure 8-10.

Successful new Web marketing approaches all involve enabling the potential customer to find information easily and to customize the depth and nature of that information, and should encourage the customer to buy. Firms should track and examine the behaviors of their Web site visitors, and then use that information to provide customized, value-added digital products and services in the marketspace. Companies that use these technology-enabled relationship management tools to improve their contact with customers will be more successful on the Web than firms that adapt advertising and promotion strategies that were successful in the physical world but that are less effective in the virtual world.

CREATING AND MAINTAINING BRANDS ON THE WEB

A known and respected brand name can present to potential customers a powerful statement of quality, value, and other desirable qualities in one recognizable element. Branded products are easier to advertise and promote, because each product carries the reputation of the brand name. Companies have developed and nurtured their branding programs in the physical marketplace for many years. Consumer brands such as Ivory soap, Walt Disney entertainment, Maytag appliances, and Ford automobiles have been developed over many years with the expenditure of tremendous amounts of money. However, the value of these and other trusted major brands far exceeds the cost of creating them.

Elements of Branding

The key elements of a brand, according to researchers at advertising agency Young & Rubicam, are differentiation, relevance, and perceived value. Product differentiation is the first condition that must be met to create a product or service brand. The company must clearly distinguish its product from all others in the market. This makes branding difficult for commodity products such as salt, nails, or plywood—difficult, but not impossible.

The classic case of branding a near-commodity product is Procter & Gamble's creation of the Ivory brand over 100 years ago. The company was experimenting with manufacturing processes and had accidentally created a bar soap that contained a high percentage of air. When one of the workers noted that the soap floated in water, the company decided to sell the soap using the differentiating characteristic in packaging and advertising by claiming "it floats." Thus was the Ivory soap brand born. **Procter & Gamble** maintains this brand differentiation on its Web site even today by listing the link to its **Ivory Soap** site under the heading "Beauty and Skin Care Products."

The second element of branding—relevance—is the degree to which the product offers utility to a potential customer. The brand will only have meaning to customers if they can visualize its place in their lives. Many people understand that **Tiffany & Co**. creates a highly differentiated line of jewelry and gift products, but very few people can see themselves purchasing and using such goods.

The third branding component, perceived value, is a key element in creating a brand that has value. Even if your product is different from others on the market and potential customers can see themselves using the product, they will not buy it unless they perceive value. Some large fast food outlets have well-established brands that actually work against them. People recognize their brands and avoid eating at their restaurants because of negative associations—such as low overall quality and high fat content menu items. Figure 8-12 summarizes the elements of a brand.

Element	Meaning to Customer
Differentiation	In what significant ways is this product or service unlike its competitors?
Relevance	How does this product or service fit into my life?
Perceived value	Is this product or service good?

Figure 8-12   Elements of a brand

If a brand has established that it is different from competing brands and that it is relevant and inspires a perception of value to potential purchasers, those purchasers will buy the product and become familiar with how it provides value. Brands only become established when they reach this level of purchaser understanding.

Emotional Branding vs. Rational Branding

Unfortunately, brands can lose their value if the environment in which they have become successful changes. A dramatic example is Digital Equipment Corporation (DEC). For years, DEC was a leading manufacturer of midrange computers. When the market for computing shifted to personal computers, DEC found that its branding did not transfer to the personal computers that it produced. The consumers in that market did not see the same perceived value or differentiation in DEC's personal computers that the buyers of midrange systems had seen for years. This is an important element of branding for Web-based firms to remember, because the Web is still evolving and changing at a rapid pace.

Companies have traditionally used emotional appeals in their advertising and promotion efforts to establish and maintain brands. One branding expert, Ted Leonhardt, has defined "brand" as "an emotional shortcut between a company and its customer." These emotional appeals work well on television, radio, billboards, and in print media, because the ad targets are in a passive mode of information acceptance. However, emotional appeals are difficult to convey on the Web because it is an active medium controlled to a great extent by the customer. Many Web users are actively engaged in such activities as finding information, buying airline tickets, making hotel reservations, and obtaining weather forecasts. These are busy people who will happily click away from emotional appeals.

Marketers are attempting to create and maintain brands on the Web by using **rational branding**. Companies that use rational branding offer to help Web users in some way in exchange for their viewing an ad. Rational branding relies on the cognitive appeal of the specific help offered, not on a broad emotional appeal. For example, Web e-mail services such as **Excite Mail**, **HotMail**, or **Yahoo! Mail** give users a valuable service—an e-mail account and storage space for messages. In exchange for this service, users see an ad on each page that provides this e-mail service. Similarly, **MasterCard** promotes its brand name online through its **Shop Smart!** program. Shop Smart! is a third-party assurance mechanism. MasterCard ensures that any Web site displaying the Shop Smart! emblem (which happens to include a large MasterCard logo) is using what Master Card defines as a "safe" method of processing transactions. In exchange for this assurance on a Web shopping site, the Web user sees the MasterCard logo.

Permission Marketing Strategies

Many businesses would like to send e-mail messages to their customers and potential customers to announce new products, new product features, or sales on existing products. However, print and broadcast journalists have severely criticized some companies for sending e-mail messages to customers or potential customers. Some companies have even faced legal action after sending out mass e-mailings. Unsolicited e-mail is often considered to be spam (see Chapter 2). However, sending e-mail messages to Web site visitors who have expressly requested the e-mail messages is a completely different story.

Many businesses are finding that they can maintain an effective dialog with their customers by using automated e-mail communications. Sending one e-mail message to a customer can cost less than one cent if the company already has the customer's e-mail address. Purchasing the e-mail addresses of persons who have asked to receive specific kinds of e-mail messages will add between a few cents and a dollar to the cost of each message sent. Another factor to consider is the conversion rate. The **conversion rate** of an advertising method is the percentage of recipients who respond to an ad or promotion. Conversion rates on requested e-mail messages range from 10% to over 30%. These are much higher than the click-through rates on banner ads, which are currently under 1% and decreasing.

The practice of sending e-mail messages to people who have requested information on a particular topic or about a specific product is called **opt-in e-mail** and is part of a marketing strategy called **permission marketing**. Seth Godin, the founder of YoYoDyne and later the vice president for direct marketing at Yahoo!, developed this marketing strategy and publicized it in a book he wrote with Don Peppers titled *Permission Marketing*.

Godin argues that, as the pace of modern life increases, time becomes a valuable commodity. Most marketing efforts that traditional businesses use to promote their products or services depend on potential customers having enough time to listen to sales pitches and pay attention to the best ones. As time becomes more precious to everyone, people no longer wish to hear and evaluate advertising and promotional appeals for products and services in which they have no interest.

Thus, a marketing strategy that only sends specific information to persons who have indicated an interest in receiving information about the product or service being promoted should be more successful than a marketing strategy that sends general promotional messages through the mass media. One Web site that offers opt-in e-mail services is **yesmail.com**.

To induce potential customers to accept, or opt in to, advertising information sent via e-mail messages, the seller must provide some incentive. This incentive could be entertainment, a chance to win a prize, or even a direct cash payment. For example, **AllAdvantage.com** is a company that pays Web users for permission to monitor their Web surfing activities. After tracking these users, AllAdvantage.com presents targeted ads to them. Advertisers are willing to pay a premium to have access to persons who have demonstrated—by their Web surfing habits—that they are interested in the products or services offered by the advertisers.

Brand-Leveraging Strategies

Rational branding is not the only way to build brands on the Web. One method that is working for well-established Web sites is to extend their dominant positions to other products and services. **Yahoo!** is an excellent example of this strategy. Yahoo! was one of the first directories on the Web. It added a search engine function early in its development and has continued to parlay its leading position by acquiring other Web businesses and expanding its existing offerings. Then, Yahoo! acquired GeoCities and Broadcast.com, and entered into an extensive cross-promotion partnership with a number of **Fox** entertainment and media companies. Yahoo! continues to lead its two nearest competitors, **Excite** and **Infoseek**, in ad revenue by adding features that Web users find useful and that increase the site's value to advertisers. Amazon.com's expansion from its original book business into CDs, videos, and auctions is another example of a Web site leveraging its dominant position by adding features useful to existing customers.

Affiliate Marketing Strategies

Of course, this leveraging approach only works for firms that already have Web sites that dominate a particular market. As the Web matures, it will be increasingly difficult for new entrants to identify unserved market segments and attain dominance. A tool that many new, low-budget Web sites are using to generate revenue is affiliate marketing. In **affiliate marketing**, one firm's Web site—the affiliate firm's—includes descriptions, reviews, ratings, or other information about a product that is linked to another firm's site that offers the item for sale. For every visitor who follows a link from the affiliate's site to the seller's site, the affiliate site receives a commission. The affiliate site also obtains the benefit of the selling site's brand in exchange for the referral.

The affiliate saves the expense of handling inventory, advertising and promoting the product, and processing the transaction. In fact, the affiliate risks no funds whatever. CDnow and Amazon.com were two of the first companies to create successful affiliate programs on the Web. CDnow's Web Buy program, which includes more than 250,000 affiliates, is one of CDnow's main sources for new customers. The Amazon.com program has over 400,000 affiliate sites. Most of these affiliate sites are devoted to a specific issue, hobby, or other interest. Affiliate sites choose books or other items that are related to their visitors' interests and include links to the seller's site on their Web pages. Books and CDs are a natural for this type of shared promotional activity, but sellers of other products and services also have successful affiliate marketing programs. **B & D Coffee**, **eToys**, and the **T-ShirtKing** all have found affiliate marketing to be a good way to attract new customers to their Web sites.

One of the more interesting marketing tactics made possible by the Web is **cause marketing**, which is an affiliate marketing program that benefits a charitable organization (and, thus, supports a "cause"). In cause marketing, the affiliate site is created to benefit the charitable organization. When visitors click a link on the affiliate's Web page, a donation is made by a sponsoring company. The page that loads after the visitor clicks the donation link carries advertising for the sponsoring companies. Many companies have found that the click-through rates on these ads are much higher than the typical banner ad click-through rates. A leading retail Web florist, **proflowers.com**, has had excellent results advertising on **The Hunger Site** page. When a visitor clicks the button on The Hunger Site page, a group of sponsoring advertisers donates food to a hungry person and a page appears in the visitor's browser with ads for the sponsors. The Hunger Site page and the sponsors' ad page are shown in Figure 8-13.

click to donate food and open page (shown below) that includes links to sponsoring advertisers' sites

links to sponsoring advertisers' sites

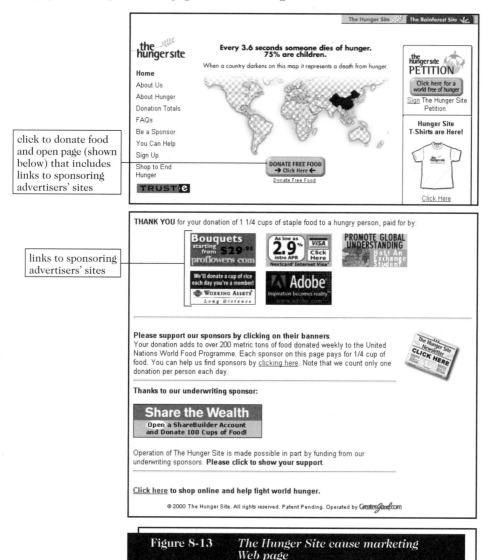

Figure 8-13 *The Hunger Site cause marketing Web page*

Viral Marketing Strategies

Traditional marketing strategies have always been developed with an assumption that the company was going to communicate with potential customers directly or through an intermediary that was acting on behalf of the company, such as a distributor, retailer, or independent sales organization. Since the Web expands the types of communication channels available, including customer-to-customer communication, another marketing approach has become popular on the Web. **Viral marketing** relies on existing customers to tell other persons—the company's prospective customers— about the products or services they have enjoyed using. Much as affiliate marketing uses Web sites to spread the word about a company, viral marketing approaches use individual customers to do the same thing. The number of customers increases much as a virus multiplies, thus the name.

Blue Mountain Arts, an electronic greeting card company, purchased very little advertising, but it has grown to become one of the most-visited sites on the Web. By late 1999, when the company was acquired by At Home Corporation for $780 million, Blue Mountain had more than 10 million people visiting its site each month. Electronic greeting cards are e-mail messages that include a link to the greeting card site. When people receive Blue Mountain Arts electronic greeting cards, they are taken to the Web site through that link and are likely to search for cards that they might like to send.

Brand Consolidation Strategies

Another way to leverage the established brands of existing Web sites was devised by **Della & James**, an online bridal registry. Although a number of national department store chains such as **Macy's** have established online registries for their own stores, Della & James offers a single registry that connects to several local and national department and gift stores, including **Crate&Barrel**, **Dillard's**, **Gump's**, **Neiman Marcus**, and **Williams-Sonoma**. The logo and branding of each participating store are featured prominently on the Della & James site. The founders of Della & James had identified an opening for a market intermediary because the average engaged couple registers at three stores. Thus, Della & James is providing a valuable consolidating activity for registering couples and their wedding guests that no store operating alone could provide. Della & James does, however, sell incidental wedding items that do not compete with the offerings of their online branding partners.

Costs of Branding

Transferring existing brands to the Web or using the Web to maintain an existing brand is much easier and less expensive than creating an entirely new brand on the Web. In 1998, a large number of companies began spending significant amounts of money to build brands on the Web. According to studies by the **Intermarket Group**, the top 100 electronic commerce sites spent an average of $8 million each that year to create and build their online brands. Two of the top spenders included the battling Web sites **Amazon.com**, which spent $133 million, and **BarnesandNoble.com**, which spent $70 million. Most of this spending was for television, radio, and print media— not for online advertising. Online brokerages E*Trade and Ameritrade Holding were also among the top five, spending $71 million and $44 million, respectively.

Promoting the company's Web presence should be an integral part of brand development and maintenance. The company's URL should always be included on product packaging and in mass media advertising on radio, television, and in print. Ensuring that the site is included in search engine databases and that the site includes appropriate META tags, as discussed earlier in this book, are both important elements in developing site awareness among the visitors that the firm wants to attract. Integrating the URL with the company logo on brochures can also be helpful in getting the word out about the Web site.

Web Site Naming Issues

Firms that have a major investment in branding a product or service must protect that investment. In Chapter 5, you learned about the security issues surrounding Web site naming. The legal and marketing aspects of Web site naming can be complicated as well. Although a variety of state and federal laws protect trademarks, the procedure for creating and using Web site names that are not trademarks can present some challenging issues. Obtaining identifiable names to use for branded products on the Web can be just as important as ensuring legal trademark protection for an existing brand investment.

In 1998, a poster art and framing company named Artuframe opened for business on the Web. With quality products and an appealing site design, the company was doing well, but it was concerned about its URL, which was www.artuframe.com. After searching for a more appropriate URL, the company's president found the Web site of Advanced Rotocraft Technology, an aerospace firm, at the URL www.art.com. After finding out that Advanced Rotocraft Technology's site was drawing 150,000 visitors each month who were looking for something art-related, Artuframe offered to buy the URL. The aerospace firm agreed to sell the URL to Artuframe for $450,000. Artuframe immediately changed its URL to **Art.com** and experienced a 30% increase in site traffic the day after implementing the name change. The newly named site did not rely on the name change alone, however. It has since entered a joint marketing agreement with Yahoo! that places an ad for Art.com on art-related search results pages. Art.com has also created an affiliate program with businesses that sell art-related products and other organizations that have Web sites devoted to art-related topics.

Another company that invested in an appropriate URL was **Cars.com**. The firm paid $100,000 to the speculator who had originally purchased the rights to the URL. Cars.com is a themed-portal site that displays ads for new cars, used cars, financing, leasing, and other car-related products and services. The major investors in this firm are newspaper publishers that wanted to retain an interest in automobile-related advertising as it moved online. Classified automobile ads are an important revenue source for many newspapers.

More recently, higher prices have prevailed in the market for domain names. Names such as Fruits.com, Question.com, Speaker.com, Tower.com, and Wisdom.com have each sold for prices over $100,000. Other names, including Cinema.com, Drugs.com, and ForSaleByOwner.com have sold for more than $500,000 each. Not long ago, eCompanies paid $7.5 million for the domain name Business.com. Figure 8-14 lists domain names that have sold for more than $1 million each.

Domain Name	Price
Business.com	$ 7.5 million
Altavista.com	$ 3.3 million
Loans.com	$ 3.0 million
Wine.com	$ 3.0 million
Autos.com	$ 2.2 million
Express.com	$ 2.0 million
WallStreet.com	$ 1.0 million

Figure 8-14 *Domain names that have sold for more than $1 million*

Although most domains that have high value are dot-com sites, the name engineering.org sold at auction to the American Society of Mechanical Engineers, a not-for-profit organization, for just under $200,000.

Several legitimate online businesses, known as **URL brokers**, are in the business of selling or auctioning domain names that they believe others will find valuable. Companies selling "good" (short and easily remembered) domain names include **Domains.com, DomainRace.com, GreatDomains.com**, and **HitDomains.com**. **Unclaimed Domains** sells a subscription to lists of recently expired domain names that it publishes periodically, and the **Netcraft** Web site has a URL search function to search for words in URLs. The **Internet Corporation for Assigned Names and Numbers (ICANN)** maintains a list of accredited **domain name registrars**, which are companies that have been authorized by **ICANN** to sell the rights to use specific domain names ending in .com, .net, and .org.

BUSINESS MODELS FOR SELLING ON THE WEB

Now that you have learned how to create an effective Web presence and use that presence to identify and reach customers, this section will present business models for selling on the Web.

Selling Goods and Services

The business model of selling goods and services on the Web is based on the mail order catalog business model that predates the Web. In this model, the seller establishes a brand image that conveys quality and uses the strength of that image to sell through catalogs mailed to prospective buyers. Buyers place orders by mail or by calling the seller's toll-free telephone number. This business model, which is often called the **catalog model**, has proven successful for a wide variety of consumer goods items, including apparel, computers, electronics, housewares, and gifts.

Expanding the mail order catalog model to the Web means the firm replaces or supplements print catalog distribution with information on its Web site. When the catalog model is expanded this way, it is often called the **Web-catalog model**. Customers can place orders through the Web site or by telephone. This flexibility has been important because many consumers are still reluctant to buy on the Web. In the first few years of consumer electronic commerce, a large number of shoppers used the Web to obtain information about products and to compare prices and features, but then made the purchase by telephone.

Many of the most successful Web catalog sales businesses are firms that were in the mail order business and have simply expanded their operations to the Web. Leading personal computer manufacturers such as **Compaq**, **Dell**, and **Gateway** have had great success selling on the Web. Dell has been a leader in allowing customers to specify exactly the configuration of computers they order on the Web. Dell has created value by designing its entire business around offering this high degree of configuration flexibility to its customers. Other personal computer manufacturers that sell directly to customers on the Web have followed Dell's lead by offering visitors different ways to access product information. These sites usually offer links to specific products and to pages designed for specific categories of customers, such as home, small business, education, or government users.

A number of apparel sellers have adapted their catalog sales model to the Web, including **Lands' End**, **L. L. Bean**, and **Talbots**. In the fast-changing world of fashion, retailers have always had to deal with the problem of overstocks—products that did not sell as well as hoped. Many retailers use outlet stores to sell their overstocks. Lands' End has found that its online overstocks page has worked so well that it is closing many of its brick-and-mortar outlet stores.

One problem that the Web presents for clothing retailers is that the color settings on computer monitors vary widely. It is difficult for customers to get a completely correct idea of what the product's color will look like when it arrives. Until technology solves this problem, most online clothing stores will send a fabric swatch on request. The swatch also gives the customer a sense of the fabric's texture—an added benefit not provided by catalogs.

Gift retailers have also successfully moved or expanded their business models to the Web. Florist **1-800-Flowers** has created an online extension to its highly successful telephone order business. **Harry and David**, the originators of the "Fruit-of-the-Month" club, opened an informational Web site to promote their existing catalog business. They were surprised by the volume of sales leads that the site generated and quickly added online ordering features to their site, which appears in Figure 8-15.

311

Figure 8-15 *Harry and David home page*

A number of retailers that did not already have an existing business have started retail operations on the Web. Some of these completely new businesses operate as Web-based deep discounters. Borrowing a concept from the physical world's Wal-Marts and discount club stores, these discounters sell merchandise such as computer equipment, software, consumer electronics, books, music CDs, and sports equipment at extremely low prices. Unlike their physical world counterparts, however, these Web discount retailers plan to sell advertising on their sites to subsidize their low product prices. This category includes firms such as **Beyond.com**, **Buy.com**, **NECX.com**, and **Outpost.com**.

Companies that have existing sales outlets and distribution networks often worry that their Web sites will take away sales from those outlets and networks. For example, Levi Strauss & Company sells its Levi's jeans and other clothing products through department stores and other retail outlets. After receiving many complaints from these stores, who had been selling Levi's products for many years, Levi Strauss decided to stop selling products on its own Web site. Such a **channel conflict** can occur whenever sales activities on a company's Web site interfere with its existing sales outlets. The **Levi's** Web site now provides product information, but directs customers who want to buy its products to physical stores that carry those products.

Selling Information or Other Digital Content

Firms that own intellectual property or rights to that property have embraced the Web as a new and highly efficient distribution mechanism. LEXIS-NEXIS is an important legal research tool that has been available as an online product for years. It provides full-text search of court cases, laws, patent databases, and tax regulations. To obtain access to this information, law firms had to subscribe and install expensive dedicated computer systems. The Web has given LEXIS-NEXIS customers much more flexibility in how they purchase information. Through the **LEXIS-NEXIS Xchange** Web site, law firms can subscribe to several versions of the service that are customized for different firm sizes and usage patterns. The Web site even offers a credit card charge option for infrequent users who do not want a subscription. LEXIS-NEXIS has used the Web to improve the delivery and variety of its existing product line and has been able to devise new products that take advantage of the Web's features.

ProQuest, a Web site that sells digital copies of published documents, has its roots in two businesses, the former Bell and Howell learning materials business and University Microfilms International (UMI). These firms had acquired reproduction rights to a variety of published and unpublished materials. For example, UMI had contracts with most North American universities to publish all doctoral dissertations and masters theses on demand. ProQuest offers digital versions of these documents for sale, along with a number of newspapers, journals, and other specialized academic publications. Many schools and libraries have subscriptions to ProQuest. **EBSCO Information Services** is a similar company that offers subscriptions to digital versions of journals, along with databases and access to electronic journals for individuals, schools, and libraries. **Dow Jones Interactive** is a more business-focused seller of subscriptions to digitized newspaper, magazine, and journal content. The Dow Jones site offers a customized digital clipping service that provides subscribers with a daily e-mail message of news on topics of interest to them.

One of the first academic organizations to make the transition to electronic distribution on the Web is (not surprisingly) the Association for Computer Machinery (ACM). The **ACM Digital Library** offers subscriptions to electronic versions of its journals to its members and to library and institutional subscribers. Academic publishing has always been a difficult business in which to make a profit, because the base of potential subscribers is so small. Even the most highly regarded academic journals often have fewer than 2000 subscribers. To break even, academic journals often must charge each subscriber hundreds or even thousands of dollars per year. Electronic publishing takes the high costs of paper, printing, and delivery out of the business model and makes dissemination of research results less expensive and more timely.

As was the case for other technologies, such as VCRs and subscription cable television, many of the early commercial adopters of Web technology were dealers in adult-themed entertainment material. Many of the first profitable sites on the Web were sellers of adult digital content. These sites pioneered the processing of credit card payment transactions and many different types of digital video technologies that are now used by all types of businesses on the Web.

Encyclopædia Britannica is an excellent example of a company that has transferred an existing brand to the Web. The Encyclopædia Britannica has developed one of the most respected brand names in research and education over its many years in print publishing. It is particularly interesting that the Encyclopædia Britannica began in 1768 as a sort of pre-computer-age frequently asked questions (FAQ) list. A group of academics collected notes they had made while conducting research and decided to publish them as a series of articles.

Encyclopædia Britannica began with two Web-based offerings. The Britannica Internet Guide was a free Web navigation aid that classified and rated information-laden Web sites. It featured reviews written by Britannica editors who also selected and indexed the sites. The company's other Web site, **Encyclopædia Britannica Online**, was available for a subscription fee or as part of its Encyclopædia Britannica CD package. Britannica used the free site to attract users to the paid subscription site.

In 1999, disappointed by low subscription sales, Britannica converted to a free, advertiser-supported site. The first day the new site became available at no cost to the public, it had over 15 million visitors, forcing Britannica to shut down for two weeks to upgrade its servers.

The Britannica.com site now offers the full content of the print edition in searchable form, plus access to the *Merriam-Webster's Collegiate Dictionary* and the *Britannica Book of the Year*. One of the most successful aspects of the new site is the way it integrates the Britannica Internet Guide Web-rating service with its print content. The Britannica Store, which also is on the Web site, still sells the CD version along with other educational and scientific products.

Britannica has gone from being a print publisher to a seller of information on the Web to an advertising-supported Web site—three major business model transitions—in just a few short years. The main value that Britannica has to sell is its reputation and the expertise of its editors, contributors, and advisors. For now, Britannica has decided that the best way to capitalize on that reputation and expertise is through an advertising-supported Web site. As you will see in the next section, other companies use this model, too.

Advertising-Supported Model

The advertising-supported business model is the one used by network television in the United States. Broadcasters provide free programming to an audience along with advertising messages. The advertising revenue is sufficient to support the operations of the network and the creation or purchase of the programs.

Many observers of the Web in its early growth period believed that the potential for Internet advertising was tremendous. However, after a few years of experience in trying to develop profitable advertising-supported business models, many of those observers are less optimistic. The success of Web advertising has been hampered by two major problems. First, as discussed earlier, no consensus has emerged on how to measure and charge for site visitor views. Since the Web allows multiple measurements, such as of number of visitors, number of unique visitors, number of click-throughs, and other attributes of visitor behavior, it has been difficult for Web advertisers to develop a standard for advertising charges, such as the CPM measure used for mass media outlets. In addition to the number of visitors or page views, stickiness is a critical element to creating a presence that will attract advertisers. Recall from Chapter 3 that the **stickiness** of a Web site is its ability to keep visitors at the site and to attract repeat visitors. People spend more time at a **sticky** Web site and are thus exposed to more advertising.

The second problem is that very few Web sites have sufficient numbers of visitors to interest large advertisers. Most successful advertising on the Web is targeted to very specific groups. However, it can be difficult to determine whether a given Web site is attracting a specific market segment unless that site collects demographic information from its visitors—information that visitors are increasingly reluctant to provide because of privacy concerns.

Only a few general-interest sites have generated sufficient traffic to be profitable based on advertising revenue alone. One of these is **Yahoo!**, which was one of the first Web directories. Because so many people use Yahoo! as a starting point for searching the Web, it has always attracted a large number of visitors. This large number of visitors made it possible for Yahoo! to expand its Web directory into one of the first portal sites. Because the Yahoo! portal's search engine presents visitors' search results on separate pages, it can include advertising on each results page that is triggered by the terms in the search. For example, when the Yahoo! search engine detects that a visitor has searched on the term "new car deals," it can place a Ford ad at the top of the search results page. Ford is willing to pay more for this ad because it is directed only at visitors who have expressed interest in new cars. This example demonstrates one attractive option for identifying a target market audience without collecting demographic information from site visitors. Unfortunately, only a few high-traffic sites are able to generate significant advertising revenues this way. Besides Yahoo!, the main portal sites in this market today are **Excite**, **Infoseek**, and **Lycos**. Smaller general-interest sites, such as the Web directory **refdesk.com**, have had much more difficulty attracting advertisers than the larger search engine sites. This may change in the future as more people use the Web.

Newspaper publishers have experimented with various ways of establishing a profitable presence on the Web. It is unclear whether a newspaper's presence on the Web helps or hurts the newspaper's business as a whole. Although it provides greater exposure for the newspaper's brand and provides a larger audience for advertising that the paper carries, it also can take away sales from the print edition, a process called **cannibalization**. Newspapers and other publishers worry about cannibalization, because it is very difficult to measure. Some publishers have conducted surveys in which they ask people whether they have stopped buying the newspaper because the contents they want to see are available online, but the results of such surveys are not very reliable.

Many leading newspapers, including **The Washington Post** and the **Los Angeles Times**, have established online presences in the hopes that they will generate enough revenue to cover the cost of creating and maintaining the Web site. The **Internet Public Library Online Newspapers** page includes links to hundreds of newspaper sites around the world.

Although attempts to create general-interest Web sites that generate sufficient advertising revenue to be profitable have met with mixed results, sites that target niche markets have been more successful. For newspapers, classified advertising is very profitable, so it is no surprise that Web sites that specialize in providing only classified advertising have profit potential if they can reach a narrow enough target market.

315

One implementation of the advertising-supported business model that appears to be successful is Web employment advertising. Firms with Web sites such as **CareerSite.com** and **JOBTRAK** offer international distribution of employment ads. As the number of people using the Web increases, these businesses will be able to move out of their current focus on technology and higher-level jobs and include advertising for all kinds of positions. These sites can use the same approach that search engine sites use to offer advertisers target markets. When a visitor specifies an interest in, for example, engineering jobs in Dallas, the results page can include a targeted banner ad for which an advertiser will pay more, because it is directed at a specific segment of the audience. Employment ad sites can also target specific categories of job seekers by including short articles on topics of interest. This will also keep qualified people who are not necessarily looking for a job coming back to the site—such people are the candidates most highly sought after by employers. The **Monster.com** page directed at mid-career executives appears in Figure 8-16. The page offers links to articles that might interest a mid-career executive and a poll customized for that audience. It also includes a banner ad at the top of the page that is directed at an audience of mid-career executives.

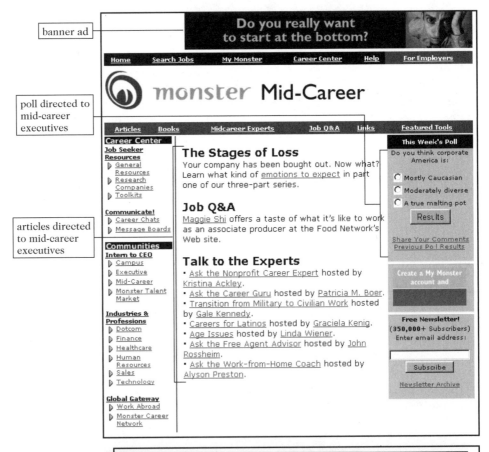

Figure 8-16 *Monster.com Web page for mid-career job seekers*

Advertising-Subscription Mixed Model

In this mixed model, which has been used for many years by newspapers and magazines, subscribers pay a fee and accept some level of advertising. In most cases, the subscribers are subjected to much less advertising than they are on advertising-supported sites. Firms have had varying levels of success in applying this model. For example, Microsoft's **Slate** e-zine returned to using an advertising-only model after failing to attract a sufficient number of subscribers. Other firms have had more success.

Two of the world's most distinguished newspapers, **The New York Times** and **The Wall Street Journal**, use a mixed advertising-subscription model. *The New York Times* version is mostly advertising-supported with a small subscription fee for visitors who want online access to the newspaper's crossword puzzle feature. This premium service originally included the bridge and chess columns, but those are now freely accessible to any site visitor. *The Wall Street Journal's* mixed model is weighted more heavily to subscription revenue. The site allows visitors to view the classified ads and certain stories from the newspaper, but most of the content is reserved for subscribers. Visitors who already subscribe to the print edition are offered a reduced rate on subscriptions to the online edition.

The **Reuters** wire service also uses a mixed model in its Web offerings. A wire service collects news reports from around the globe, consolidates them, and sells them to newspapers, radio and television stations, governments, and large companies. The value added by a wire service is consolidation and filtering. For example, a company might want a wire service to provide every story it collects on the company and its competitors. Reuters provides some news headlines on its site, but it refers Web visitors to its subscribers, including Yahoo!, Lycos, Infoseek, and **C-NET**, through links on its headlines page.

Sports fans visit the **ESPN** site for all types of sports-related information. Leveraging its brand name from its cable television businesses, ESPN is one of the most-visited sites on the Web. It sells advertising and offers a vast amount of free information, but die-hard fans can subscribe to its Insider service to obtain access to even more sports information.

Northern Light is a search engine with a twist. In addition to searching the Web, it searches its own database of journal articles and other publications to which it has acquired reproduction rights. Northern Light's business model extends the usual Web search engine to include services such as those offered by ProQuest or EBSCO. A Northern Light search results page appears in Figure 8-17.

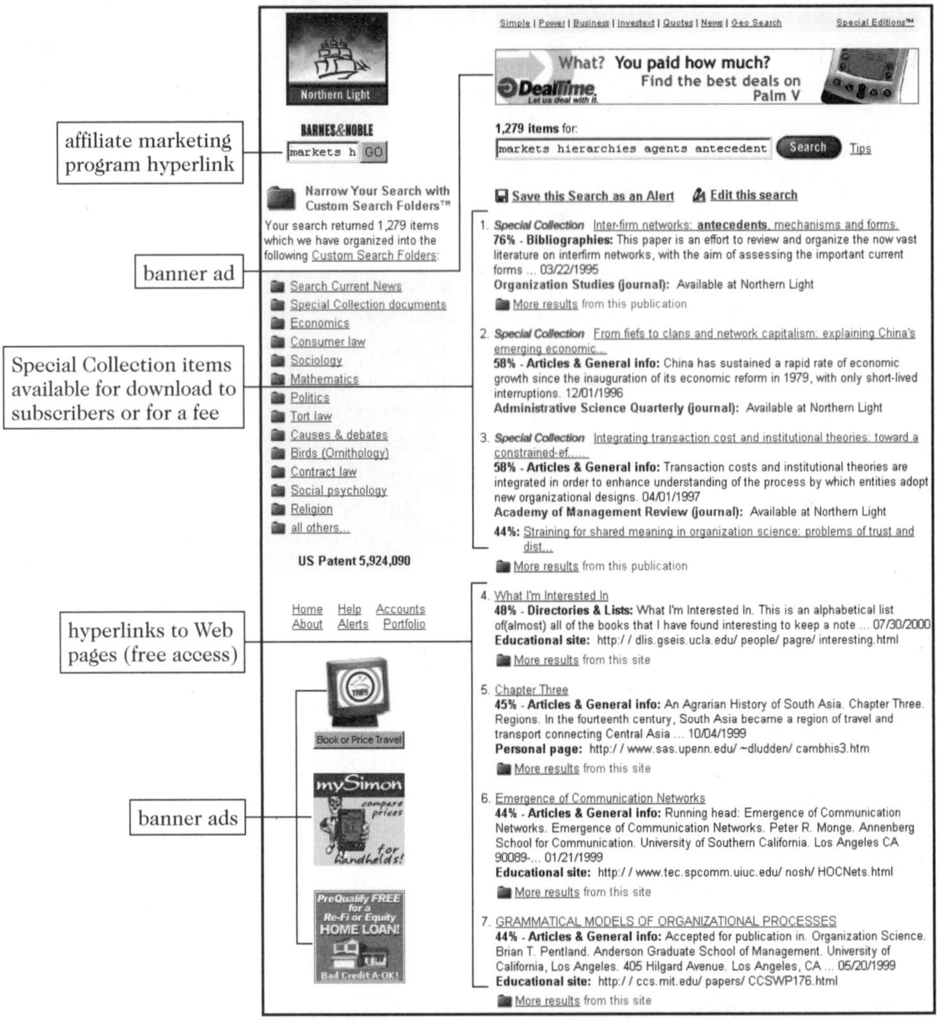
affiliate marketing program hyperlink
banner ad
Special Collection items available for download to subscribers or for a fee
hyperlinks to Web pages (free access)
banner ads
Simple | Power | Business | Investext | Quotes | News | Geo Search
Special Editions™
Northern Light
What? You paid how much?
Find the best deals on Palm V
DealTime
Let us deal with it.
BARNES&NOBLE
markets h GO
1,279 items for:
markets hierarchies agents antecedent
Search
Tips
Narrow Your Search with Custom Search Folders™
Your search returned 1,279 items which we have organized into the following Custom Search Folders:
Search Current News
Special Collection documents
Economics
Consumer law
Sociology
Mathematics
Politics
Tort law
Causes & debates
Birds (Ornithology)
Contract law
Social psychology
Religion
all others...
US Patent 5,924,090
Home Help Accounts
About Alerts Portfolio
Save this Search as an Alert
Edit this search
1. Special Collection Inter-firm networks: antecedents, mechanisms and forms.
76% - Bibliographies: This paper is an effort to review and organize the now vast literature on interfirm networks, with the aim of assessing the important current forms ... 03/22/1995
Organization Studies (journal): Available at Northern Light
More results from this publication
2. Special Collection From fiefs to clans and network capitalism: explaining China's emerging economic...
58% - Articles & General info: China has sustained a rapid rate of economic growth since the inauguration of its economic reform in 1979, with only short-lived interruptions. 12/01/1996
Administrative Science Quarterly (journal): Available at Northern Light
3. Special Collection Integrating transaction cost and institutional theories: toward a constrained-ef......
58% - Articles & General info: Transaction costs and institutional theories are integrated in order to enhance understanding of the process by which entities adopt new organizational designs. 04/01/1997
Academy of Management Review (journal): Available at Northern Light
44%: Straining for shared meaning in organization science: problems of trust and dist...
More results from this publication
4. What I'm Interested In
48% - Directories & Lists: What I'm Interested In. This is an alphabetical list of(almost) all of the books that I have found interesting to keep a note ... 07/30/2000
Educational site: http://dlis.gseis.ucla.edu/people/pagre/interesting.html
More results from this site
5. Chapter Three
45% - Articles & General info: An Agrarian History of South Asia. Chapter Three. Regions. In the fourteenth century, South Asia became a region of travel and transport connecting Central Asia ... 10/04/1999
Personal page: http://www.sas.upenn.edu/~dludden/cambhis3.htm
More results from this site
6. Emergence of Communication Networks
44% - Articles & General info: Running head: Emergence of Communication Networks. Emergence of Communication Networks. Peter R. Monge. Annenberg School for Communication. University of Southern California. Los Angeles CA 90089-... 01/21/1999
Educational site: http://www.tec.spcomm.uiuc.edu/nosh/HOCNets.html
More results from this site
7. GRAMMATICAL MODELS OF ORGANIZATIONAL PROCESSES
44% - Articles & General info: Accepted for publication in. Organization Science. Brian T. Pentland. Anderson Graduate School of Management. University of California, Los Angeles. 405 Hilgard Avenue. Los Angeles, CA ... 05/20/1999
Educational site: http://ccs.mit.edu/papers/CCSWP176.html
More results from this site
Book or Price Travel
mySimon
compare prices
for handhelds!
PreQualify FREE for a Re-Fi or Equity HOME LOAN!
Bad Credit A-OK!

Figure 8-17 *Northern Light search results page*

The search results marked "Special Collection" are included in the Northern Light proprietary database. Users can read the bibliographic information for these entries, which sometimes includes an abstract, at no charge. Users can download the full text of any special collection item for a small fee, which varies depending on where the article appeared. Alternatively, users can subscribe to several different plans that offer free or reduced-price downloads.

Fee-for-Transaction Models

Travel agents earn commissions on each airplane ticket, hotel reservation, auto rental, or vacation that they book. These commissions are paid to the travel agent by the transportation or lodging provider. The travel agency business model involves receiving a fee for facilitating a transaction. The value added by a travel agent is that of information consolidation and filtering. A good travel agent knows many things about the traveler's destination and knows enough about the traveler to select the information elements that will be useful and valuable to the traveler. Computers, particularly computers networked to large databases, are very good at information consolidation and filtering. In fact, travel agents have used networked computers, such as the **SABRE Group** systems, for many years to make reservations for their customers.

When the Internet emerged as a new way to network computers and then became available to commercial users, a number of online travel agencies began doing business on the Web. Existing travel agencies did not, in general, rush to the new medium. They believed that the key value they added, personal customer service, could not be replaced with a Web site. Therefore, the first Web-based travel agencies were new entrants. One of these sites, **Travelocity**, is based on the same SABRE system that traditional travel agents use. Microsoft has entered the online travel agency business with its **Expedia** site.

In addition to earning commissions from the transportation and lodging providers, these sites generate advertising revenue from ads placed on travel information pages. These ads are similar to those on search engine results pages, because advertisers can target them without obtaining demographic details about the site visitor. For example, if you are booking a flight to Chicago, the page that lists airline ticket options may also carry a banner ad for a hotel in Chicago or a car rental company that is running a promotion in the Chicago area.

Auto dealers buy cars from the manufacturer and sell them to consumers. They provide showrooms and salespeople to help customers learn about product features, arrange financing, and make a purchase decision. Most auto dealers negotiate the prices at which they sell their cars; thus, the salesperson's job also includes extracting the highest possible price from the consumer. Many people do not like negotiating car prices. **Autobytel** and other firms offer the consumer who has learned about product features, arranged financing, and is ready to purchase a new car an option that removes the salesperson from the process. Autobytel will locate dealers in the buyer's area that are willing to sell the car specified by the buyer (including make, model, options, and color) for a small premium over the dealer's nominal cost. The buyer can purchase the car from the dealer without negotiating with a salesperson. Autobytel charges participating dealers a fee for this service. In effect, Autobytel is taking the salesperson out of the value chain. To the extent that the salesperson provides little or no value to the consumer, Autobytel is reducing the transaction costs in the process. The removal of an intermediary, such as the salesperson in this case, from a value chain is called **disintermediation**.

Stock brokerage firms use a fee-for-transaction model. They charge their customers a commission for each trade executed. In the introduction to this chapter, you learned how Merrill Lynch had to face competition from online and discount brokers. To the extent that Merrill Lynch's stockbrokers fail to provide value to their customers, they are susceptible to the same kind of disintermediation as the auto salespeople being replaced by Autobytel. The online brokers are replacing the traditional broker services with an alternate service that has greater perceived value by investors today—trading over the Web.

Other sales agency businesses may move to the Web soon. Although insurance companies currently have little interest in selling policies and investments on the Web, the fee-for-transactions model appears to be a good fit and might reduce operating costs by disintermediating salespeople in that industry. Other candidates for increased movement to the Web are the real estate brokerage industry and online banking. In both cases, a Web presence could significantly disintermediate sales and branch management operations.

Summary

By understanding how the Web differs from other media and designing a Web site to capitalize on those differences, firms can create an effective Web presence that delivers value to visitors. Every organization must realize that visitors to its Web site arrive with a variety of expectations, prior knowledge, and skill levels, and are connected to the Internet through different technologies. Knowing how these factors can affect the visitor's ability to navigate the site and extract information from the site can help organizations design better Web sites. Enlisting the help of users when building test versions of the Web site is also a good way to create a Web site that represents the organization well.

Firms must understand the nature of communication on the Web so they can use it to identify and reach the largest possible number of qualified customers. Technology-enabled relationship management can provide better returns for businesses on the Web than the traditional approaches of market segmentation and micromarketing.

Firms on the Web can use rational branding instead of the emotional branding techniques that work well in mass media advertising. Some businesses on the Web are sharing and transferring brand benefits through affiliate marketing and cooperative efforts among brand owners. Businesses are using four different revenue-generating models as they sell goods and services on the Web: the Web-catalog model, the advertising-supported model, the advertising-subscription mixed model, and the fee-for-transaction model.

Key Terms

Ad View

Addressable Media

Affiliate Marketing

Cannibalization

Cause Marketing

Channel Conflict

Click

Click-Through

Conversion Rate

Cost Per Thousand (CPM)

Customer Relationship Management (CRM)

Disintermediation

E-Zine

Electronic Customer Relationship Management (eCRM)

Impression

Many-to-Many Communications

Many-to-One Communication Model

Market Segmentation

Marketspace

Mass Media

Micromarketing

One-to-Many Communication Model

One-to-One Communication Model

Opt-In E-Mail

Page View

Permission Marketing

Personal Contact

Presence

Prospecting

Rational Branding

Repeat Visit

Segments

Stakeholders

Stickiness

Sticky

Technology-Enabled Relationship Management

Trial Visit

URL Brokers

Usability Testing

Viral Marketing

Visit

Web-Catalog Model

Review Questions

1. In one paragraph, define the term *presence*. Write an additional paragraph in which you explain why firms that do business on the Web should be more concerned about presence than firms that operate only in the physical world.

2. Write a brief paragraph in which you describe two Web design mistakes that firms often made when they created their first Web sites in the mid-1990s.

3. In 500 words or fewer, explain how promoting products on the Web is different from using mass media promotion or personal contact.

4. What are the key elements of technology-enabled relationship management? What advantages does technology-enabled relationship management have over traditional seller-customer interactions? Keep your answer under 250 words.

5. Write a paragraph in which you describe the conditions under which a Web site can hope to become profitable if it relies exclusively on advertising revenue.

Exercises

1. Page 291 includes a list of six things that Web sites can do to meet the needs of visitors. Find one Web site that includes three or more of the items on the list. Explain how the Web site that you choose accomplishes each. You may want to use the **Webby Awards** site as a starting point in your search, but do not use any of the award nominees or winners as your example site.

2. You have been employed by Bob Drudge, the owner of **refdesk.com**, to sell space on his site to advertisers. Create a promotional press release of no more than 300 words in which you describe the advantages of advertising on refdesk.com. You may decide to promote space on the main page, specific other pages, or all pages. Be prepared to explain why your promotional strategy should work. You may find the

Art of Web Site Promotion, **Promotion World**, **Sitelaunch**, or **Seltzer's "How to Publicize Your Web Site over the Internet"** Online Companion links helpful in your task.

3. Identify four occupations, other than *auto salesperson* and *travel agent*, that you believe are exposed to the risk of disintermediation posed by firms moving activities to the Web.

Describe what specific activities might be moved to the Web that would cause this risk. For each occupation, explain whether the employees would be able to find jobs in the new business activity and explain what those jobs would be. You may want to use **About.com** or your favorite Web search engine to find information on this topic.

For Further Study and Research

Andrews, W. 1999. "Competing by Tweaking Old Methods," *Internet World*, March 22, 13–14.

Barron, K. 1999. "Realtors 1, Web 0," *Forbes*, 163(3), February 8, 64–65.

Berner, R. 2000. "Going that Extra Inch," *Business Week*, September 18, 84.

Biggs, M. 2000. "E-Commerce Success Requires Commitment to Building a Proper Customer Community," *InfoWorld*, 22(14), April 3, 68.

Brandt, R. 1998. "Internet Kamikazes: CNET's Halsey Minor," *Upside*, 10(1), January, 100, 136.

Brown, E. and J. Fox. 1999. "Nine Ways to Win on the Web," *Fortune*, 139(10), May 24, 112–121.

Buckman, R. 1999. "Wall Street Is Rocked by Merrill's Online Plans," *The Wall Street Journal*, June 2. C1.

Business Week. 1999. "Learn E-Business or Risk E-Limination," March 22, 122.

Byrnes, N. 2000. "Avon: The New Calling," *Business Week*, September 18, 136–143.

Carr, N. 2000. "Hypermediation: Commerce as Clickstream," *Harvard Business Review*, 78(1), January–February, 46–47.

Champy, J. 1998. "Direct Sales or Electronic Channels? Give the Customer a Choice," *Computerworld*, 32(47), November 23, 64.

Chuck, L. 1999. "On Being 'Consumer-ed': Marketing the User," *Searcher*, 7(5), May, 10–12.

Christensen, C. and M. Overdorf. 2000. "Meeting the Challenge of Disruptive Change," *Harvard Business Review*, 78(2), March–April, 66–75.

Cuneo, A. 1999. "Lands' End Ads Pitch 'Ultimate Direct Merchant,'" *Advertising Age*, 70(17), April 19, 85.

Daly, J. 2000. "Sage Advice: Interview with Peter Drucker," *Business 2.0*, August 22, 134–144.

Dodge, J. 1998. "Big Companies Think Small on E-Commerce," *PC Week*, 15(37), September 14, 3.

The Economist. 2000. "Shopping Around the Web," 354(8159), February 26, 5–6.

Evans, P. and T. Wurster. 1997. "Strategy and the New Economics of Information," *Harvard Business Review*, 75(5), September–October, 71–83.

Financial Executive. 1999. "CFOs Say Company Internet Involvement to Jump," 15(3), May–June, 57.

Gardner, E. 1999. "Art.com," *Internet World*, March 15, 13. Also available online at: (http://www.iw.com/print/1999/03/15/).

Godin, S. and D. Peppers. 1999. *Permission Marketing: Turning Strangers into Friends, and Friends into Customers*. New York: Simon & Schuster.

Gunderson, A. 2000. "Double, Double...Toil and Trouble: Online Shopping is Still a Muddle," *Fortune*, September 4, 374.

Hicks, M. 1999. "Sold on the Simplicity of Web Sites," *PC Week*, June 7, 77.

Hodges, M. 1997. "Is Web Business Good Business?" *Technology Review*, 100(6), August–September, 22–28.

Hof, R. 2000. "Creative Coddling, Great Word of Mouth," *Business Week*, September 19, 57.

Hoffman, D. and T. Novak. 2000. "How to Acquire Customers on the Web," *Harvard Business Review*, 78(3), May–June, 179–188.

Hogan, M. 1999. "The Internet Business 100," *PC Computing*, 12(7), July, 145–161.

Kelsey, D. 2000. "Top E-Tail Web Sites' Customer Service Not Good," *BizReport*, August 8. Available online at: (http://www.bizreport.com/research/2000/08/20000809-1.htm).

Kling, J. 1999. "Seven Great Ways to Use Your Company's Web Site," *Harvard Management Update*, January, 3–4.

Komenar, M. 1997. *Electronic Marketing*. New York: John Wiley & Sons.

Koprowski, G. 1998. "The (New) Hidden Persuaders: What Marketers Have Learned About How Consumers Buy on the Web," *The Wall Street Journal*, December 7, R10.

Lee, L. 2000. "An E-Commerce Cautionary Tale," *Business Week*, March 20, 46–47.

Levine, R., C. Locke, D. Searle, and D. Weinberger. 2000. *The Cluetrain Manifesto: The End of Business as Usual*. Cambridge, MA: Perseus.

Locke, C. 1998. "Fear and Loathing on the Web," *The Industry Standard*, July 9.

Lytel, J. 2000. "Domain-Name Disputes Get Personal," *BizReport*, September 22. Available online at: (http://www.bizreport.com/marketing/2000/09/2000922-1.htm).

Magid, L. 1999. "Web Ads Get Smart," *Upside Today*, June 9. Available online at: (http://www.upside.com/texis/mvm/larry_magid?id=375d863b0).

Managing Intellectual Property. 1999. "Developments," June, 90.

Mowry, M. 2000. "Net Retail's Grim Reality," *The Industry Standard*, August 7, 154–155.

Narisetti, R. 1998. "New and Improved: Ad Experts Talk About How Their Business Will Be Transformed by Technology," *The Wall Street Journal*, November 16, R33.

Nash, K. 2000. "Little E-Engines That Could," *Computerworld*, 34(16), April 17, 46–47.

Neilsen, J. 1999. *Designing Websites With Authority: Secrets of an Information Architect*. Indianapolis: New Riders.

Neilsen, J. 2000. "End of Web Design," *Alertbox*, July 23. Available online at: (http://www.useit.com/alertbox/20000723.html).

Neuborne, E. and R. Hof. 1998. "Branding on the Net," *Business Week*, November 9, 76–81.

Notess, G. 2000. "Internet Search Engine Update," *Online*, January–February, 13–14.

Nunes, P., D. Wilson, and A. Kambil. 2000. " The All-in-One Market," *Harvard Business Review*, 78(3), May–June, 19–20.

Ojala, M. 2000. "The Business of Domain Names," *Online*, May 1, 78.

Oliva, R. 2000. "Brainstorm Your E-business," *Marketing Management*, 9(1), Spring, 55–57.

Orenstein, S. 2000. "Boo.com: A Cautionary Tale," *The Industry Standard*, June 5, 106–113.

Ramsey, C. 2000. "Managing Web Sites as Dynamic Business Applications," *Intranet Design Magazine*, June. Available online at: (http://idm.internet.com/articles/200006/wm_index.html).

Rayport, J. and J. Sviokla. 1995. "Exploiting the Virtual Value Chain," *Harvard Business Review*, 73(6), November–December, 75–85.

Reed, S. 1999. "Can't Get No Satisfaction: Online Shopping Can Be as Bad as the Real Thing," *InfoWorld*, 21(21), May 24, 83.

Reichheld, F. and P. Schefter. 2000. "E-Loyalty: Your Secret Weapon on the Web," *Harvard Business Review*, 78(4), July–August, 105–113.

Reyes, E. 2000. "Customer Service: The Lost Element," *Business 2.0*, July 11, 191–195.

Ross, P. 1999. "Web Winners," *Forbes*, 163(8), April 19, 226–228.

Schaffer, E. 2000. "A Better Way For Web Design," *InformationWeek*, May 1, 194.

Schwartz, E. 1997. *Webonomics*. New York: Broadway Books.

Schwartz, E. 1999. *Digital Darwinism*. New York: Broadway Books.

Shelton, B. 1999. "Building Customer Loyalty on the Web," 9(3), March 1, 24–29.

323

Shneiderman, B. 1997. *Designing the User Interface: Strategies for Effective Human-Computer Interaction*. Reading, MA: Addison-Wesley.

Shrake, S. 2000. "Start With the Right Brand Name," *Target Marketing*, April 1, 52.

Sklar, J. 2000. *Principles of Web Design*. Cambridge, MA: Course Technology.

Slatalla, M. 2000. "Wonder How You'd Look in Yellow Eye Shadow?" *The New York Times*, August 18, G-4.

Technology Review. 1998. "That Mess on Your Web Site," 101(5), September–October, 72–75.

Tedeschi, B. 2000. "Charitable Groups Discover New Revenue in Retailing Goods via Their Own Web Sites," *The New York Times*, March 27, C-11.

Tedeschi, B. 2000. "Retailers Are Letting Their Right Hand Know What The Left Hand is Up To for Better Customer Service," *The New York Times*, July 24, C-9.

Tweney, D. "Market Pressures Will Change the Shape of Online Advertising: Only the Clever Will Survive," *InfoWorld*, 20(36), September 7, 49.

Walsh, J. 1999. "Is Your Site Really Working?" *InfoWorld*, 21(10), March 8, 53, 56.

Walsh, T. 2000. "Four Wheel Drive," *Business 2.0*, August 22, 151–164.

Weinberger, D. 1999. "0:1 Marketing," *Journal of Hyperlinked Organizations*, May 20. Available online at:
(http://www.hyperorg.com/backissues
/joho-may20-99.html#01).

Zeff, R. and B. Aronson. 1997. *Advertising on the Internet*. New York: John Wiley & Sons.

Zellner, W. 2000. "The Trick to Selling Airline Tickets Online? Minimalism," *Business Week*, September 18, 90.

STRATEGIES FOR PURCHASING AND SUPPORT ACTIVITIES: FROM ELECTRONIC DATA INTERCHANGE TO ELECTRONIC COMMERCE

INTRODUCTION

General Electric (GE) is one of the largest and most successful companies in the United States. It engages in a wide range of businesses around the world, including production of appliances, electrical and electronic products, broadcasting, and a variety of financial and insurance activities. One of its oldest lines of business is **GE Lighting**, which produces over 30,000 different kinds of lightbulbs in its 28 North American plants and other locations around the world. The raw materials used in making lightbulbs are fairly standard items: glass, aluminum, various insulating plastics, and filament materials. However, a major portion of each lightbulb's cost is the money that GE Lighting spends on indirect materials and parts for the machines used to fabricate and assemble the bulbs. These indirect materials and parts must conform to detailed GE specifications laid out on more than 3 million individual blueprints and other design drawings.

Since the technologies for making lightbulbs are mature and well known, GE Lighting can solicit bids from a variety of suppliers for indirect materials and machinery replacement parts. This is a high volume, low item-value business. Unfortunately, the bidding process at GE Lighting had become very slow and inefficient. Each transaction required the purchasing department to request the relevant blueprints, photocopy them, attach them to other material specification documents, and mail the whole package to suppliers who might be interested in bidding on the item. It would often take purchasing personnel more than four weeks to gather the information, send it to potential suppliers, obtain and evaluate suppliers' bids, negotiate with the chosen suppliers, and place an order. These long delays were limiting GE Lighting's flexibility and ability to respond to requests from its customers.

By applying the tools of electronic commerce to these purchase transactions, GE Lighting was able to make major improvements to the entire parts acquisition process. Today, purchasing personnel have access to a procurement system through their desktop computers. When they need to buy replacement parts for a machine, they create a new purchase file that includes basic quantity, delivery date, and delivery location information. Then, from a list generated by a continuously updated supplier database, they select suppliers from whom they will request quotes. Finally, they attach electronic copies of all necessary blueprints and engineering drawings, which are now digitized and stored in another database, and with a mouse click send the entire bid package off in an encrypted format to all the selected suppliers. Assembling the bid package now takes hours instead of a week or more. Suppliers are asked to respond within a short time period—usually a week—through the Internet. The purchasing staff member can evaluate the returned bids and award a contract online, completing the entire process in about 10 days.

The most significant savings for GE Lighting were in process time reduction, from four weeks or more to 10 days, and in the elimination of paper and the costs of paper handling. However, the company also realized other benefits. Because the online system made it easier to send out bid packages, purchasing was able to send out more bids to a wider range of suppliers. In particular, many foreign suppliers who had been difficult to reach with mailed bid packages were now being included in the solicitation for quotes. The increased competition drove prices down; GE Lighting has saved up to 20% on many of these items since moving the bid process online. Suppliers have welcomed the reduced time lag between submitting the bid and finding out whether GE Lighting would award them the contract; this has made their production planning easier.

- Strategies that businesses use to improve purchasing, logistics, and other support activities
- The ways that firms are creating network organizations that extend beyond traditional enterprise limits
- Electronic data interchange, how it works, and how businesses are moving it to the Internet
- Supply chain management and how businesses are using the Internet and Web technologies to improve it
- The software packages that companies are using to implement business-to-business electronic commerce and supply chain management

PURCHASING, LOGISTICS, AND SUPPORT ACTIVITIES

In the previous chapter, you learned about strategy issues that arise when businesses and other organizations provide information to potential customers and other stakeholders. In terms of the value chain model described in Figure 1-12, you learned about the primary activities *identify customers*, *market and sell*, and *deliver*. You also became familiar with several business models for selling on the Web. Although those business models apply to both business-to-business and business-to-consumer electronic commerce, the emphasis in Chapter 8 was on business-to-consumer advertising, promotion, and sales activities.

In this chapter, you will learn how businesses use electronic commerce to improve their primary activities of *purchasing* and *logistics*, and all of the support activities shown in Figure 1-12. At first glance, none of these business elements seems as exciting as creating a Web presence and selling to new customers on the Web; however, the potential for cost reduction and business process improvement in purchasing, logistics, and support activities is tremendous.

An emerging characteristic of purchasing, logistics, and support activities is that they need to be flexible. A purchasing or logistics strategy that works this year may not work next year. Fortunately, economic organizations are evolving from the hierarchical structures they have used since the Industrial Revolution to new, more flexible network structures. These network structures are made possible by the reductions in transaction costs caused by the emergence of the Internet and the Web.

Purchasing Activities

Purchasing activities include identifying vendors, evaluating vendors, selecting specific products, placing orders, and resolving any issues that arise after receiving the ordered goods or services. These issues might include late deliveries, incorrect quantities shipped, incorrect items shipped, and defective items. By monitoring all relevant elements of purchase transactions, purchasing managers can play an important

327

role in maintaining and improving product quality and reducing cost. Many managers call this function procurement instead of purchasing to distinguish the broader range of responsibilities. The term **procurement** generally includes all purchasing activities, plus the monitoring of all elements of purchase transactions. It also includes the job of managing and developing relationships with key suppliers.

In many cases, procurement staff must have high levels of product knowledge. Specialized Web purchasing sites can be particularly useful in these situations. For example, the task of outfitting a hospital emergency room or operating room can take six months. Firms such as **Neoforma** gather a wide array of related medical fixtures and supplies in one virtual space to make procurement for this activity much more efficient and effective. The Neoforma product sourcing Web site includes pages that offer digital movies of equipped hospital rooms showing the fixtures, furniture, and equipment that the company sells. One of these pages, displaying a room that Neoforma equipped at a radiology imaging center, appears in Figure 9-1.

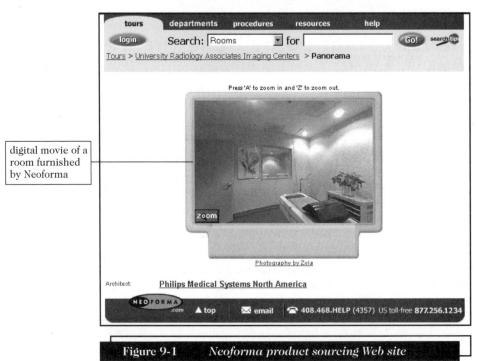

digital movie of a room furnished by Neoforma

Figure 9-1   Neoforma product sourcing Web site

The Neoforma site offers digital movies of a number of rooms the company has outfitted for various medical practices and hospitals. The site visitor can zoom the movie in or out to obtain a better view of the room's design and fixtures.

A number of companies that manufacture general industrial merchandise and standard machine tools that are used in a variety of industries have created Web sites through which businesses can purchase items to fulfill recurring needs. Many of the products that companies buy on a recurring basis are commodities—that is, standard items that buyers usually select using price as their main criterion. These products are often called **maintenance, repair, and operating (MRO)** supplies. By using a Web site to process orders, the vendors in this market can save the cost of printing and shipping catalogs, and the cost of handling telephone orders. Some industry analysts estimate that the cost to process an MRO order through a Web site can be less than one-tenth of the cost of handling the same order by telephone.

One of the largest MRO suppliers in the world is **W.W. Grainger**. Its Web site offers over 220,000 products for sale. Grainger's Web store, which appears in Figure 9-2, offers visitors a variety of ways to access information about and order Grainger products.

Figure 9-2 *Grainger.com Web store*

A visitor can enter the online catalog, use the product search box at the top of the page, or search by clicking a hyperlink to one of the categories listed in the middle of the page.

McMaster-Carr is another major MRO supplier that has established a Web site for business-to-business sales. In 1999, **Milacron** opened a Web site to sell MRO items such as cutting tools, grinding wheels, and manufacturing fluids. Milacron, a diversified machinery manufacturer, had experienced difficulty in effectively reaching one of its key markets—over 100,000 widely scattered machine and job shops—and is planning to reach that market through its Web presence. Milacron is even accepting credit cards in payment for purchases on its Web site, which is highly unusual for an industrial products vendor. By accepting credit cards, Milacron gives both its smaller customers and its new customers (who do not have an established credit record with Milacron) a convenient way to buy.

Office equipment and supplies are also items that are used by a wide variety of businesses. Market leaders **Office Depot** and **Staples** each have well-designed Web sites devoted to helping business purchasing departments buy these routine items as easily as possible. On their business-to-business Web sites, **Digi-Key** and **Newark Electronics** sell electronic parts, and **Global Computer Supplies** sells computers and related items

Logistics Activities

The classic objective of logistics has always been to provide the right goods in the right quantities in the right place at the right time. Businesses have been increasing their use of information technology to achieve this objective. Indeed, major transportation companies such as **Schneider Logistics**, **Ryder System**, and **J.B. Hunt** now want to be seen by their customers as information management firms as well as freight carriers. For example, the recently introduced Schneider Track and Trace system delivers real-time shipment information to Web browsers on its customers' computers. This system uses a Java applet to show the customer which freight carrier is transporting a shipment, where the shipment is, and when it should arrive at its destination. As mentioned in Chapter 2, **FedEx** has freight tracking Web pages available to its customers, as does **UPS**. Firms that run their own trucking operations have implemented tracking systems that use global positioning satellite technology to monitor vehicle movements.

Logistics activities include managing the inbound movements of materials and supplies and the outbound movements of finished goods and services. Thus, receiving, warehousing, controlling inventory, scheduling and controlling vehicles, and distributing finished goods are all logistics activities. The Web and the Internet are providing an increasing number of opportunities to manage these activities better as they lower transaction costs and provide constant connectivity between firms engaged in logistics management.

Support Activities

A 1999 article in *Inc.* magazine described the troubles facing Allegiance Telecom. The company was growing very rapidly and hiring over 100 people each month to staff its sales offices throughout the United States. Each new hire had to receive a full briefing on medical, dental, and retirement benefits plans and then he or she had to select from among several options for each. Since Allegiance was growing so rapidly, its human resources staff was spread thin and could not be in every sales office for every hire. The company turned to **Online Benefits**, a firm that duplicates its clients' human resource functions on a password-protected Web site that is accessible to clients' employees. The employees can then access their employers' benefits information, find the answers to frequently asked questions, and even perform complex benefit option calculations.

Other firms that offer support activities services include the following: **TheTrip.com**, employee travel policies; **CyLex Systems**, document storage; **PayMaxx**, payroll processing; **Atrieva**, electronic file storage; and **HotOffice**, group calendar and communications. Larger firms are building these types of functions into their intranet systems. For smaller firms, **DigitalWork** offers a number of business support activities, including debt collection, employee recruitment tools, and press release assistance.

Support activities include the general categories that appear in Figure 1-12: *finance and administration, human resources*, and *technology development*. Finance and administration includes activities such as making payments, processing payments received from customers, planning capital expenditures, and budgeting and planning to ensure that sufficient funds will be available to meet the organization's obligations as they come due. The operation of the computing infrastructure of the organization is also an administration activity. Human resource activities include hiring, training, and evaluating employees, administering benefits, and complying with government record-keeping regulations. Developing technology can include a wide variety of activities, depending on the nature of the business or organization. It can include networking research scientists into virtual collaborative workgroups, posting research results, publishing research papers online, and providing connections to outside sources of research and development services.

One common activity that supports multiple primary activities is training. In many companies, the human resources department handles training. Other companies may decentralize this function and have individual departments administer it. For example, insurance firms expend large amounts of resources on sales training. In most insurance companies, the sales and marketing department administers this training.

In 1999, the Swedish telecommunications giant Ericsson launched a Web site for current and former employees, families of those employees, and employees of approved business partners. This extranet was designed to facilitate knowledge management. **Knowledge management** is the intentional collection, classification, and dissemination of information about a company, its products, and its processes. This type of knowledge is developed over time by individuals working for or with a company and is often difficult to gather and distill. Ericsson has more than 100,000 employees scattered across the globe.

Managers hope that this knowledge network will generate new ideas, help solve problems, and improve business processes throughout the international organization. Designers of the system have identified their biggest challenge: to direct the information they collect in the extranet to projects and product development activities that will benefit from that information.

BroadVision has installed an internal system called K-Net, or Knowledge Network, that organizes all information sources that its employees use regularly in their jobs. It found that many of its employees were visiting between 10 and 20 Web sites each day in the course of doing their jobs. K-Net brings together all of the information that each employee needs and combines it into one dashboard-style interface presented on a Web browser. Much of the interface is customized for employees, although some parts of the interface—such as health insurance, vacations days, and other human resources information—is standardized for all employees. BroadVision has found the K-Net system to be so useful that it is partnering with Bank of

America, Hewlett-Packard, and a European travel services company, Amadeus, to develop a version of K-Net that it will sell to other companies.

Network Model of Economic Organization

In Chapter 1, you learned about the three different forms of economic organization: markets, hierarchies, and networks. One trend that is becoming clear in purchasing, logistics, and support activities is the shift away from hierarchical structures toward network structures. The traditional purchasing model had one hierarchically structured firm negotiating purchase terms with several similarly structured supplier firms and playing each supplier against the others. As is typical in a network organization, more businesses are using their procurement departments to negotiate using a variety of tools and in a variety of economic forms. For example, a buying firm might enter into a partnership with a supplier to develop a new technology that will reduce product costs overall. The technology development might be done by a third firm using research conducted by a fourth firm.

While reading the previous Support Activities section, you may have noticed that companies can have other firms perform various support activities. Again, these are examples of firms moving to a network model of economic organization. Imagine a business that uses one supplier to manage its payroll, another to administer its employee benefits plans, and a third to handle its document storage needs. The document storage service supplier might store the documents of the payroll service supplier and the benefits administration firm. The payroll service supplier might handle the payroll for the benefits administration firm. A fourth firm might provide online backup storage for the files of the other three companies. Of course, the payroll firm and the employee benefits firm might form a marketing partnership to sell both of their services to particular market segments. The document storage firm and the online backup storage firm might form a similar strategic alliance.

Highly specialized firms can now exist and trade services very efficiently on the Web. The Web is enabling this shift from hierarchical forms of economic organization to network forms. These emerging networks of firms are more flexible and can respond to changes in the economic environment much more quickly than hierarchically structured businesses ever could. You can learn more about the economics of networked organizations at the **Network Economics** Web site maintained by the University of California, Berkeley. The roots of Web technology for business-to-business transactions, however, lie in a very hierarchically structured approach to interfirm information transfer: electronic data interchange.

ELECTRONIC DATA INTERCHANGE

You learned in Chapter 1 that electronic data interchange (EDI) is a computer-to-computer transfer of business information between two businesses that uses a standard format of some kind. The two businesses that are exchanging information are called **trading partners**. Firms that exchange data in specific standard formats are said to be **EDI-compatible**. The business information exchanged is often transaction data; however, it can also include other information related to transactions, such as price quotes and order status inquiries. Transaction data in business-to-business transactions includes the information traditionally included on paper invoices, purchase orders, requests for quotations, bills of lading, and receiving reports. The data on these five types of forms accounts for over 75% of all information exchanged by trading partners in the United States. Thus, EDI was the first form of electronic commerce to be widely used in business—some 20 years before anyone used the term "electronic commerce" to describe anything!

Early Business Information Interchange Efforts

The emergence of large business organizations that occurred in the late 1800s and early 1900s brought with it the need to create formal records of business transactions. In the 1950s, companies began to use computers to store and process internal transaction records, but the information flows between businesses continued to be printed on paper; purchase orders, invoices, bills of lading, checks, remittance advices, and other standard forms were used to document transactions.

The process of using a person or a computer to generate a paper form, mailing that form, and then having another person enter the data into the trading partner's computer was slow, inefficient, expensive, redundant, and unreliable. By the 1960s, businesses that engaged in large volumes of transactions with each other had begun exchanging transaction information on punched cards or magnetic tape. Advances in data communications technology eventually allowed trading partners to transfer data over telephone lines instead of shipping punched cards or magnetic tapes to each other.

Although these information transfer agreements between trading partners increased efficiency and reduced errors, they were not an ideal solution. Since the translation programs that one trading partner wrote usually would not work for other trading partners, each company participating in this information exchange had to make a substantial investment in computing infrastructure. Only large trading partners could afford this investment, and even those companies had to have a significant number of transactions to justify the cost. Smaller or lower-volume trading partners could not afford to participate in the benefits of these paper-free exchanges.

In 1968, a number of freight and shipping companies joined together to form the Transportation Data Coordinating Committee (TDCC), which was charged with exploring ways to reduce the paperwork burden that shippers and carriers faced. The TDCC created a standardized information set that included all the data elements that shippers commonly included on bills of lading, freight invoices, shipping manifests, and other paper forms. Instead of printing a paper form, shippers could transform information about shipments into a computer file that conformed to the TDCC standard format. The shipper could electronically transmit that computer file to any freight company that had adopted the TDCC format. The freight company translated the TDCC format into data it could use in its own information systems. The savings from not printing and handling forms, not entering the data twice, and not having to worry about error-correction procedures was significant for most shippers and freight carriers.

Although these early industry-specific data interchange efforts were very helpful, their benefits were limited to members of the industries that created standard-setting groups. In addition, most businesses that are in a particular industry buy goods and services from businesses that are in other industries. For example, a machinery manufacturer might buy from steel mills, paint distributors, electrical assembly contractors, and container manufacturers. Almost every business needs to buy office supplies and the services of freight and transportation companies. Thus, full realization of EDI's economies and efficiencies required standards that could be used by companies in all industries.

Emergence of Broader Standards

After a decade of fragmented attempts at setting broader EDI standards, a number of industry groups and several large companies decided to mount a major effort to create a set of cross-industry standards for electronic components, mechanical equipment, and other widely used items. The **American National Standards Institute** (**ANSI**) has been the coordinating body for standards in the United States since 1918. ANSI does not set standards, but it has created a set of procedures and organizational standards for the development of national standards and it accredits committees that follow those procedures.

In 1979, ANSI chartered a new committee to develop uniform EDI standards. This committee is called the **Accredited Standards Committee X12** (**ASC X12**). The committee meets three times each year to develop and maintain EDI standards. The committee and its subcommittees include information systems professionals from over 800 businesses and other organizations. Membership is open to organizations and individuals who have an interest in the standards. The administrative body that coordinates ASC X12 activities is the **Data Interchange Standards Association** (DISA).

The ASC X12 standard has benefited from the participation of members from a wide variety of industries. The standard currently includes specifications for several hundred **transaction sets**, which are the names of the formats for specific business data interchanges. Figure 9-3 lists some of the more commonly used ASC X12 transaction sets.

104 - Air Shipment Information	829 - Payment Cancellation Request
110 - Air Freight Details and Invoice	840 - Request for Quotation
125 - Multilevel Railcar Load Details	841 - Specifications/Technical Information
151 - Electronic Filing of Tax Return Data Acknowledgement	842 - Nonconformance Report
170 - Revenue Receipts Statement	843 - Response to Request for Quotation
180 - Return Merchandise Authorization and Notification	846 - Inventory Inquiry/Advice
	847 - Material Claim
204 - Motor Carrier Shipment Information	850 - Purchase Order
210 - Motor Carrier Freight Details and Invoice	853 - Routing and Carrier Instruction
213 - Motor Carrier Shipment Status Inquiry	854 - Shipment Delivery Discrepancy Information
214 - Transportation Carrier Shipment Status Message	855 - Purchase Order Acknowledgment
	856 - Ship Notice/Manifest
304 - Shipping Instructions	857 - Shipment and Billing Notice
317 - Delivery/Pickup Order	859 - Freight Invoice
325 - Consolidation of Goods in Container	860 - Purchase Order Change Request–Buyer-Initiated
350 - U.S. Customs Release Information	861 - Receiving Advice/Acceptance Certificate
404 - Rail Carrier Shipment Information	865 - Purchase Order Change Acknowledgment/Request–Seller-Initiated
410 - Rail Carrier Freight Details and Invoice	
421 - Estimated Time of Arrival and Car Scheduling	867 - Product Transfer and Resale Report
	869 - Order Status Inquiry
440 - Shipment Weights	870 - Order Status Report
466 - Rate Request	879 - Price Change
511 - Requisition	893 - Item Information Request
810 - Invoice	920 - Loss or Damage Claim–General Commodities
812 - Credit/Debit Adjustment	
813 - Electronic Filing of Tax Return Data	924 - Loss or Damage Claim–Motor Vehicle
820 - Payment Order/Remittance Advice	997 - Functional Acknowledgment
828 - Debit Authorization	998 - Set Cancellation

Figure 9-3   *Commonly used ASC X12 transaction sets*

Detailed specifications for the ASC X12 transaction sets are available at the **Harbinger.net** and **Extol** Web sites. Although the X12 standards were quickly adopted by major firms in the United States, businesses in other countries continued to use their own national standards in many cases. In the mid-1980s, the United Nations Economic Commission for Europe invited North American and European EDI experts together to build a common set of EDI standards based on the successful experiences of U.S. firms in using the ASC X12 standards. In 1987, the United Nations published its first standards under the title **EDI for Administration, Commerce, and Transport (EDIFACT**, or **UN/EDIFACT**). As you can see from Figure 9-4, a number of the commonly used UN/EDIFACT standard transaction sets are similar to those in the ASC X12 standard.

AUTHOR	-	Authorization		
BOPCUS	-	Balance of Payment Customer Transaction Report	IFTCCA	-	Forwarding/Transport Shipment Charge Calculation
BOPDIR	-	Direct Balance of Payment Declaration	IFTDGN	-	Dangerous Goods Notification
BOPINF	-	Balance of Payment Information from Customer	IFTFCC	-	International Transport Freight Costs/Other Charges
COARRI	-	Container Discharge/Loading Report	IFTMAN	-	Arrival Notice
COHAOR	-	Container Special Handling Order	INVOIC	-	Invoice
CONAPW	-	Advice on Pending Works	INVRPT	-	Inventory Report
CONDPV	-	Direct Payment Valuation	ORDCHG	-	Purchase Order Change Request
CONITT	-	Invitation to Tender	ORDERS	-	Purchase Order
CONPVA	-	Payment Valuation	ORDRSP	-	Purchase Order Response
CONQVA	-	Quantity Valuation	PAXLST	-	Passenger List
COPRAR	-	Container Discharge/Loading Order	PAYMUL	-	Multiple Payment Order
COREOR	-	Container Release Order	PAYORD	-	Payment Order
COSTCO	-	Container Stuffing/Stripping Confirmation	PRODEX	-	Product Exchange Reconciliation
			QALITY	-	Quality Data
			QUOTES	-	Quote
COSTOR	-	Container Stuffing/Stripping Order	RECADV	-	Receiving Advice
CREADV	-	Credit Advice	REMADV	-	Remittance Advice
CUSDEC	-	Customs Declaration	REQDOC	-	Request for Document
CUSRES	-	Customs Response	REQOTE	-	Request for Quote
DEBADV	-	Debit Advice	SSREGW	-	Notification of Registration of a Worker
DELFOR	-	Delivery Schedule			
HANMOV	-	Cargo/Goods Handling and Movement	STATAC	-	Statement of Account
IFCSUM	-	Forwarding and Consolidation Summary	SUPRES	-	Supplier Response

Figure 9-4 *Commonly used UN/EDIFACT transaction sets*

The ASC X12 organization has voted to move its U.S. standards toward the UN/EDIFACT international standards; however, no date for the final migration has been set. Both organizations created their transaction sets by extracting the information items from the paper forms used to document business transactions. Some critics of the current EDI standards argue that this reliance on forms has made it difficult for businesses to integrate EDI data flows into their business process–oriented information systems. Unfortunately, changing EDI transaction sets to follow business processes instead of paper transaction forms would require a complete redesign of standards that have become part of many organizations' computing infrastructures over the past 30 years.

How EDI Works

Although the basic idea behind EDI is straightforward, its implementation can be complicated, even in fairly simple business situations. For example, consider a company that needs a replacement for one of its metal cutting machines. This section will describe the steps involved in making this purchase using a paper-based system, and then explain how the process would change using EDI. In both of these examples, assume that the vendor uses its own vehicles instead of a common carrier to deliver the purchased machine.

Paper-Based Purchasing Process

The buyer and vendor in this example are not using any integrated software for business processes internally; thus, each information processing step results in the production of a paper document that must be delivered to the department handling the next step. Information transfer between the buyer and vendor is also paper-based and can be delivered by mail, courier, or fax. The information flows that will occur in the paper-based version of the purchasing process example are shown in Figure 9-5.

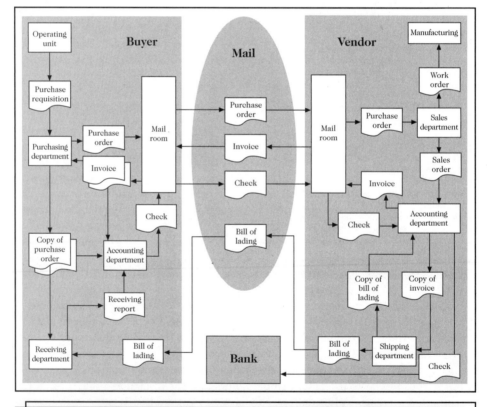

Figure 9-5   *Information flows in the paper-based purchasing process*

Once the production manager in the operating unit decides that the metal cutting machine needs to be replaced, the following process begins:

- The production manager completes a purchase requisition form and sends it to purchasing. This requisition will describe the machine that is needed to perform the metal cutting operation.
- Purchasing contacts vendors to negotiate price and terms of delivery. When purchasing has selected a vendor, it prepares a purchase order and forwards it to the mail room.
- Purchasing also sends one copy of the purchase order to receiving so that it can plan to accept delivery when scheduled; purchasing sends another copy to accounting to advise it of the financial implications of the order.
- The mail room sends the purchase order it received from purchasing to the selected vendor via mail or courier.
- The vendor's mail room receives the purchase order and forwards it to its sales department.
- The vendor's sales department prepares a sales order that it sends to its accounting department and a work order that it sends to manufacturing. The work order describes the machine's specifications and authorizes manufacturing to begin work on it.
- When the machine is completed, manufacturing notifies accounting and sends the machine to shipping.
- The accounting department sends the original invoice to the mail room and a copy of the invoice to the shipping department.
- The mail room sends the invoice to the buyer via mail or courier.
- The vendor's shipping department uses its copy of the invoice to create a bill of lading and sends it with the machine to the buyer.
- The buyer's mail room receives the invoice at about the same time as its receiving department receives the machine with its bill of lading.
- The buyer's mail room sends one copy of the invoice to purchasing so it knows that the machine has been shipped, and sends the original invoice to accounting.
- The buyer's receiving department checks the machine against the bill of lading and its copy of the purchase order. If the machine is in good condition and matches the specifications on the bill of lading and the purchase order, receiving completes a receiving report and delivers the machine to the operating unit.
- Receiving sends a completed receiving report to accounting.
- Accounting makes sure that all details on its copy of the purchase order, the receiving report, and the original invoice match. If they do, accounting issues a check and forwards it to the mail room.
- The buyer's mail room sends the check via mail or courier to the vendor.
- The vendor's mail room receives the check and sends it to accounting.
- Accounting compares the check to its copies of the invoice, bill of lading, and sales order. If all details match, accounting deposits the check in the vendor's bank and records the payment received.

EDI Purchasing Process

The information flows that will occur in the EDI version of this example purchasing process are shown in Figure 9-6. The mail service has been replaced with the data communications of an EDI network, and the flows of paper within the buyer's and vendor's organizations have been replaced with computers running EDI translation software.

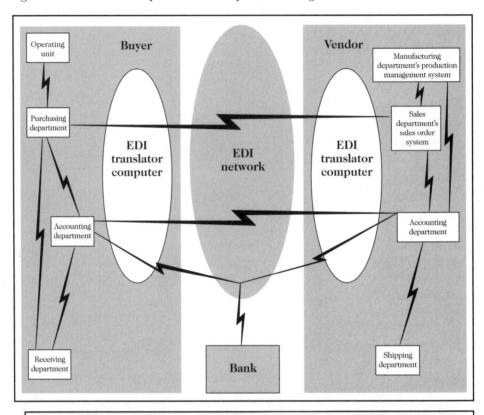

Figure 9-6   *Information flows in the EDI purchasing process*

In this version, when the operating unit manager decides that the metal cutting machine needs to be replaced, the following process begins:

- Operating unit manager sends an electronic message to its purchasing department. This message will describe the machine that is needed to perform the metal cutting operation.
- Purchasing contacts vendors by telephone or e-mail or through their Web sites to negotiate price and terms of delivery. After selecting a vendor, purchasing sends a message to the sales department announcing the selection.
- The buyer's EDI translator computer converts this message to a standard format purchase order transaction set, then forwards the message through an EDI network to the vendor.

- Purchasing also sends one electronic message to the buyer's receiving department so it can plan to accept delivery when it is scheduled; purchasing sends to the buyer's accounting department another electronic message that includes details such as the agreed purchase price.
- The vendor's EDI translator computer receives the purchase order transaction set message and converts it to the file format that the vendor's information systems use.
- The converted purchase order details appear in the sales department's sales order system and are automatically forwarded to the production management system in manufacturing and accounting's system.
- The information that was automatically forwarded to manufacturing describes the machine's specifications and authorizes manufacturing to begin work on it.
- When the machine is completed, manufacturing notifies accounting and sends the machine to the vendor's shipping department.
- The vendor's shipping department sends an electronic message to its accounting department indicating that the machine is ready to ship.
- The vendor's accounting department sends a message to its EDI translator computer, which converts the message to the standard invoice transaction set and forwards it through the EDI network to the buyer.
- The buyer's EDI translator computer receives the invoice transaction set before its receiving department receives the machine. The computer then converts the invoice data to a format that the buyer's information systems can use. The invoice data becomes immediately available to both the buyer's accounting department and its receiving department.
- When the machine arrives, the buyer's receiving department checks the machine against the invoice information on its computer system. If the machine is in good condition and matches the specifications shown in the buyer's system, receiving sends a message to accounting confirming that the machine has been received in good order. It then delivers the machine to the operating unit.
- The buyer's accounting department system compares all details in the purchase order data, receiving data, and the decoded invoice transaction set from the vendor. If the details all match, the accounting system notifies its bank to reduce the buyer's account and increase the vendor's account by the amount of the invoice. The EDI network may provide services that perform this task.

Value-Added Networks

As you can see by comparing the paper-based purchasing process in Figure 9-5 to the EDI purchasing process in Figure 9-6, the departments are exchanging the same messages among themselves, but EDI reduces paper flow and streamlines the interchange of information among departments within a company and between companies. These efficiencies were responsible for the benefits described in the GE Lighting example presented in the introduction to this chapter. The three key elements shown in Figure 9-6 that alter the process so dramatically are the EDI network (instead of the mail service) that connects the two companies and the two EDI translator computers that handle the conversion of data from the formats used internally by the buyer and the vendor to standard EDI transaction sets. Trading partners can implement the EDI network and EDI translation processes in several ways. Each of these uses one of two basic approaches: direct connection or indirect connection.

Direct Connection between Trading Partners

The first approach, called **direct connection EDI**, requires each business in the network to operate its own on-site EDI translator computer (as shown in Figure 9-6). These EDI translator computers are then connected directly to each other using modems and dial-up telephone lines or dedicated leased lines. The dial-up option becomes troublesome when customers or vendors are located in different time zones and transactions are time-sensitive or high in volume. The dedicated leased line option can become very expensive for businesses that must maintain many connections with customers or vendors. Trading partners that use different communications protocols can make both direct connection options difficult to implement.

Indirect Connection between Trading Partners

Instead of connecting directly to each of its trading partners, a company might decide to use the services of a value-added network. As you learned in Chapter 1, a value-added network (VAN) is a company that provides communications equipment, software, and skills needed to receive, store, and forward electronic messages that contain EDI transaction sets. To use the services of a VAN, a company must install EDI translator software that is compatible with the VAN. Often, the VAN will supply this software as part of its operating agreement. Figures 9-7 and 9-8 show the differences between direct connection EDI and indirect connection EDI that uses a VAN.

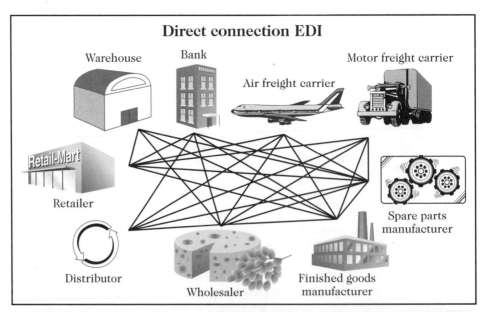

Figure 9-7   *Direct connection EDI*

To send an EDI transaction set to a trading partner, the VAN customer connects to the VAN using a dedicated or dial-up telephone line and then forwards the EDI-formatted message to the VAN. The VAN logs the message and delivers it to the trading partner's mailbox on the VAN computer. The trading partner then dials in to the VAN and retrieves its EDI-formatted messages from that mailbox. This approach is called **indirect connection EDI** because the trading partners pass messages through the VAN instead of connecting their computers directly to each other.

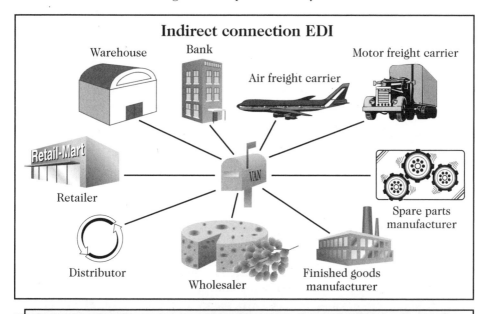

Indirect connection EDI

Warehouse · Bank · Motor freight carrier · Air freight carrier · Retail-Mart · VAN · Retailer · Spare parts manufacturer · Distributor · Wholesaler · Finished goods manufacturer

Figure 9-8   *Indirect connection EDI through a VAN*

Companies that provide VAN services include **General Electric Information Services**, **GPAS**, **Harbinger Corporation**, **IBM Global Services**, **IMS Network**, **Kleinschmidt**, and **Sterling Software**. Advantages of using a VAN include the following:

- Users only need to support the VAN's one communications protocol instead of many possible protocols used by trading partners.
- The VAN records message activity in an audit log. This VAN audit log becomes an independent record of transactions, and this record can be helpful in resolving disputes between trading partners.
- The VAN can provide translation between different transaction sets used by trading partners (for example, the VAN can translate an ASC X12 set into a UN/EDIFACT set).
- The VAN can perform automatic compliance checking to ensure that the transaction set is in the specified EDI format.

VANs do have a number of disadvantages, however. One major issue is cost. Most VANs require an enrollment fee, a monthly maintenance fee, and a transaction fee. The transaction fee can be based on transaction volume, transaction length, or both. Trading partners that have few transactions often find it difficult to justify the high fixed costs of the enrollment fee and the monthly maintenance fee. For example, the up-front cost of implementing EDI, including software, enrollment fee, and hardware, can exceed $50,000. Other trading partners that have high transaction volumes find the ongoing transaction-based fees to be prohibitive.

In the past, many vendors were forced into bearing the high costs of participating in EDI to satisfy the needs of one or two large customers. This happened frequently to suppliers of the auto industry and the retail merchandising industry. Using VANs can become cumbersome and expensive for companies that want to do business with a number of trading partners that each use different VANs. Although some VANs do offer the service of exchanging messages with other VANs, the cost of this service can be unpredictable. Also, inter-VAN transfers do not always provide a clear audit trail for use in dispute resolution. Recently, firms began to look to the Internet as a communications medium that might hold promise for overcoming some of the disadvantages of traditional EDI.

EDI on the Internet

As the Internet gained prominence as a tool for conducting business, trading partners who had been using EDI began to view the Internet as a potential replacement for the expensive leased lines and dial-up connections they had been using to support both direct and VAN-aided EDI. Companies that had been unable to afford EDI began to look at the Internet as an enabling technology that might get them back in the game of selling to large customers who demanded EDI capabilities of their suppliers.

The major roadblocks to conducting EDI over the Internet initially were general concerns about security and the Internet's general inability to provide audit logs and third-party verification of message transmission and delivery. As the basic TCP/IP structure of the Internet was enhanced with secure protocols such as SHTTP and various encryption schemes, businesses worried less about security issues, although concerns still existed. The lack of third-party verification continues to be an issue, since the Internet has no built-in facility for that. Because EDI transactions are business contracts and often involve large amounts of money, the issue of nonrepudiation is significant. Recall from Chapter 6 that **nonrepudiation** is the ability to establish that a particular transaction actually occurred. It prevents each party from repudiating, or denying, the transaction's validity or existence. In the past, the nonrepudiation function was provided either by a VAN's audit logs for indirect connection EDI or a comparison of the trading partners' message logs for direct connection EDI.

Open Architecture of the Internet

A number of new firms, such as **Commerce One**, **DynamicWeb Enterprises**, the **EC Company**, **IPNet**, and **VanTree**, have begun providing EDI services on the Internet. Firms that had been providing traditional VAN services, such as **AT&T IP Services**, are also beginning to offer EDI on the Internet. EDI on the Internet is also called **open EDI** because the Internet is an open architecture network, as you learned in Chapter 2. Many of the new EDI offerings go beyond traditional EDI and help trading partners accomplish information interchanges that have evolved faster than the EDI standard transaction sets.

The open architecture of the Internet allows trading partners virtually unlimited opportunities for customizing their information interchanges. New tools such as XML are helping trading partners be even more flexible in exchanging detailed information. Three EDI groups have recently met and charged a new ASC X12 task group with these broad objectives:

- Converting the ASC X12 EDI data elements and transaction set structures to XML in a way that retains a one-to-one mapping between the existing ASC X12 and the new XML data elements
- Developing XML data element names that allow people with existing ASC X12 transaction set expertise to continue to use that expertise when working with the new XML transaction sets
- Meeting the needs of application-to-application and human-to-application interfaces

Other firms are extending their internal networks (intranets) to their trading partners; this makes the intranets extranets. Technologies such as virtual private networks (VPNs) are providing the security that makes such extranets increasingly attractive. For example, Nintendo USA has been using an EDI-based product registration system to prevent fraudulent returns. The system allows retailers to send directly to Nintendo USA the serial numbers of Nintendo systems that they have sold. This worked well for large retailers, but the benefits did not offset the costs for smaller toy stores. In 1998, Nintendo expanded the registration system to include non-EDI adopters. Using an IPNet software package that captures serial number and other warranty information at the cash register and then sends it over the Internet to Nintendo, smaller retailers now enjoy all the benefits of EDI at a much lower cost than traditional EDI.

Financial EDI

Although Internet EDI is growing and offering new, flexible information interchange solutions for many trading partners, some elements of EDI remain difficult to transfer to the Internet. The EDI transaction sets that provide instructions to a trading partner's bank are called financial EDI (FEDI). All banks have the ability to perform electronic funds transfers (EFTs), which are the movements of money from one bank account to another. You learned about these in Chapter 1. These bank accounts may be customer accounts or the accounts that banks keep on their own behalf with each other. When EFTs involve two banks, they are executed using a clearinghouse. In the United States, most EFTs are handled through the **Automated Clearing House (ACH)**. **EDI-capable banks** are those banks that are equipped to exchange payment and remittance data through VANs. Some banks also offer VAN services for nonfinancial transactions, too. These banks are called **value-added banks (VABs)**. Nonbank VANs that can translate financial transaction sets into ACH formats and transmit them to banks that are not EDI-capable are sometimes called **financial VANs (FVANs)**.

Many companies are reluctant to send over the Internet FEDI transaction sets that contain transfer instructions for large amounts of money—in some cases, millions of dollars—because of the perceived low level of security on the Internet. FEDI transaction sets are negotiable instruments—the electronic equivalent of checks.

The reliability of FEDI itself is an issue, too. Since FEDI uses the Internet, it can be exposed to problems that are less likely to occur on the dedicated leased telephone lines used for connecting to a VAN. For example, if an Internet router outage delays an instruction to transfer $10 million, a trading partner could easily lose a day's interest on the funds. Thus, companies that have established indirect connection EDI through a VAN are likely to continue doing so to ensure added security for FEDI transaction sets, even though the cost is much higher than the cost of using the Internet for FEDI.

Hybrid EDI Solutions

Some firms are offering hybrid EDI solutions that use the Internet for part of the transaction. For example, **Bottomline Technologies'** PayBase package allows trading partners to send payment instructions over the secure ACH network and the detailed payment description information directly to the trading partner over the Internet. Since the payment description information is not a negotiable instrument, the concerns about security are less critical than with FEDI transactions. Other hybrid solutions include EDI-HTML translation services. These are offered by banks and other EDI service firms. One example of this kind of service is Northern Trust's **NetTransact** service, which provides an interface for the smaller business that is connected to the Internet but does not have EDI translation capability. Figure 9-9 shows how a service such as NetTransact works.

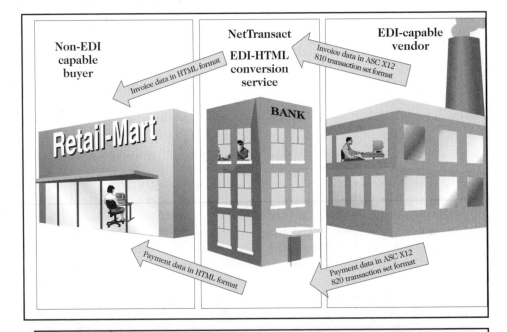

Figure 9-9   *NetTransact EDI-HTML conversion service*

For example, if an EDI-enabled vendor sends an invoice transaction set to a trading partner using NetTransact, that trading partner (the buyer) receives an HTML document instead of the ASC X12 810-formatted invoice data. The buyer can read the HTML document in a Web browser and can download the HTML document into a local accounting or purchase tracking system. NetTransact converts the buyer's response to the invoice into an ASC X12 820-formatted payment transaction set and forwards the set to the vendor or the vendor's VAN. To the vendor, this transaction appears to be all EDI. To the buyer, the transaction is no more difficult than using browser software to surf the Web. Increasingly, businesses will explore methods such as these that use the connectivity of the Internet to reduce the costs of EDI.

The four largest EDI service vendors—GE Information Services, Harbinger, IBM Global Services, and Sterling Commerce—have all recognized that the Internet, or at least its TCP/IP communications protocol, is becoming the standard way that companies' systems talk to each other. According to a 1999 article in *Upside* magazine, these EDI giants are also working on ways to incorporate XML as a part of EDI transaction sets or as a replacement for them.

EDI was the original form of electronic commerce and it appears that it will continue to evolve and be a part of the electronic commerce boom on the Internet. Total EDI transaction volume is estimated to be about $3.3 billion in 2000 and is expected to reach $3.8 billion by 2002. The share of EDI carried by VANs has been declining. In 1997, VANs carried more than 95% of all EDI traffic. Experts predict that VANs' share of the transaction volume will decline to about 50% by 2002. This reflects a large expected increase in the number of smaller businesses that are willing to engage in EDI now that the use of an expensive VAN is no longer absolutely required.

SUPPLY CHAIN MANAGEMENT

In Chapter 1, you examined how companies can organize their strategic business unit activities using an industry value chain. The part of an industry value chain that precedes a particular strategic business unit is often called a **supply chain**. A company's supply chain for a particular product or service includes all the activities undertaken by every predecessor in the value chain to design, produce, promote, market, deliver, and support each individual component of that product or service.

For example, the supply chain of an automobile manufacturer would include every activity undertaken by every component supplier, including engine manufacturers, steel fabricators, glass manufacturers, wiring harness assemblers, and thousands of others. The purchasing department has traditionally been charged with buying all of these components at the lowest price possible. Usually, purchasing staff did this by identifying qualified vendors and asking them to prepare bids that described what they would supply and how much they would charge. The purchasing staff would then select the lowest bid that still met the quality standards for the component. This bidding process led to a very competitive environment with a large number of suppliers that focused excessively on the cost of individual components and ignored the total supply chain costs, including the cost to the manufacturing organization of dealing with such a large number of suppliers.

Value Creation in the Supply Chain

In recent years, businesses have realized that they can save money and increase product quality by taking a more active role in negotiations with suppliers. By engaging suppliers in cooperative, long-term relationships, companies have found that they can work together with these suppliers to identify new ways to provide their own customers faster, cheaper, and better service. This process of taking an active role in working with suppliers to improve products and processes is called **supply chain management**. By coordinating the efforts of supply chain participants, firms that engage in supply chain management are reaching beyond the limits of their own organization's hierarchical structure and creating a new network form of organization among the members of the supply chain.

Supply chain management was originally developed as a way to reduce costs. It focused on very specific elements in the supply chain and tried to identify opportunities for process efficiency. Today, supply chain management is used to add value in the form of benefits to the ultimate consumer at the end of the supply chain. This requires a more holistic view of the entire supply chain than had been common in the early days of supply chain management.

Businesses that engage in supply chain management work to establish long-term relationships with a small number of very capable suppliers. These suppliers, called **tier one suppliers**, in turn develop long-term relationships with a larger number of suppliers that provide components and raw materials to them. These **tier two suppliers** manage relationships with the next level of suppliers, called **tier three suppliers**, who provide them with components and raw materials. A key element of these relationships is trust between the parties. The long-term relationships that are created among participants in the supply chain are called **supply alliances**. The level of information sharing that must take place among the supply chain participants can be a major barrier to entering into these alliances. Firms are not used to disclosing detailed operating information and often perceive that information disclosure might hurt the firm by placing it at a competitive disadvantage.

In exchange for the stability of the closer, long-term relationships, buyers expect annual price reductions and quality improvements from suppliers at each stage of the supply chain. However, all supply chain participants share information and work together to create value. Ideally, the supply chain coordination creates enough value that each level of supplier can share the benefits of reduced cost and more efficient operations. Supply chain management has been gaining momentum during the past decade and is supported by major purchasing groups such as the **National Association for Purchasing Management** and the **Supply Chain Council**.

By working together, supply chain members can reduce costs and increase the value of the product or service to the ultimate consumer. The advantage of coordinated planning and design can be especially significant when the lower-level suppliers (those in tiers two and three) would not have otherwise realized what factors are important to the ultimate consumer or what problems and bottlenecks exist in the processes at other levels in the industry value chain.

One area in which differences in organizational goals often arise is described by Marshall Fisher in his 1997 *Harvard Business Review* article. He explains that firms often organize themselves to achieve either efficient process goals or market-responsiveness flexibility goals. Some companies structure themselves to be efficient producers, whereas others structure themselves to be flexible producers. The kinds of things that allow a firm to be an efficient, low-cost producer are exactly the things that prevent a firm from being flexible enough to respond to market changes. The efficient producer will invest in expensive machines that can stamp out large numbers of low-cost items. This investment drives the cost of production down, but makes it difficult for the producer to be flexible. The large investment in specialized machinery prevents that producer from reconfiguring the plant layout, for example. If even one member of the supply chain for a product that requires flexible production operates as an efficient producer (instead of as a flexible producer), every other firm in the supply chain will suffer. The efficient producer will create bottlenecks that will hamper the best efforts of all other supply chain members. Clear communication up and down the supply chain can keep each participant informed of what the ultimate consumer is demanding. The participants then can plot a strategy that will meet those demands.

Technology in the Supply Chain

Clear communications, and quick responses to those communications, are a key element of successful supply chain management. Technologies, and especially the technologies of the Internet and the Web, can be very effective communication enhancers. For the first time, firms can effectively manage the details of their own internal processes and the processes of other members of their supply chain. Software that uses the Internet can help all members of the supply chain review past performance, monitor current performance, and predict when and how much of certain products need to be produced. Figure 9-10 lists the advantages of using Internet and Web technologies in supply chain management.

Suppliers can:
- Share information about customer demand fluctuations
- Receive rapid notification of product design changes and adjustments
- Provide specifications and drawings more efficiently
- Increase the speed of processing transactions
- Reduce the cost of handling transactions
- Reduce errors in entering transaction data
- Share information about defect rates and types

Figure 9-10   *Advantages of using Internet and Web technologies in supply chain management*

Boeing, the largest producer of commercial aircraft in the world, faces a huge task in keeping its production on schedule. Each airplane requires more than 1 million individual parts and assemblies, and each airplane is custom configured to meet the purchasing airline's exact specifications. These parts and assemblies must be completed and delivered on schedule or the production process comes to a halt.

In 1997, production and scheduling errors required Boeing to shut down two entire assembly operations for several weeks, costing the company over $1.5 billion. To prevent this from ever happening again, Boeing has invested in a number of new information systems that increase production efficiency by providing planning and control over logistics in every element of its supply chain. Using EDI and Internet links, Boeing is working with suppliers so that they can provide exactly the right part or assembly at exactly the right time. Even before starting an airplane into production,

Boeing makes the engineering specifications and drawings available to its suppliers through secure Internet connections. As work on the airplane progresses, Boeing keeps every member of the supply chain continually informed of completion milestones achieved and necessary schedule changes.

By its second year of using these new systems, Boeing had cut in half the time needed to complete individual assembly processes. It has realized similar reductions in part defect costs. The combined effects of these increased efficiencies are helping Boeing do a much better job of meeting its customers' needs. Instead of waiting 36 months for delivery, its customers can now have their new airplanes in 10 to 12 months.

To further benefit customers, Boeing launched a spare parts Web site, **Boeing PART** (part analysis and requirements tracking). Over 500 airlines that are Boeing customers do not use EDI to order replacement parts. Boeing PART lets those customers register and then order parts using their Web browsers. The site is processing over 5000 transactions per day at a significantly lower cost to Boeing than if it were handling faxes, telephone calls, and mailed purchase orders. Boeing can deliver most parts ordered through Boeing PART on the same or next day.

Although **Dell Computer** has become famous for its use of the Web to sell custom-configured computers to individuals and businesses, it has also used technology-enabled supply chain management to give customers exactly what they want. It has reduced the amount of inventory it keeps on hand from three weeks' sales to six days' sales. Ultimately, Dell would like to see inventory levels measured in minutes. By increasing the amount of information it has about its customers, Dell has been able to dramatically reduce the amount of inventory it must hold. Dell has also shared this information with members of its supply chain.

Dell's top suppliers have access to a secure Web site that shows them Dell's latest sales forecasts along with other information about planned product changes, defect rates, and warranty claims. In addition, the Web site tells suppliers who Dell's customers are and what they are buying. All of this information helps these tier one suppliers plan their production much better than they could otherwise. The information sharing goes in both directions in Dell's supply chain: Tier one suppliers are required to provide Dell with current information on their defect rates and production problems. As a result, all members of the supply chain work together to reduce inventories, increase quality, and provide high value to the ultimate consumer. Much of this cooperative work requires a high level of trust. To enhance this trust and develop a sense of community, Dell maintains bulletin boards as an open forum in which its supply chain members can share their experiences in dealing with Dell and each other.

For Boeing, Dell, and other firms such as **PPG Industries** and **BOC Gases**, the use of Internet and Web technologies in managing supply chains has yielded significantly increased process speed, reduced costs, and increased flexibility. All of these attributes combine to allow a coordinated supply chain to produce products and services that better meet the needs of the ultimate consumer.

The major issue that most companies must deal with in forming supply chain alliances is developing trust. Continual communication and information sharing are key elements in building trust. Since Internet and Web technologies are tools that improve communications at a very low cost, they are ideal aids for enhancing the creation of a highly coordinated and effective supply chain. A 1999 joint *Information Week* and *Business Week* poll found that over 40% of information technology managers believed that information technology was improving both relationships with suppliers and supply chain management initiatives overall.

SOFTWARE FOR PURCHASING, LOGISTICS, AND SUPPORT ACTIVITIES

Software for conducting purchasing, logistics, and support activities on the Web can be classified into two general categories: business-to-business commerce software and supply chain management software. Both of these types of software are sold primarily to large companies and are very expensive. Many of the software packages currently available are in the early stages of development and have been installed only at a few customer sites.

Enterprise Resource Planning Software

Many of the companies that are candidates for these software packages have already implemented enterprise resource planning software. You learned in Chapter 4 that enterprise resource planning (ERP) software is designed to help a company integrate all of its manufacturing, finance, distribution, and other internal business functions into one information system. Major ERP vendors include **J. D. Edwards**, **Oracle**, **PeopleSoft**, and **SAP**. Although ERP software has a strong internal focus, some of the major ERP vendors have begun to introduce add-on software that enables their ERP systems to perform business-to-business commerce and supply chain management tasks. However, most companies are still buying separate software from specialized vendors to perform these functions and then are connecting it to their ERP software themselves.

Business-to-Business Commerce Software

Business-to-business commerce software is designed to help companies build Web sites that host catalog and other commercial sales activities. Some of these products, such as Netscape's **SellerXpert** (which includes Netscape's **ECXpert** communications-handling software) and Open Market's **LiveCommerce-Transact** combination, are full-featured products that help companies put catalogs online and include searching and order management tools. **BroadVision** offers several business-to-business software products, including BroadVision Business Commerce and BroadVision Procurement, that provide similar functions. Although you can customize these products, they also will operate as delivered. All of these products provide programming interfaces that you can use to connect them to existing ERP systems and inventory databases.

The other business-to-business commerce software packages are toolkits that help the customer custom configure catalog and order management systems. Described in more detail in Chapter 4, these products include IBM's **WebSphere Commerce Suite** and Microsoft's **Commerce Server 2000**. The advantage of these products is that they provide an excellent environment for crafting exactly the catalog and order management site desired. Their main drawback is that they do not include any ready-to-run applications. In addition, each package has limitations. WebSphere Commerce Suite is optimized to run with IBM's DB2 database system, and Commerce Server 2000 requires a Windows NT operating system.

All of these packages are targeted at mid-sized and larger businesses with substantial in-house technology support staff. The cost of developing a working business-to-business Web site with any of these packages can easily exceed $100,000, including the cost of programming modifications and basic hardware.

For very large companies that want to automate routine purchasing decisions, **Ariba** is currently the leading choice. Ariba provides a way for companies to standardize purchase requisitions for office supplies, as mentioned in the introduction to Chapter 3, but Ariba also provides firms with standardized access to MRO products and other routinely purchased items that are used throughout their organizations. This type of software is called an **operating resource management system** (ORMS). Ariba defines **operating resources** as including information technology, telecommunications equipment, professional services, MRO supplies, travel and entertainment expenses, and office equipment.

Ariba uses a Web browser interface that allows any employee to enter the system and complete an online requisition. The requisition is automatically routed to the appropriate person for electronic authorization, and then sent directly to the supplier. Ariba automatically sends information about the purchase to the firm's accounting system. Ariba is written in Java, so it works on most platforms. This is important in large companies that may be running a variety of operating systems. Ariba also includes programming interfaces to all major ERP software packages, which helps ensure that the accounting and logistics for each transaction are properly tracked. Ariba is a very expensive software product—a typical installation costs between $1 million and $3 million—but the large firms that use it find that they save tremendous amounts of money by lowering the cost of handling so many mundane transactions.

Supply Chain Management Software

Although Ariba automates a number of routine purchasing processes and provides links to suppliers, it is a procurement solution, not a full supply chain management product. Supply chain management software includes demand forecasting tools and planning capabilities that allow all supply chain members to coordinate their activities and continually adjust their production levels.

Currently, the two major firms offering supply chain management software are i2 Technologies and Manugistics. The i2 Technologies product, **RHYTHM**, includes components that manage demand planning, supply planning, and demand fulfillment. The demand planning module includes proprietary algorithms customized for specific industry markets that examine customers' buying patterns and generate continuously updated forecasts. The supply planning module coordinates distribution logistics, inventory level forecasting, collaborative procurement, and supply allocations. The demand fulfillment module handles the execution elements, including order management, customer verification, backlog control, and order fulfillment.

The **Manugistics** supply chain management product includes a constraint-based master planning module that controls the other elements of the system. These other elements include modules for transportation management, replenishment management, manufacturing planning, scheduling, purchase planning, and materials control.

As ERP vendors continue to introduce their own supply management tools, competitive pressure on companies such as i2 and Manugistics will increase. It is likely that this competition will lead to even better and more comprehensive products that incorporate a full range of supply chain management principles in new and more effective ways.

Summary

In this chapter, you learned that companies are using Internet and Web technologies in a variety of ways to improve their purchasing and logistics primary activities. Businesses also are making similar improvements in a wide range of support activities, such as human resources, accounting, and technology development. Firms are finding it more important than ever to extend the reach of their enterprise planning and control activities beyond their organization's legal definition and to include parts of other organizations. This emerging network model of organization was introduced in Chapter 1 and is used in this chapter to describe the growth in interorganizational communication and coordination.

EDI, the first example of electronic commerce, was first developed by freight companies to reduce the paperwork burden of processing repetitive transactions. The spread of EDI to virtually all large companies over the past 30 years has led smaller businesses to seek an affordable way of participating in EDI. The Internet is now providing the inexpensive communications channel that EDI lacked for so many years.

The increase in communications capabilities offered by the Internet and the Web is, and will continue to be, an important force driving the adoption of supply chain management techniques in a variety of industries. Supply chain management incorporates several elements that can be implemented and enhanced through the use of the Internet and the Web. Increasingly, firms are using integrated software tools to connect with their supply chain alliance partners to become more efficient and provide more value to the ultimate consumer of their value chain's products and services.

Key Terms

Accredited Standards Committee X12 (ASC X12)
American National Standards Institute (ANSI)
Automated Clearing House (ACH)
Direct Connection EDI
EDI for Administration, Commerce, and Transport (EDIFACT)
EDI-Capable Banks
EDI-Compatible
Financial VANS (FVANS)
Indirect Connection EDI
Knowledge Management
Maintenance, Repair, and Operating (MRO)
Nonrepudiation

Open EDI
Operating Resource Management System (ORMS)
Operating Resources
Procurement
Supply Alliances
Supply Chain
Supply Chain Management
Tier One Suppliers
Tier Three Suppliers
Tier Two Suppliers
Trading Partners
Transaction Sets
Value-Added Banks

Review Questions

1. In what ways can the Internet and the Web improve logistics management? Provide a list of at least five suggestions.

2. Which industries were the first to establish standard EDI transaction sets? State why, in your opinion, these industries were more interested in setting standards than other industries.

3. Define ASC X12 and briefly describe its operations.

4. Provide two examples of how a supply chain participant that has a different business strategy than other members of the supply chain can reduce the efficiency and value of the supply chain as a whole.

5. In under 300 words, explain how enterprise resource planning software and supply chain management software differ in their goals and implementations.

Exercises

1. Use the **Thomas Register of American Manufacturers** Web site (note: free registration is required to access this site) to locate an industrial product with which you are completely unfamiliar. Note how many companies offer that product, how many of those companies have catalogs or Web sites, how many offer online ordering, and how many offer literature by fax. Summarize what you have learned about the product and its availability on the Web in a report of approximately 400 words.

2. Your boss, Andrew Wheeler, is the president of a small plastic parts fabricator. He wants you to look into improving client interactions through EDI. You know that a number of companies that provide VAN services for companies engaging in EDI, such as **General Electric Information Services**, **GPAS**, **Harbinger Corporation**, **IBM Global Services**, **IMS Network**, **Kleinschmidt**, and **Sterling Software**, have begun offering Internet EDI services too. Some of these VAN operators are targeting smaller businesses that, in the past, would not have been able to afford to implement EDI. Choose three of the VAN providers listed and examine their Web sites.

For the three VAN providers that you choose, determine whether they are offering Internet EDI services. Also decide whether, in your opinion, their Web sites are targeting smaller businesses. In a memo to Andrew of approximately 200 words, summarize your findings for the three VAN providers you chose.

3. A number of standard-setting organizations offer memberships to business firms. You are working for Grace Crowley, chief information officer (CIO) of a medium-sized company that manufactures commercial seating products. These products include furniture for waiting rooms made in standard sizes and a wide range of customer-specified chair designs for auditoriums, music halls, and theaters. Grace has asked you to investigate the benefits of joining two standard-setting organizations, the **Open Buying on the Internet Consortium** and **RosettaNet**. Prepare two memos for Grace, one for each organization. In your memos, outline the purposes of each organization, and the costs and benefits of becoming a member. Then present your recommendation regarding whether your company should join.

For Further Study and Research

Adams, E. 2000. "Goodbye EDI, Hello XML?" *World Trade*, 13(2), February, 54–55.

Ariba, Inc. 2000. *B2B Marketplaces in the New Economy*. Mountain View, CA: Ariba, Inc. Available online at: (http://www.ariba.com/com_plat/white_paper_form.cfm).

Ayers, J. 1999. "Supply Chain Strategies," *Information Systems Management*, 16(2), Spring, 72–80.

Baatz, E. 1999. "How Tech Tools Unlock Supply Value," *Purchasing*, 126(6), April 22, 28–34.

Barnes, H. 1999. "Getting Past the Hype: Internet Opportunities for B-to-B Marketers," *Marketing News*, 33(3), February 1, 11–12.

Booker, E. 1999. "XML Applications Stand Up To EDI," *Internetweek*, April 19, 8.

Bort, J. 2000. "Checking the B2B Foundation, *Network World*, 17(37), September 11, 78–80.

Bovel, D. and M. Joseph. 2000. "From Supply Chain to Value Net," *Journal of Business Strategy*, 21(4), July–August, 24–28.

Brook, J. 1998. "Monitoring Purchases Via the Web," *Internetweek*, November 23, 9.

Busch, J. and L. Reisman. 2000. "B-to-B Exchanges: Know Your Domain," *The Industry Standard*, July 24, 96.

Clark, P. 2000. "Payment Options Proliferate Online," *B to B*, 85(14), September 11, 36.

Colberg, T., N. Gardner, K. Horan, D. McGinnis, P. McLauchlin, and Y-H. So. 1995. *The Price Waterhouse EDI Handbook*. New York: John Wiley & Sons.

Dalton, G. 1999. "Tighter Supply Chains," *Information Week*, February 22, 34.

Dalton, G., B. Violino, and J. Mateyaschuk. 1999. "E-Business Evolution," *Information Week*, June 7, 50–57.

Discount Store News. 1999. "EDI: Internet Revolutionizes EDI," 38 (10), P3–P6.

Dobbs, J. 1999. *Competition's New Battleground: The Integrated Value Chain*. Cambridge, MA: Cambridge Technology Partners.

The Economist. 1997. "Big, Boring, Booming," 343(8016), May 10, 16–18.

Elliot, M. 2000. "Buyer's Guide: Supply Chain Software," *IIE Solutions*, 32(6), June, 39–44.

Esichaikul, V. and C. Chaichotiranant. 1999. "Selecting an EDI Third-Party Network," *Information Systems Management*, 16(1), Winter, 26–32.

Fickel, L. 1998. "MicroAge's Internet-Based Training Program," *CIO Magazine*, July 15, 72.

Financial Executive. 1999. "CFOs Say Company Internet Involvement to Jump," 3(57), May–June, 57.

Fisher, M. 1997. "What Is the Right Supply Chain for Your Product?" *Harvard Business Review*, 75(2), March–April, 105–116.

Fisher, S. 1999. "EDI Vendors Spin a New Web," *Upside*, 11(5), May, 62–64.

Forbes. 2000. "B2B…to Be?" 166(5), August 22, 124–128.

Frook, J. 1998. "Wal-Mart Opens its Arms to Internet EDI," *Internetweek*, June 22, 16.

Gamble, R. 1998. "Electronic Billing Gets a Test," *Treasury & Risk Management*, 8(8), October, 53–54.

Hamel, G. 2000. "Waking Up IBM," *Harvard Business Review*, 78(4), July–August, 137–144.

Hart, P. and C. Saunders. 1998. "Emerging Electronic Partnerships: Antecedents and Dimensions of EDI Use from the Supplier's Perspective," *Journal of Management Information Systems*, 14(4), Spring, 87–112.

Henriott, L. 1999. "Transforming Supply Chains Into E-Chains," *Supply Chain Management Review Global Supplement*, Spring, 15–18.

Higgins, K. 2000. "Dot-Org To Dot-Com: The Shift Is On," *InformationWeek*, July 3, 66–68.

Hill, S. 1999. "Supply Chain Management in the Age of E-Commerce," *Apparel Industry Magazine*, 60(3), March, 60–63.

Hoffman, C. 2000. "Run XBRL Right Now," *Journal of Accountancy*, 190(2), August, 28–29.

Hoffman, R. 1998. "Four Solutions to Rev Up your E-Commerce Business," *Network Computing*, December 15, 75–86.

Holt, S. 1998. "Back on the Supply-Chain Gang," *InfoWorld*, 20(46), November 9, 78–79.

353

Ince, J. 2000. "To B2B, or not to B2B," *Upside*, August, 130–136.

Kaplan, S. and M. Sawhney. 2000. "E-Hubs: The New B2B Marketplaces," *Harvard Business Review*, 78(3), May–June, 97–103.

Karpinski, R. 1999. "Web Links Supply Chain to Storefront," *InternetWeek*, June 21, 8–9.

Kay, E. 2000. "From EDI to XML," *Computerworld*, 34(25), June 19, 84–85.

Kerwin, K., M. Stepanek, and D. Welch. 2000. "At Ford, E-Commerce Is Job 1," *Business Week*, February 28, 74–77.

King, J. 2000. "Quietly, Private E-Markets Rule," *Computerworld*, 34(36), September, 1, 16.

Lewis, W. 2000. "Pillar of the Community: XML Is Becoming the Standard Platform," *Intelligent Enterprise*, 3(13), August 18, 32–38.

Lin, B. and C. Hsieh. 2000. "Online Procurement: Implementation and Managerial Implications," *Human Systems Management*, 19(2), 105–110.

Marcella, A. and S. Chan. 1993. *EDI Security, Control, and Audit*, Norwood, MA: Artech House.

Massetti, B. and R. Zmud. 1996. "Measuring the Extent of EDI Usage in Complex Organizations: Strategies and Illustrative Examples," *MIS Quarterly*, 20(3), September, 331–345.

Mastony, C. and G. Button. 2000. "Maintenance, Repair & Operations," *Forbes*, 166(2), July 17, 174–176.

McNatt, R. and D. Welch. 2000. "Oh, What a Feeling: B2B," *Business Week*, May 15, 14.

Mickahail, A. 1999. "Understand Supply Chain Products and Players," *Automatic I.D. News*, 15(4), April, 48–50.

Ovans, A. 2000. "E-Procurement at Schlumberger," *Harvard Business Review*, 78(3), 21–22.

Power, C. 1999. "Internet Systems Imperil EDI for Corporate Buying," *American Banker*, 164(73), April 19, 15.

Premkumar, G. 2000. "Interorganization Systems and Supply Chain Management: An Information Processing Perspective," *Information Systems Management*, 17(3), Summer, 56–69.

Pushkin, A. and B. Morris. 1997. "Understanding Financial EDI," *Management Accounting*, 70(5), November, 42–46.

Radosevich, L. 1997. "The Once and Future EDI," *CIO Magazine*, January 1. Available online at: (http://www.cio.com/archive/ec_future_edi.html).

Raths, D. 1999. "Let the Net Do It," *Inc.*, 21(4), March 16, 52–58.

Roberts, B. 1998. "Portals, You Say? This One's Private: Ericsson's Intranet Is a Give-and-Take Affair with Employees," *Intranet Design Magazine*, December 14. Available online at: (http://idm.internet.com/articles/200003/pt_03 _15_00f.html).

Roberts, B. 1999. "24 Little Hours," *Internet World*, 5(26), August 1, 40–41.

Rosencrance, L. 2000. "Transporters Move to Deliver on E-Commerce," *Computerworld*, 34(27), July 3, 22.

Schenecker, M., G. Desai, J. Patel, and J. Levitt. 1998. "Goodbye To Old-Fashioned EDI," *Information Week*, December 14, 73–80.

Schonfeld, E. 1999. "A Site Where Hospitals Can Click to Shop," *Fortune*, 139(7), April 12, 150.

Schwartz, E. 2000. "Exchange Evolution Points to Higher Savings," *InfoWorld*, 22(26), June 26, 12.

Senn, J. 1992. "Electronic Data Interchange," *Information Systems Management*, 9(1), Winter, 45–53.

Shinal, J. 2000. "A Web-Profit Prophet Spreads the Word," *Business Week*, September 18, 68.

Stein, T. and J. Sweat. 1998. "Killer Supply Chains," *Information Week*, November 9, 36–42.

Stundza, T. 1999. "A Tale of Two Commodities: Metals, Chemicals Buyers Eye E-Commerce," *Purchasing*, 126(6), April 22, S20–S27.

Sweat, J. 1999. "Integrated Enterprise," *Information Week*, June 14, 18–19.

Szygenda, R. 1999. "Executive Report: IT Value Chain," *Information Week*, February 8, 4–6.

Teschler, L. 2000. "New Role for B-to-B Exchanges: Helping Developers Collaborate," *Machine Design*, 72(19), October 5, 52–58.

Tie, R. 2000. "Comments Encouraged on Newly Named XBRL," *Journal of Accountancy*, 189(6), June, 14–15.

Trommer, D. and S. Scheck. 1999. "Making the Connection: Automating Supply Chain Processes," *Electronic Buyer's News*, May 3, 50.

354

Tweney, D. 1999. "Microsoft Takes Aim at Language Barriers to Business Information," *InfoWorld*, 21(11), March 15, 59.

Upin, E. 2000. "On the Verge of the Next Great Wave," *Upside*, August, 170–172.

Wagner, M. 2000. "Vendor, Dotcom Thyself: Sun Turns Net Focus Inward," *InternetWeek Online*, September 5. Available online at: (http://www.internetwk.com/lead/lead090500.htm).

Waugh, R. and S. Elliff. 1998. "Using the Internet to Achieve Purchasing Improvements at General Electric," *Hospital Material Management Quarterly*, 20(2), November, 81–83.

Weintraub, A. 2000. "The Be-All and End-All of B2B Sites?" *Business Week*, June 5, 56.

Wilson, T. 2000. "EDI Is Alive and Kicking, Study Says," *InternetWeek*, February 21, 15.

Zerega, B. 1998. "Extranet Lights a Fire Under Gas-Ordering Business," *InfoWorld*, 20(33), August 17, 46.

Ziemke, P. and P. Meadows. 2000. "Bridging Work and Home," *Upside*, 12(9), September, 105–110.

STRATEGIES FOR WEB AUCTIONS, VIRTUAL COMMUNITIES, AND WEB PORTALS

In 1995, Pierre Omidyar was working as a software developer and, in his spare time, operating a small Web site that provided updates on the Ebola virus. His girlfriend collected Pez candy dispensers but had trouble finding other people who shared her interest. Omidyar decided to help her out by adding a small auction function to his Web site so that people could trade Pez dispensers and other items. Interest in the site's auctions grew so rapidly that within a year, Omidyar had quit his job to devote his full energies to the Web auction business he had created. By the end of its second year in operation, Omidyar's Web site, which he called **eBay**, had auctioned over $95 million worth of goods.

Inspired by this success, Omidyar obtained $5 million in funding from **Benchmark Capital** in 1997. Benchmark also helped him recruit a top-notch management team for the business. In September 1998, eBay offered its stock to the public and raised $63 million. Like many Internet companies, eBay experienced extremely rapid growth. In two years, sales grew from $372,000 to $47 million. Unlike many Internet companies, eBay was profitable from its inception; its net income in 1998 was over $2 million. Because of its high growth rate and solid profitability, eBay was able to return to the stock market and raise an additional $765 million in April 1999.

In just three years, eBay established itself as the dominant Web site for general consumer auctions with over 2 million registered buyers and sellers.

By 2000, eBay had increased its number of registered users to more than 15 million and it was running 63 million auctions annually for goods valued at more than $1.3 billion. These auctions were earning eBay a net income of more than $30 million on revenue that was growing at a rate of approximately 100% per year.

Since eBay was one of the first auction Web sites and because it has pursued an aggressive promotion strategy, it has become the first choice site for many people who want to participate in auctions. Both buyers and sellers benefit from a large market-place such as the one eBay has created. EBay's early advantage in the online auction business will be very difficult for competitors to overcome.

In Chapter 8, you learned how businesses are using the Web to create online identities, reach customers, and sell to them. In Chapter 9, you learned how businesses are using the Web to purchase goods and work with their suppliers more effectively. In both of these chapters, the focus was on how companies can use the Web to do better the things that they have been doing for years: buying and selling. In this chapter, you will learn how new companies are forming to take advantage of the Web and how existing companies are using the Web to do things that they have never done before. These new things include running auctions, creating virtual communities, and operating Web portals.

357

LEARNING OBJECTIVES

In this chapter, you will learn about:
- The key characteristics of the six major auction types
- Strategies for general and specific consumer Web auction sites
- Strategies for business-to-business Web auction sites
- How businesses can use virtual communities to increase brand awareness and sales
- Strategies for Web portal sites

AUCTION BASICS

In many ways, online auctions provide a business opportunity that is perfect for the Web. An auction site can charge both buyers and sellers to participate, and it can sell advertising on its pages. People interested in trading specific items can form a market segment that advertisers will pay extra to reach. Thus, the same kind of targeted advertising opportunities that search engine sites generate with their results pages are available to advertisers on auction sites. This combination of revenue-generating characteristics makes it relatively easy to develop Web auction business models that yield profits early in the life of the project.

One of the Internet's strengths is that it can bring together people who share narrow interests but who are geographically dispersed. Web auctions can capitalize on that ability by either catering to a narrow interest or providing a general auction site that has sections devoted to specific interests.

Origins of Auctions

The earliest auctions for which we have written records are from Babylon in about 500 B.C. In those auctions, men bid against each other for the women they wished to marry. Roman soldiers used auctions to liquidate the property they took from their vanquished foes. In 193 A.D., the Praetorian Guard auctioned off the entire Roman Empire after killing the Emperor Pertinax. In later years, Buddhist temples held auctions to sell off the possessions of deceased monks.

Auctions became common activities in 17th-century England, where taverns held regular auctions of art and furniture. The 18th century saw the birth of two British auction houses—**Sotheby's** in 1744 and **Christie's** in 1766—that continue to be major auction firms today. The British settlers of the colonies that would become the United States brought auctions with them. Colonial auctions were used to sell farm equipment, animals, tobacco, and, sadly, human beings.

In an auction, a seller offers an item or items for sale, but does not establish a price. This is called "putting an item up for bid" or "putting an item on the (auction) block." Potential buyers are given information about the item or some opportunity to examine the item and then offer **bids**, which are the prices they are willing to pay for the item. The potential buyers, or **bidders**, each have developed **private valuations**, or amounts they are willing to pay for the item. The whole auction process is managed by an **auctioneer**. In some auctions, people employed by the seller or the auctioneer can make bids on behalf of the seller. These people are called **shill bidders**. Shill bidders can artificially inflate the price of an item and may be prohibited from bidding by the rules of a particular auction.

English Auctions

Many different kinds of auctions exist. Most people who have attended or seen an auction on television have experienced only one type of auction, the **English auction**, in which bidders publicly announce their successive higher bids until no higher bid is forthcoming. At that point, the auctioneer pronounces the item sold to the highest bidder at that bidder's price. This type of auction is also called an **ascending-price auction**. An English auction is an **open auction** (or **open-outcry auction**) because the bids are publicly announced; however, there are other types of auctions that include this feature that are also called open auctions.

In some cases, an English auction has a minimum bid or reserve price. A **minimum bid** is the price at which an auction begins. If no bidders are willing to pay that price, the item is removed from the auction and not sold. In some auctions, a minimum bid is not announced, but if the seller's **reserve price**, or simply **reserve**, is not exceeded, the item is withdrawn from the auction and not sold.

English auctions that offer multiple units of an item for sale and that allow bidders to specify the quantity they want to buy are called **Yankee auctions**. When the bidding in a Yankee auction concludes, the highest bidder is allotted the quantity bid. If items remain after satisfying the highest bidder, those remaining items are allocated to successive lower (next-highest) bidders until all items are distributed. Although all successful bidders receive the quantity of items on which they bid, they only pay the price bid by the *lowest* successful bidder.

English auctions have drawbacks for both sellers and bidders. Since the winning bidder is only required to bid a small amount more than the next-highest bidder, winning bidders tend not to bid their full private valuations. Also, bidders risk becoming caught up in the excitement of competitive bidding and then bidding more than their private valuations. This psychological phenomenon, called the **winner's curse**, has been extensively documented by William Thaler and other behavioral economists.

Dutch Auctions

The **Dutch auction** is a form of open auction in which bidding starts at a high price and drops until a bidder accepts the price. Because the price drops until a bidder claims the item, Dutch auctions are also called **descending-price auctions**. Farmer's cooperatives in the Netherlands use this type of auction to sell perishable goods such as produce and flowers, which is how it came to be known as a "Dutch" auction.

In most Dutch auctions, the seller offers a number of similar items for sale. One common implementation of a Dutch auction uses a clock that drops the price with each tick. The first bidder to call out "stop," which stops the clock, becomes the winning bidder. The winning bidder can take all or any part of the auctioned items at that price. If any items remain, the clock is restarted and continues until all the items are taken by successive lower bidders. A Dutch auction is often better for the seller because the bidder with the highest private valuation will not let the bid drop much below that valuation for fear of losing the item to another bidder. Dutch auctions are particularly good for moving large numbers of commodity items quickly.

Sealed-Bid Auctions

In **sealed-bid auctions**, bidders submit their bids independently and are usually prohibited from sharing information with each other. In a **first-price sealed-bid auction**, the highest bidder wins. If multiple items are being auctioned, successive lower (next-highest) bidders are awarded the remaining items at the prices they bid. The **second-price sealed-bid auction** is the same as the first-price sealed-bid auction except that the highest bidder is awarded the item at the price bid by the *second*-highest bidder.

At first glance, one might wonder why a seller would even consider such an auction, because it gives the item to the winning bidder at a lower price. William Vickrey won the **1996 Nobel Prize in Economics** for his studies of the properties of this auction type. He concluded that it yields higher returns for the seller, encourages all bidders to bid the amounts of their private valuations, and reduces the tendency for bidders to collude. Since the winning bidder is protected from an erroneously high bid, all bidders tend to bid higher than they would in a first-price sealed-bid auction. Second-price sealed-bid auctions are commonly called **Vickrey auctions**.

Double Auctions

In a **double auction**, buyers and sellers each submit combined price-quantity bids to an auctioneer. The auctioneer matches the sellers' offers (starting with the lowest price and then going up) to the buyers' offers (starting with the highest price and then going down) until all the quantities offered for sale are sold to buyers. This type of auction works well only for items of known quality, such as securities or graded agricultural products, that are regularly traded in large quantities. Double auctions can be operated in either sealed-bid or open-outcry formats. The **New York Stock Exchange** conducts sealed-bid double auctions of stocks and bonds in which the auctioneer, called a **specialist**, manages the market using its own funds when necessary. The **Chicago Board of Trade** conducts open-outcry double auctions of commodity futures and stock options.

The six auction types described in this section are the most commonly used in business today. Figure 10-1 summarizes the key characteristics of each of these six major auction types.

Auction type	Key characteristics
English auction	Starting from a low price, bidding increases until no bidder is willing to bid higher.
Dutch auction	Starting from a high price, bidding automatically decreases until the bidder accepts the price.
First-price sealed-bid auction	Secret bidding process; the highest bidder pays the amount of the highest bid.
Second-price sealed bid auction (Vickrey auction)	Secret bidding process; the highest bidder pays the amount of the *second*-highest bid.
Double auction open-outcry auction	Buyers and sellers declare combined price-quantity bids. The auctioneer matches seller offers (lowest to highest) with buyer offers (highest to lowest). Buyers and sellers can modify bids based on knowledge gained from other bids.
Double auction sealed-bid auction	Buyers and sellers declare combined price-quantity bids. The auctioneer (specialist) matches seller offers (lowest to highest) with buyer offers (highest to lowest). Buyers and sellers cannot modify their bids.

Figure 10-1   *Six major auction types*

WEB AUCTION STRATEGIES

A 1999 online survey conducted by *PC Computing* magazine on its Web site asked visitors about their experiences with Web auctions. The magazine found that 37% of people responding to the survey had made purchases and 12% had sold items using a Web auction site. However, 15% of respondents stated that they would never use a Web auction site. Despite these detractors, Web auctions are one of the fastest-growing segments of online business today. In a 2000 poll of experienced Internet users, the market research firm Greenfield Online found that 44% of respondents had participated in an online auction—35% had purchased items and 14% had sold items. Business analysts and researchers predict that Web auctions will account for 30% of all electronic commerce by 2002. Although the online auction business is changing rapidly as it grows, three broad categories of auction Web sites are emerging: general consumer auctions, specialty consumer auctions, and business-to-business auctions.

General Consumer Auctions

One of the most successful consumer auction Web sites is eBay, the company described in the introduction to this chapter. The eBay home page appears in Figure 10-2.

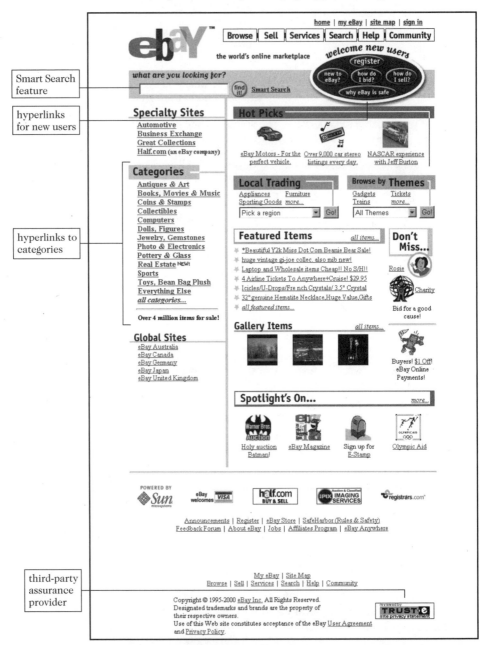

Smart Search feature

hyperlinks for new users

hyperlinks to categories

third-party assurance provider

Figure 10-2   EBay home page

The eBay home page includes links to categories of items. Alternatively, the potential bidder can use the Smart Search feature to find a specific item by entering descriptive terms. The bottom of the page includes a link to the third-party assurance provider **TRUSTe**.

Sellers and buyers must register with eBay and agree to the site's basic terms of doing business. Sellers pay eBay a listing fee and a sliding percentage of the final selling price. Buyers pay nothing to eBay. In addition to paying the basic fees, sellers can choose from a variety of enhanced and extra-cost services, including having their auctions listed in boldface type and included in lists of preferred auctions.

In an attempt to address buyer concerns about seller reliability, eBay has instituted a rating system. Buyers can submit ratings of sellers after doing business with them. These ratings are converted into graphics that appear with the seller's nickname on each auction in which that seller participates. Although this system is not without flaws, many eBay bidders feel that it affords them some level of protection from unscrupulous sellers. The converse is true also—sellers rate buyers, which provides some protection for sellers from unscrupulous buyers.

The most common format used on eBay is a computerized version of the English auction. The eBay English auction allows the seller to set a reserve price. In eBay English auctions, the bidders are listed, but the bid amounts are not disclosed until after the auction is over. This is a slight variation on the in-person English auction, but since eBay always shows a continually updated high bid amount, a bidder who monitors the auction can see the bidding pattern as it occurs. The main difference between eBay and a live English auction is that bidders do not know who placed which bid until the auction is over. The eBay English auction also allows sellers to specify that an auction be made private. In an eBay private auction, the site never discloses bidders' identities and the prices they bid. At the conclusion of the auction, eBay notifies only the seller and the highest bidder.

Another auction type offered by eBay is an increasing-price format for multiple item auctions that eBay calls a Dutch auction. This format is not a true Dutch auction, but is instead what economists call a Yankee auction, which is a multiple item form of the English auction.

363

In either type of eBay auction, bidders must constantly monitor the bidding activity. All eBay auctions have a **minimum bid increment**, the amount by which one bid must exceed the previous bid, which is about 3% of the bid amount. To make bidding easier, eBay allows bidders to make a proxy bid. In a **proxy bid**, the bidder specifies a maximum bid. If that maximum bid exceeds the current bid, the eBay site will automatically enter a bid that is one minimum bid increment higher than the current bid. As new bidders enter the auction, the eBay site software continually enters higher bids for all bidders who have placed proxy bids. Although this feature is designed to make bidding require less bidder attention, if a number of bidders enter proxy bids on one item, the bidding rises rapidly to the highest proxy bid offered. This rapid rise in the current bid often occurs in the closing hours of an eBay auction.

EBay has been so successful because it was the first major Web auction site for consumers that did not cater to a specific audience and because it has advertised widely. In 2000, eBay spent more than $100 million to market and promote its Web site. A significant portion of this promotional budget was spent on traditional mass media, such as television advertising. For eBay, such advertising has proven to be the best way to reach its main market: people who have a hobby or very specific interest in items that are not locally available. Whether those items are jewelry, antique furniture, coins, first-edition books, or stuffed animals, eBay has created a place where people can become collectors, dispose of their collections, or simply trade out of their collections.

Other firms have tried to enter the general consumer auction market that eBay dominates. One of the most promising new entrants was Auction Universe. Times Mirror, the parent company of the *Los Angeles Times* newspaper, started Auction Universe in 1997 and sold it in 1998 to the **Classified Ventures** partnership of eight major newspaper companies (including Times Mirror itself). These companies were concerned that classified advertising on the Web posed a threat to newspaper classified advertising, which is one of the most profitable elements in the newspaper business, and decided to start their own Web sites for classified ads such as **Apartments.com**, **Cars.com**, and **NewHomeNetwork.com**. These sites earn revenue by charging for running ads, by selling advertising on their pages, or both. Classified Ventures believed that the Auction Universe site could become an important and profitable part of its Web presence.

Auction Universe closed in August 2000. Classified Ventures' classified ad sites continue to operate. The Auction Universe site had been modeled on eBay and offered similar types of auctions and services for buyers and sellers. Some critics believed that the Auction Universe interface was more intuitive than eBay's and included a better search engine; however, the site failed to mount a sustained challenge to eBay's dominance. Even with major corporate sponsorship and a $10 million advertising campaign behind it, Auction Universe was unable to displace the advantage eBay obtained as the first Web auction site for general consumers.

Since one of the major determinants of Web auction site success is attracting enough buyers and sellers to create markets in enough items, some Web sites that already have a large number of visitors are entering this business. Portal sites such as **Yahoo!** and **Excite** have created auctions, again using the eBay model. The Excite Auctions home page appears in Figure 10-3.

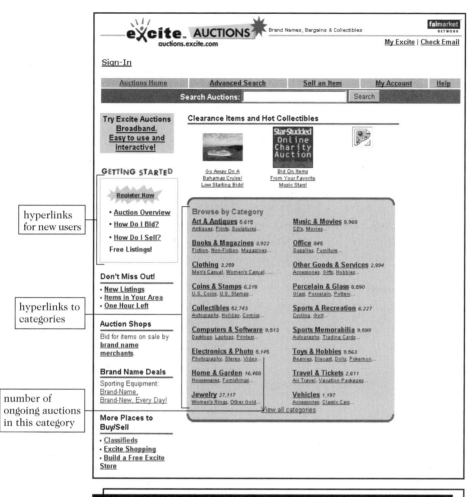

hyperlinks for new users

hyperlinks to categories

number of ongoing auctions in this category

Figure 10-3 *Excite Auctions home page*

As you examine the Excite page in Figure 10-3, you will notice that it includes many of the same features as the eBay home page. It has links to categories of auction items and a search function. Excite differs in that it has a drastically smaller number of ongoing auctions than eBay. Even though Excite has a large number of visitors, it has been unsuccessful in directing those visitors to its auction sites. The CNET auction site, which primarily auctions computing and consumer electronics items, has similarly small numbers of auctions.

Yahoo! has had greater success in attracting large numbers of auction participants, in part because it offers its auction service to sellers at no charge. However, Yahoo! has been less successful in attracting buyers, resulting in less bidding action in each auction than generally occurs on eBay. Since Yahoo! draws a large number of visitors every month, it is possible that Yahoo! will be able to increase participation in its auctions. Yahoo! is one of the most-visited sites on the Web—in mid-2000 Media Metrix reported that 49 million Web users visit Yahoo! at least once each month.

Amazon.com, the pioneering Web bookseller, has also recently expanded its business to include auctions. Although its number of auctions is still small, Amazon is aggressively marketing its new business. Unlike eBay, Amazon has not yet earned profits. Some industry observers note that Amazon might earn more by charging a commission on the auction of a used book than it could earn by selling the same title as a new book. With over 20 million registered users of its existing book, music, and video sales site, Amazon is well-positioned to challenge eBay. Unlike visitors to portal sites such as Excite, visitors to Amazon are already accustomed to purchasing goods on the Web. This may give Amazon an edge over portal sites in converting its visitors to auction participants.

One of the aggressive marketing positions that Amazon has already taken is its "Auctions Guarantee." This guarantee directly addresses concerns raised in the media by eBay customers about being cheated by unscrupulous sellers. When Amazon opened its Auctions site, it agreed to reimburse any buyer for merchandise purchased in an auction that was not delivered or that was "materially different" from the seller's representations. Amazon limited its guarantee to items costing $250 or less; however, buyers of more expensive items generally protect themselves by using a third-party **escrow service**, which holds the buyer's payment until he or she receives and is satisfied with the purchased item. EBay responded immediately by offering its customers a similar guarantee, but not before Amazon had established itself as a serious competitor in the Web auction business.

In June 1999, Amazon announced a joint online venture with Sotheby's, the famous British auction house, to hold Web auctions of fine art, antiques, jewelry, and other high-value collectibles. In general, it is difficult to sell these types of items on the Web because of the importance of direct, in-person inspection. Such inspections help establish the item's authenticity and condition. Sotheby's and its international network of dealers obtain the items for their online auctions and guarantee the authenticity and condition of items, just as at a Sotheby's in-person auction. Again, Amazon is addressing a serious concern of some of eBay's most prized customers, those who participate in auctions of high-value items. The Sotheby's joint venture and Amazon's recent press releases suggest that Amazon is trying very hard to differentiate its auction site from eBay's as a more attractive home for the upper end of the auction market.

Most general consumer Web auction sites invite independent sellers to auction their items on the site. Only a few general consumer Web auction sites offer items for sale from their own inventory. One of the more unusual and interesting of these sites is the **Klik-Klok Dutch Auction**. Klik-Klok sells general consumer items from its inventory using a true Dutch auction format. Figure 10-4 shows an in-progress Klik-Klok auction of a table clock.

Table Clock

Suggested retail price: $ 90.00
Starting price: $ 75
Shipping and handling: 9.50

Polished brass case featuring single columns, floating dial on a thick mineral glass panel. This handsome clock also has an alarm. Size 6 7/8" x 7 3/8"

quantity selector

price drops as timer counts down to zero

Dutch Auction Home
Department Store
Home

Current Price: $68.80
Time Remaining: 0:31
Pieces Remaining: 5

1

Buy Now!!

Figure 10-4 *Klik-Klok Dutch auction in progress*

These Dutch auctions occur in a very short time period, usually a few minutes. The price begins at a stated starting price and drops every few seconds as the timer counts down to zero. Bidders can enter a quantity up to the number in the "Pieces Remaining" indicator and click the "Buy Now!" button to enter a bid at the price shown in the "Current Price" indicator.

Most of these general consumer Web auction sites are recent entries. The premier Web auction site in this category is still eBay. Although it appears that portal sites or well-established Web merchants, such as Amazon.com, might be able to establish successful auction sites, they must overcome the strong advantage that eBay has built. One serious issue that any challenger to eBay will face is that the economic structure of markets is biased against new entrants. Since markets become more efficient (yielding fairer prices to both buyers and sellers) as the number of buyers and sellers increases, new auction participants are inclined to patronize established marketplaces. Thus, existing auction sites, such as eBay, are inherently more valuable to customers than new auction sites. This basic economic fact will make the task of creating other successful general consumer Web auction sites even more difficult in the future.

Specialty Consumer Auctions

Rather than struggle to compete with a well-established rival such as eBay in the general consumer auction market, a number of firms have decided to identify special interest market targets and create specialized Web auction sites that meet the needs of those market segments. Several early Web auction sites started by featuring technology items such as computers, computer parts, photographic equipment, and consumer electronics.

Doug Salot started an auction site, **Haggle Online**, in September 1996. Salot had been buying and selling computer equipment on the Internet's Usenet newsgroups before the Web existed. He saw the potential for the Web's graphical user interface in creating auctions. Haggle has officially branched out to include items unrelated to computers; however, technology products continue to be its mainstay. The site even includes an online computer museum! Other sites devoted to computers and related products include the **CNET.com** technology portal site and the Ziff-Davis computer publishing companies' **ZDNet** site.

Unlike the CNET.com and ZDNet sites, which only auction items provided by independent sellers, **Onsale** uses its Web auction site to sell its own inventory of close-outs and refurbished computers and computer-related items. Onsale offers warranties on some of its refurbished products and has established itself as a specialized Web auction site in its chosen market segment. Creative Computers, a direct-mail catalog marketer of computer hardware and software, is the owner of the **uBid** auction site, which targets the same technology market as Onsale. Both Onsale and uBid have added other auction categories to their sites, but their true strengths are in the computer and technology markets. A Canadian company that sells auction software to other firms that want to conduct specialty auctions, **Bid.com**, has also had some success in the refurbished computer market.

Although computers and technology were obvious early market segments that would find Web auctions appealing in the first wave of electronic commerce, a number of other specialized Web auction sites have emerged recently. Golfers in search of bargains on new and used clubs can find them at the **Golf Club Exchange** Web auction shown in Figure 10-5.

Figure 10-5   *Golf Club Exchange Web auction*

Coin collectors are attracted to sites such as **Coin Universe**. Wine enthusiasts can bid on their favorite vintages at **Winebid**. Tobacco fanciers can find the perfect smoke at **Cigar-Bid.com**. Antique and art lovers can bid on dealer's offerings at

eHammer. Investors can bid on the rates they wish to earn on USABancShares certificates of deposit at its Web auction site, **CDenergy**.

All of these sites gain an advantage by identifying a strong market segment that has readily identifiable products that are desired by persons with relatively high levels of disposable income. Golf clubs, wine, and technology products all meet these requirements very well. As other Web auction site developers identify similar market segments, these specialized consumer auctions may become profitable niches that can successfully coexist with large general consumer sites, such as eBay. One company that has identified a completely different way to segment the Web auction market is **CityAuction**. CityAuction lists its auctions by the location of an item's seller. The site's assumption is that people are more willing to bid on items offered for sale by a local person.

Business-to-Business Auctions

Unlike consumer Web auctions, business-to-business Web auctions evolved to meet a very specific need. Many manufacturing companies periodically need to dispose of unusable or excess inventory. Despite the best efforts of procurement and production management, businesses occasionally buy more raw materials than they need. Many times, unforeseen changes in customer demand for a product can saddle manufacturers with excess finished goods or spare parts.

Depending on its size, a firm will typically use one of two methods to distribute excess inventory. Large companies sometimes have liquidation specialists that find buyers for these unusable inventory items. Smaller businesses often sell their unusable and excess inventory to **liquidation brokers**, which are firms that find buyers for these items. Web auctions are the logical extension of these inventory liquidation activities to a new and more efficient channel, the Internet.

Two of the three main business-to-business Web auction models that are emerging are direct descendants of these two traditional methods for handling excess inventory. In the large company model, the business creates its own auction site that sells excess inventory. In the small company model, a third-party Web auction site takes the place of the liquidation broker and auctions excess inventory listed on the site by a number of smaller sellers. The third business-to-business Web auction model resembles consumer Web auctions. In this model, a new business entity enters a market that lacked efficiency and creates a site at which buyers and sellers that have not historically done business with each other can participate in auctions. An alternative implementation of this model occurs when a Web auction replaces an existing sales channel.

One of the earliest examples of the large company model is Ingram Micro's Auction Block site, which Ingram Micro started in 1997. Ingram Micro is a major distributor of computers and related equipment to value-added resellers (VARs), which are companies that configure computer hardware and software, such as network servers, for business users. Because computer technology changes rapidly, Ingram Micro often finds itself with outdated disk drives, computer chips, and other items that it formerly turned over to liquidation brokers.

Ingram Micro now auctions those items to its established customers through the Auction Block site. In 2000, Ingram Micro auctioned over $6 million worth of obsolete inventory through its Auction Block site and is expanding the site to include automated ordering, automatic bidding, and previews of future auctions. The VARs that are Ingram Micro's main customers also now have the option of putting the Auction Block program on their own sites, which will allow their customers to participate in the bidding.

Ingram Micro estimates that the auction prices it receives on the site average about 60% of the items' cost. This percentage compares favorably to the average of 10% to 25% of cost that Ingram Micro had been obtaining from liquidation brokers. In effect, large companies such as Ingram Micro are removing the liquidation brokers from the value chain and claiming the brokers' intermediary profits. You learned in Chapter 8 that this process is called disintermediation.

Another large computer technology company that decided to build its own auction site to dispose of obsolete inventory is CompUSA. Although CompUSA sells to individuals, a significant portion of its sales are to corporate customers. Instead of selling through liquidation brokers, CompUSA decided to let midsized and smaller businesses bid directly on its technology inventory. Its Web auction site, **CompUSA Auctions**, appears in Figure 10-6.

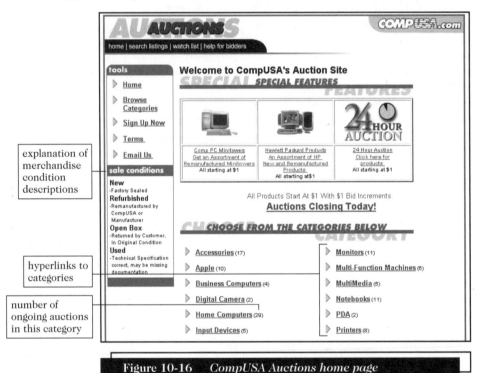

Figure 10-16   CompUSA Auctions home page

In the second business-to-business model, smaller firms sell their obsolete inventory through an independent third-party auction site. In some cases, these Web auctions are conducted by the same liquidation brokers that have always handled the disposition of obsolete inventory. These brokers have adapted to the changed environment and have implemented electronic commerce to stay in business. One example is the **DoveBid** site established by the Ross-Dove Company, which has been a traditional liquidation broker for many years.

Other third-party auction sites have been started by newcomers or even by companies that want to liquidate their inventory and are willing to do the same for other companies in their industry. Examples of third-party Web auction sites are **Auction IT** for computer equipment, **Going, Going...Sold!** for laboratory equipment, **FastParts.Com** for electronic components, and **J.R. Metals Quick Bid Auction** for steel.

Examples of completely new businesses that have created opportunities for business-to-business trade on Web auction sites include **NECX Online Exchange** for computer and electronics parts and **SciQuest** for laboratory instruments and supplies. The $1.6 trillion per year chemicals market has seen a number of new Web auction sites appear recently. Sites such as **CheMatch** for bulk petrochemicals and **Chemdex** for research laboratory chemicals are targeting specific market segments, whereas the World Chemical Exchange is targeting the entire chemicals market. Click **ChemConnect** on the Online Companion to see the World Chemical Exchange's auction.

In other industries, new auction markets on the Web are replacing their older ways of doing business. In the financial markets, mortgage brokers now have access to the **OpenClose Connection**, a Web auction site that allows brokers to put mortgage loan packages up for bids from lenders. On the **Band-X** Web auction site, telecommunications companies can buy or sell time on their networks to each other. Sellers list the number of minutes that they have available, and the price of airtime minutes fluctuates in response to buyers' bids on those minutes.

Established securities trading organizations such as the New York Stock Exchange (NYSE) and the Chicago Board of Trade (CBOT) are facing an electronic challenge to their time-honored ways of doing business. In 1998, a new venture was funded by electronic brokers E*Trade and Ameritrade Holdings, with contributions from several other brokerage firms called the **International Securities Exchange** (ISE). This new exchange is the first to be registered in the United States since 1973. In May 2000, the ISE began trading 82 of the most actively traded stock options contracts. The ISE's goal is to provide a low-cost trading forum in which the exchange will ultimately be able to trade 600 different stock options.

This new electronic securities exchange poses a threat to all existing securities exchanges, because its lower participation fees might attract the most lucrative large trades of active issues from existing exchanges. Once the ISE becomes better established and expands its list of securities traded, industry analysts question whether traditional exchanges such as the NYSE and the CBOT can continue to exist.

One very interesting innovation is the new approach to bidding pioneered by **FreeMarkets Online**. Instead of using a public Web auction site, FreeMarkets uses its software and hardware tools to coordinate private online auctions that allow businesses to solicit bids from suppliers. Instead of undergoing the laborious process of sending out request for proposal packages to many suppliers, a business can list its request for proposals with FreeMarkets. Companies that have used FreeMarkets report savings of 10% to 20% in their procurement costs. FreeMarkets gives firms a way of beginning to procure electronically without first investing in their own integrated system. In effect, FreeMarkets has moved the traditional first-price sealed-bid auction form onto the Internet.

Auction-Related Services

A common concern among people bidding in Web auctions is the reliability of the sellers. Recent surveys indicate that as many as 15% of all Web auction buyers have either not received the items they purchased, or found the items to be different from the seller's representation. About half of those buyers were unable to resolve their dispute to their satisfaction. When purchasing high-value items, buyers can use an escrow service to protect their interests.

You learned earlier in this chapter that an escrow service is an independent party that holds a buyer's payment until the buyer receives the purchased item and is satisfied that the item is what the seller represented it to be. Some escrow services will take delivery of the item from the seller and inspect it for the buyer. Escrow services do, however, charge fees ranging from 1% to 5% of the item's cost, subject to a minimum fee. The minimum fee provision can make escrow services uneconomical for small purchases. Escrow services that will handle Web auction transactions include **i-Escrow**, **SecureTrades**, and **TradeSafe Online**. Some of these escrow firms also sell auction buyer's insurance that can protect buyers from nondelivery risks and some quality risks.

Another service offered by some firms on the Web is a directory of auctions. Sites such as **Auction Guide** offer guidance for new auction participants and helpful tips on bidding strategies and other auction fine points for more experienced buyers and sellers. The **AuctionInsider** site, shown in Figure 10-7, is an auction directory site that provides links to auctions sorted by category, advice for auction participants, and an e-mail newsletter devoted to auction topics.

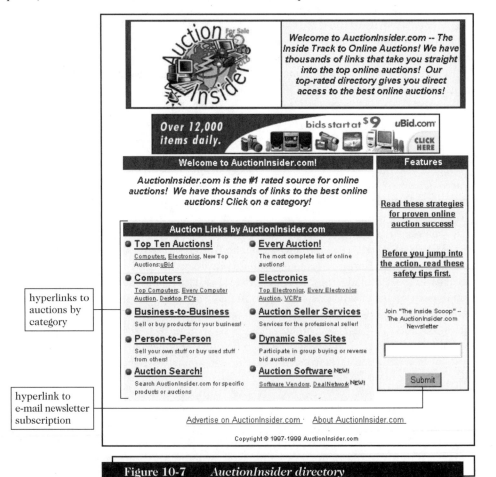

hyperlinks to auctions by category

hyperlink to e-mail newsletter subscription

Figure 10-7 *AuctionInsider directory*

Web sites that offer auction search and price-monitoring services include **Bidder's Edge**, **Price Watch,** and **PriceSCAN**. The Bidder's Edge site allows users to monitor bidding action on specific items or categories of items as it occurs on some of the more popular Web auction sites. Price Watch is an advertiser-supported site on which those advertisers post their current selling prices for computer hardware, software, and consumer electronics items. Although this monitoring is a retail pricing service designed to help shoppers find the best price on new items, Web auction-goers find it can help them with their bidding strategies. PriceSCAN is a similar price-monitoring service that also includes prices on books, movies, music, and sporting goods, in addition to the types of items monitored by Price Watch.

Seller-Bid Auctions and Group Purchasing Sites

An interesting variation on the traditional auction model has come to the Web on sites such as **TravelBids**: Sellers bid the prices at which they are willing to sell. These types of auctions are sometimes called **reverse auctions**, because the role of the bidder as buyer is reversed to bidder as seller. For example, at the TravelBids site, a person who has booked an airline or cruise reservation can post that reservation. Travel agents then bid for the right to ticket that booking. The bids decrease until no more travel agents bid or the buyer accepts the low bid.

Another new phenomenon on the Web is the group purchasing site. On a **group purchasing site**, the seller posts an item with a price. As individual buyers enter bids on an item (these bids are agreements to buy one unit of that item, but no price is specified), the site can negotiate a better price with the item's provider. The posted price ultimately decreases as the number of bids increases, but only if the number of bids increases.

Mercata is the first major group purchasing site on the Web for consumers. It offers group purchasing services, which it calls PowerBuys, on a variety of merchandise, including electronics, appliances, luggage, and jewelry. The Mercata site encourages potential buyers to send e-mail messages to their friends who might be interested in the same item. When more people bid on a particular item, its price decreases. The Mercata home page appears in Figure 10-8.

Demandline.com is a group purchasing site for businesses. It pools demand from small businesses for core services such as credit card processing, telecommunications, payroll services, and software support. It takes the combined demand of many smaller firms to large providers of these services and negotiates lower rates than the businesses could obtain on their own.

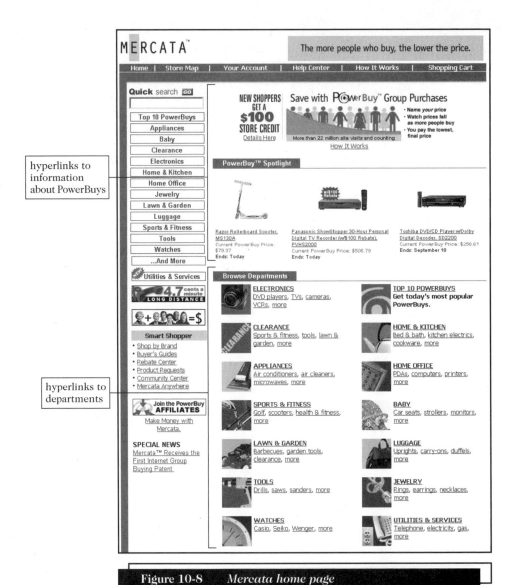

hyperlinks to information about PowerBuys

hyperlinks to departments

Figure 10-8 *Mercata home page*

VIRTUAL COMMUNITY STRATEGIES

Consider the following scenario:

> Fran Dennison has arrived in Paris one day early for a series of business meetings. She is hoping to recover from her jet lag and enjoy a little French food before her work begins. She has found a lovely café and, using her basic knowledge of French, has successfully ordered lunch.

Fran is reading the business section of *Le Monde*, a local newspaper. She begins reading an article about one of the business partners she will meet tomorrow, but her French is not good enough to completely understand the article. Fran opens her notebook computer and enters a request for translation services. She specifies that she needs immediate real-time translation of up to 500 words and is willing to pay up to 20 cents per word. She notes that the material to be translated is an article in today's *Le Monde*; she also enters the title of the article. Her computer, which contains a cellular link to her office network, launches an immediate search of online communities and marketplaces for this exact service. Two minutes later, a message appears on her computer from a French graduate student in the United States, Philippe Desmarest. His message indicates that he is willing to provide an immediate translation at Fran's quoted rate and that his computer has found the article on the *Le Monde* Web site. Five minutes later, an English translation appears in Fran's mailbox and $94.20 has been moved from her checking account to Philippe's. Fran has time to read the article and think about how she will adjust her presentation at tomorrow's meeting before her salad arrives.

This scenario could not occur today, but it is very close to becoming possible. Three key elements are required to make things such as Fran's on-demand translation a reality: cellular-satellite communications technology, electronic marketplaces, and software agents. All three of these elements exist today, but they have yet to be completely integrated.

Cellular-satellite communications technology capable of linking Fran to the Internet already exists and can be packaged with notebook computers. In 1999, electronic auction company **eBay** and cellular-satellite communications company **SkyTel Communications** announced a wireless person-to-person online trading service that uses two-way pagers, smart phones, and other communications devices.

Electronic marketplaces are growing out of virtual online communities such as **GeoCities** and **Tripod**. By integrating the bidding technologies developed by Web auction sites, these communities can become gathering places for people who want to buy and sell a wide range of products and services. The **MIT Media Lab** and other researchers have developed a number of intelligent software agents (also called software robots, or bots, which you learned about in Chapter 3). These agents are programs that traverse the Web and find items for sale that meet a buyer's specifications. In addition to finding product item matches, software agents such as Simon can find the lowest price for an item. Simon is one of the best shopping agents currently available. You can find Simon at the **mySimon** Web site, which appears in Figure 10-9.

375

Figure 10-9   *MySimon home page*

Some software agents are focused on a particular category of product, such as **Best Book Buys**, which searches more than 20 online bookstores for the best prices on books. The **MIT Software Agents Group** and other researchers are developing other software agents that track ratings of buyer and seller reputations in addition to obtaining price information. In much the same way that eBay makes reputation reports available to its bidders and sellers about each other, more general software agents can create and search databases of all kinds of buy-sell transactions on the Web. A Web site called the **BotSpot**, which is maintained by internet.com, contains information about a number of software agents, including a list of more than 40 shopping bot sites.

In the next section, you will learn how companies are creating virtual communities on the Web for people like Fran who engage in the business processes you learned about earlier in this book: identifying customers, promoting and selling goods and services to those customers, purchasing raw materials and supplies, and managing both the firm's internal operations and its relationships with other participants in the firm's supply chains.

Virtual Communities

A **virtual community**, also called a **Web community** or an **online community**, is a gathering place for people and businesses that does not have a physical existence. Howard Rheingold described the characteristics of these communities in his 1993 book, *The Virtual Community*, which has become recognized as the definitive book on the subject. Virtual communities exist on the Internet today in various forms, including Usenet newsgroups, chat rooms, and Web sites. These virtual communities help companies, their customers, and their suppliers plan, collaborate, transact business, and interact in ways that benefit all of them.

Although most Web communities are business-to-consumer strategy implementations, some successful business-to-business virtual communities have emerged. Milacron's **Milpro** site is a good example of a business-to-business virtual community. Milacron manufactures machine tools and sells them to a wide variety of industrial customers. Some of these customers are large manufacturers that Milacron can easily service. However, many of Milacron's customers are small, geographically isolated job shops. Job shops manufacture custom metal parts and specialized machinery.

Milacron decided that the best way to maintain contact with these customers was to create a virtual community in which customers could interact with Milacron and with each other. The Milpro site, shown in Figure 10-10, includes a Milacron product catalog, machine ordering and quotation services, and links to customer account information.

link to the Milpro Wizard help feature

377

Figure 10-10 *Milpro business-to-business virtual community site*

In addition to providing these Milacron-focused services, the site also includes a number of links to interactive features that are focused on customers and their interactions with each other. For example, the site offers a machinery flea market in which customers can trade used machinery and spare parts with each other. The site also features the Milpro Wizard, which offers free help to job shops. The Wizard can recommend specific products for specialized needs and offers suggestions for solving process problems. By providing this kind of free assistance, Milacron is using a rational branding strategy to attract customers and potential customers to the site and entice them into becoming regular visitors.

The lower border of the Web page even includes drawings of drill bits and grinding wheels. Milacron uses these graphics and the trademarked subtitle "Your Internet Tool Crib" to convey a feeling that this is "the" site for people who know machine tools. In this way, Milacron combines a small bit of emotional branding appeal with its rational branding strategy. The sense of community that Milacron hopes its Milpro site will provide for its customers has great value. For example, if the site encourages customers to check in regularly and stay longer, those customers will gradually develop a high opinion of Milacron and its products. This highly interactive method of using a sticky Web site to improve a firm's image with its customer can be a powerful brand-building strategy. For business-to-consumer sites, high levels of stickiness are even more important, since they can be used to justify higher advertising rates.

Early Web Communities

One of the first Web communities was the **WELL**. The WELL, which is an acronym for "whole earth 'lectronic link," predates the Web. It began as a series of dialogs among the authors and readers of the *Whole Earth Review* in 1985. Most WELL members were originally from the San Francisco Bay area, and the influence of that area's counterculture heritage is a significant part of the WELL's ambiance. Members of the WELL pay a monthly fee to participate in its forums and conferences. The WELL has been home to many important researchers and participants in the growth of the Internet and the Web. Its membership also includes noted writers and artists. In 1999, the e-zine publisher **Salon.com** bought the WELL and has maintained the sense of community that has existed there for 15 years.

As the Web emerged in the mid-1990s, its potential for creating new virtual communities was quickly exploited. In 1995, Beverly Hills Internet opened a virtual community site that featured two Webcams aimed down Hollywood streets and had links to entertainment information Web sites. The theme of this community was the formation of digital cities around the focus of the Webcams. The founders of Beverly Hills Internet wanted to create a sense of community and felt that the Webcams would help accomplish that goal. Their hope was that people would be attracted by the Webcam images and want to add their own contributions, thus becoming members of a virtual neighborhood. Members were given free space on the Web site to create pages within these virtual cities to add their contributions. The first of these digital cities were created around Webcams in the Los Angeles area and therefore were named for Los Angeles-area communities. As the site grew to include more geographic areas, it changed its name to **GeoCities**. GeoCities earned revenue by selling advertising on members' Web pages and on pop-up pages that appeared whenever a visitor accessed a member's site. GeoCities grew rapidly and was purchased in 1999 by Yahoo! for $5 billion.

Other similar sites became virtual communities. Tripod was founded in 1995 in Massachusetts and offered its participants free Web page space, chat rooms, news and weather updates, and health information pages. Like GeoCities, Tripod sold advertising on its main pages and on participants' Web pages. The search engine site **Lycos** purchased Tripod in 1998 for $58 million.

Theglobe.com, also started in 1995, was the outgrowth of a class project at Cornell University. The students who created the site built in bulletin boards, chat rooms, discussion areas, and personal ads. They then sold advertising to support the site's operation. Later additions included news feeds, an online art gallery, and shopping pages. Although Theglobe.com offers free Web page space, it does not emphasize that feature to the same extent that competing virtual communities do.

Web Portal Strategies

By the late 1990s, virtual communities were selling advertising to generate revenue. Search engine sites and Web directories also were selling advertising to generate revenue. Beginning in 1998, a wave of purchases and mergers occurred among these sites. Still using an advertising-only revenue generation model, the new sites that emerged included all the features offered by virtual community sites, search engine sites, Web directories, and other information-providing and entertainment sites. These portals, which you first learned about in Chapter 3, are so named because their goal is to be every Web surfer's entry to the Web.

Many Web observers believe that Web portal sites will be the great revenue-generating businesses of the future and that adding portal features to their existing sites or converting those sites to portals is a wise business strategy. They think that incorporating the sense of belonging developed in Web communities with the handy tools of search engine and Web directory sites will yield Web sites with high degrees of stickiness that will be extremely attractive to advertisers.

One rough measure of stickiness is how long each user spends at the site. Figure 10-11 lists 10 of the most popular sites on the Web (based on the number of home users who accessed the sites during August 2000). The information in Figure 10-11 is adapted from a **Nielsen//NetRatings** report and shows sites grouped by owner. For example, the numbers for Yahoo! include activity on all Yahoo! sites. Nielsen//NetRatings measures the number of unique visitors to determine site popularity. These leading sites had between 12 and 53 million unique visitors during the month measured. In Figure 10-11, the sites are not ranked by popularity, but by the average number of minutes that users spent on the sites.

Name	Minutes per unique vistor	Millions of unique visitors
Yahoo!	68	48
MSN	51	36
AOL-Netcenter	33	53
Excite@Home	28	22
Disney	24	19
Lycos Network	17	24
Time Warner	16	15
AltaVista	16	12
Microsoft	12	34
About.com	9	15

Adapted from reports for August 2000 published by
Nielsen//NetRatings at http://www.nielsen-netratings.com/

Figure 10-11 *Stickiness of popular Web sites*

Since Web portals ask their members to provide demographic information about themselves, the potential for targeted marketing is very high. Industry observers predicting success for Web portals may be correct. The top 10 most-visited Web sites shown in Figure 10-11 include nine Web portals. High visitor counts can yield high advertising rates for these sites. Web portals have been able to obtain up-front cash payments from advertisers, which is very unusual for any kind of advertising sale. Excite paid Netscape (now a part of AOL) a $70 million advance fee for two-year rights to a prominent advertising location on its Netcenter Web portal site. Other portal sites have negotiated advertising deals that include a percentage of sales generated from sales leads on the portal site.

The companies that run Web portals certainly believe in the power of portals. They have been aggressively adding sticky features such as chat rooms, e-mail, and calendar functions—often by purchasing the companies that create those features. In addition to buying the virtual community site Tripod, Lycos purchased the online directory WhoWhere? for $133 million. In 1999 alone, Yahoo! spent over $10 billion in cash and stock to expand the range of services available on its Web portal site.

As you learned earlier, Milacron's Milpro site provided a number of virtual community features that would enable that site to be expanded from a virtual community site to a Web portal site. Another example of a smaller business that is creating a portal site is **homebid.com**. This company has been conducting online auctions of residential real estate, but plans to become an all-purpose real estate Web portal by allowing customers to buy a home, finance it, and register for electricity, telephone, trash collection, and other necessary services all through its site.

Although the major Web portals will continue to compete with each other for large numbers of Web visitors in the coming years, most companies should consider whether a focused portal strategy makes sense for them.

Summary

Companies are now using the Web to do things that they have never done before, such as operating auction sites, creating virtual communities, and serving as Web portals. You learned about the key characteristics of the six major auction types (English, Dutch, first-price sealed-bid, second-price sealed-bid, double auction open-outcry, and double auction sealed-bid) and learned how firms are using Web auction sites to sell goods to their customers and generate advertising revenue. Newspaper companies and other businesses have created Web sites that sell classified advertising. Escrow services have begun providing protection to Web auction participants by acting as independent, trusted third parties that hold both goods and payments until the buyer is satisfied with the transaction. Buyers have become sellers by participating in reverse auctions, in which service providers bid for the right to provide lowest-cost services. Businesses are also conducting Web auctions to liquidate excess inventories instead of selling those inventories to liquidation brokers.

New companies have formed to take advantage of the Web's ability to bring together people and organizations that share narrow interests but that are geographically dispersed. Businesses are exploiting this characteristic of the Web by creating virtual communities with their customers and suppliers and by using these communities to sell goods and services. The major Web search engine sites have evolved into Web portals, and smaller businesses are beginning to use similar Web portal strategies to improve brand awareness, increase sales, keep visitors coming back, and generate advertising revenue.

Key Terms

Ascending-Price Auction
Auctioneer
Bid
Bidder
Descending-Price Auction
Double Auction
Dutch Auction
English Auction
Escrow Service
First-Price Sealed-Bid Auction
Group Purchasing Site
Liquidation Broker
Minimum Bid
Minimum Bid Increment
Online Community
Open Auction

Open-Outcry Auction
Private Valuation
Proxy Bid
Reserve
Reserve Price
Reverse Auction
Sealed-Bid Auction
Second-Price Sealed-Bid Auction
Shill Bidder
Specialist
Vickrey Auction
Virtual Community
Web Community
Winner's Curse
Yankee Auction

Review Questions

1. In one paragraph, name and briefly describe the key characteristics of the English auction format. In a second paragraph, explain how those characteristics are implemented in an English format Web auction.

2. In approximately 200 words, define the term *reserve price* and explain how the use of a reserve price can affect the progress and outcome of an auction.

3. Write three paragraphs in which you name and briefly describe each of the three models that are emerging for business-to-business Web auctions.

4. Describe the services that an escrow service company offers. Name one advantage and one disadvantage of using an escrow service. Limit your answer to 300 words.

5. Describe at least four sticky features that Web sites use to attract and keep visitors. For each feature, explain what makes it sticky. Explain why stickiness is important to companies operating Web sites. Limit your answer to 500 words.

Exercises

1. Use the Online Companion to examine the projects list at the **MIT Software Agents Group Projects** site. Choose a software agent technology used in one of these projects and, in approximately 200 words, describe how you could use it in an electronic commerce application.

2. Both **CNET.com** and **ZDNet** are technology-oriented Web portal sites that include specialty consumer Web auctions devoted to computers and computer-related items. Examine the auctions at these two sites and compare them to at least one other specialty consumer Web auction site devoted to similar items, such as **Haggle Online**, **Onsale**, or **uBid**. In your comparison, consider the number of auctions, the range of auctioned items, and the bidding activity. Summarize your findings in a report of approximately 200 words.

3. You have been hired as an electronic commerce consultant to Oyster Bay, Inc., a dealer in ocean-going yachts. Oyster Bay maintains offices and marinas in major U.S. East Coast ports. The typical purchaser of an Oyster Bay yacht is a high-income business executive, a retiree, or a person of significant inherited wealth. Oyster Bay salespersons have noted that their customers are increasingly aware of the Web. Prepare a proposal for an Oyster Bay Web portal site. You do not need to design the Web pages, but your proposal should include a detailed list of features that should be included in the site design. Describe each feature in detail and explain why you believe it should be included. For each feature, note whether it will be supplied by Oyster Bay personnel or purchased from an outside supplier. To learn more about existing yacht sales sites, you may wish to consult the **Grand Yachts Northwest**, **Hansen Yacht Sales**, and **Yacht Sales International** links in the Online Companion.

For Further Study and Research

Anders, G. 1998. "Search for Pez Launched Web's Newest Mogul," *The Wall Street Journal*, September 25, B1.

Anders, G. 1999. "Amazon.com Will Go Head to Head With eBay, OnSale in Online Auctions," *The Wall Street Journal*, March 29, A3.

Atwood, B. 1999. "Net Companies Get their Bids on Auction Biz," *Billboard*, 111(16), April 17, 70.

Banaghan, M. 1998. "Bargain Hunters Find a New Trail of Web-Site Auctions," *Business Review Weekly: BRW*, 20(29), August 3, 66.

Barron, J. 2000. "For 1776 Copy of Declaration, A Record in an Online Auction," *The New York Times*, June 30, B1.

Beales, N. 1999. "CommonPlaces' Web Strategy Targets Students: CollegeBytes.com Site Aims to Be the Internet Hub Among Undergraduates," *The Wall Street Journal*, May 10, B13D.

Bensinger, K. and A. Peers. 2000. "Baseball's Most Valuable Card Goes to Auction," *The Wall Street Journal*, June 5, B1.

Borden, M. 1999. "And Perhaps a Furby to Go with your Fuhrer?" *Fortune*, 139(9), May 10, 28.

Borrego, A. 2000. "How I Learned to Stop Worrying and (Almost) Love EBay," *Inc.*, May 16, 32–36.

Campbell, S. and J. Hagendorf. 1999. "Distributors Spruce Up Online Auction Sites," *Computer Reseller News*, November 22, 57–58.

Carrns, A. 1999. "www.doctorsmedicinesdiseases-galore.com—Today's Cybercraze Is Any Web Site Devoted To Health or Maladies," *The Wall Street Journal*, June 10, B1.

Cassady, R. 1967. *Auctions and Auctioneering*, Berkeley, CA: University of California Press.

Collett, S. 1999. "Thin Clients Pull in Car Customers," *Computerworld*, 33(14), April 5, 41.

Consumer's Research Magazine. 1999. "Fraud Surge," 82(3), March, 7–9.

Dalton, G. "Online Auctions Pick Up," *Information Week*, February 2, 37.

DeCaro, F. 1999. "To the Flea Market in Pajamas on EBay," *The New York Times*, May 9, Section 9, 6.

Dobrzynski, J. 2000. "F.B.I. Opens Investigation of EBay Bids," *The New York Times*, June 7, 1.

Dodge, J. 1999. "Making a Web Site Work: Three Companies Illustrate How Comfortable Internet Homes Can Be," *The Wall Street Journal*, June 14, 16.

Draenos, S. 2000. "Bidding for Auction Success," *Upside*, May 1, 41–44.

The Economist. 1997. "Going, Going..." May 31, 61.

Fisher, S. 2000. "Web Auctions Open Doors for Art Collectors," *InfoWorld.com*, September 15, Available online at: (http://www2.infoworld.com/articles/hn/xml/00/09/18/000918hnetend.xml).

Freeman, L. 2000. "Blue Mountain Arts: Mark Rinella," *Advertising Age*, June 26, S20.

Furchgott, R. and R. McNatt. 1999. "When Bidders Are Ringers," *Business Week*, May 17, 8.

Fusaro, R. 1999. "Internet Auction Sites Need Sharp Customer Support," *Computerworld*, 33(4), January 25, 28.

Gilbert, J. and A. Kerwin. 1999. "Newspapers Carve Slice of Auction Pie," *Advertising Age*, 70(26), June 21, 32–34.

Gomes, L. 2000. "EBay's Earnings Beat Forecasts as Sales, Number of Users Surge," *The Wall Street Journal*, July 26, B6.

Green, H. 2000. "Letting the Masses Name Their Price," *Business Week*, September 18, 44.

Grimes, A. 2000. "Hoaxes on EBay Raise Questions on Auction Web Site's Safeguards," *The Wall Street Journal*, January 7, B6.

Gross, N. 1999. "Building Global Communities: How Business Is Partnering with Sites that Draw Together Like-Minded Consumers," *Business Week*, March 22, EB42.

Guernsey, L. 2000. "A New Caveat for eBay Users: Seller Beware," *New York Times*, August 3, G1.

Habal, H. 1999. "An Auction-Style Web Site Starting a Marketing Push," *American Banker*, 164(29), February 12, 8.

Hansell, S. and J. Dobrzynski. 2000. "EBay Cancels Art Sale and Suspends Seller," *The New York Times*, May 11, A1.

Hochstein, M. 1999. "Web Site Plan Revised To Discourage 'Auction' of Fannie-Approved Loans," *American Banker*, 164(82), April 30, 8.

Hof, R. 2000. "Will Auction Frenzy Cool?" *Business Week*, September 18, 140.

InfoWorld. 1999. "Focus on I-Commerce," 21(6), February 8, 59.

Jones, K. 1999. "Two B-to-B Auction Models Emerge," *Inter@ctive Week*, 5(11), March 23, 44.

Kennedy, J. 1998. "Radio Daze," *MIT's Technology Review*, 101(6), November–December, 68–71.

Kollock, P. 1998. "Design Principles for Online Communities," *PC Update*, 15(5), June, 58–60.

Krauss, M. 1999. "Net Sites Favored by Sports Nuts Point Way to Future of the Web," *Marketing News*, April 12, 14.

Kuchinskas, S. 1999. "Eh, What's Up, EBay?" *Brandweek*, 40(13), March 29, 40.

Ma, M. 1999. "Agents in E-Commerce," *Communications of the ACM*, 42(3), March, 78–80.

Machlis, S. 1999. "Should My Company Care About Portals?" *Computerworld*, 33(9), March 1, 44–45.

McGuire, D. 2000. "Auction Sites Stay Popular Despite Fraud Warnings—Study," *Newsbytes*, September 25. Available online at: (http://www.nbnn.com/).

McLean, B. 1999. "Sothebys.com," *Fortune*, 139(3), February 15, 200.

Merlino, L. 2000. "Auction Anxiety," *Upside*, October. Available online at: (http://www.upside.com/texis/mvm/story?id=39ac0ec70).

Methvin, D. 1999. "Online Auction Action," *Windows Magazine*, 10(6), June, 159.

Nardi, B. and V. O'Day. 1999. *Information Ecologies: Using Technology with Heart*, Cambridge, MA: MIT Press.

Newsweek. 1999. "Spend Time, Get a Dime," March 15, 84.

Peers, A., G. Anders, and K. Bensinger. 1999. "Web Auctions Get Haute," *The Wall Street Journal*, June 17, B1.

Petersen, A. 1999. "Some Places to Go When You Want to Feel Right at Home: Communities Focus on People Who Need People," *The Wall Street Journal*, January 6, B6.

Petrecca, L. and B. Snyder. 1998. "Auction Universe Puts in $10 Mil Bid for Customers," *Advertising Age*, 43(8), October 26, 8.

Plotkin, H. 1998. "Art Net," *Forbes*, 161(7), April 6, 29–31.

Priluck, J. 1999. "MIT: E-Commerce Just Beginning," *Wired News*, February 24, Available online at: (http://www.wired.com/news/news/technology/story/18104.html).

Quan, J. 1999. "Risky Business," *Rolling Stone*, March 4, 91–92.

Razzi, E. 1999. "Online Lenders: Bidding for You," *Kiplinger's Personal Finance Magazine*, 53(5), May, 56–58.

Rheingold, H. 1993. *The Virtual Community: Homesteading on the Electronic Frontier*, New York: HarperCollins.

Robins, W. 2000. "Auctions.com Now a Dot-Goner," *Editor & Publisher*, August 28, 6.

Ross, P. 1999. "Web Winners," *Forbes*, 163(8), April 19, 226–228.

Roth, D. 2000. "Fraud's Booming in Online Auctions," *Fortune*, May 29, 276.

Roth, D. 2000. "Meet EBay's Worst Nightmare," *Fortune*, June 26, 199–202.

Schuyler, N. 2000. "Going… Going… Gotcha!" *PC World*, October 1, 181.

Shellenbarger, S. 1999. "New Technologies Can Help the Elderly Stay More Connected," *The Wall Street Journal*, April 21, B1.

Sliwa, C. 1999. "EBay Tops in Auction Niche," *Computerworld*, 33(19), May 10, 38.

Solomon, S. 1999. "Go Sell it on the Mountain," *Inc.*, December, 48–62.

Tedeschi, B. 2000. "Creating Marketplaces for Business-to-Business Transactions," *The New York Times*, January 24, C10.

Thaler, R. 1994. *The Winner's Curse: Paradoxes and Anomalies of Economic Life*, Princeton, NJ: Princeton University Press.

Turban, E. 1997. "Auctions and Bidding on the Internet: An Assessment," *International Journal of Electronic Markets*, 7(4), 7–11.

Turner, R. 1999. "Winning Bid," *Money*, 28(3), 201–203.

Useem, J. 1999. "For Sale Online: You," *Fortune*, 140(1), July 5, 66–73.

Vickrey, W. 1961. "Counterspeculation, Auctions, and Competitive Sealed Tenders," *Journal of Finance*, 16(1), March, 8–37.

The Wall Street Journal. 2000. "Art of Antiquing Turned on Its Head by Online Auctions," August 8, B12B.

Wamsley, D. 1999. "Online Ad Auctions Offer Sites More than Bargains," *Advertising Age*, 70(13), March 29, 43.

Wang, S. 1999. "Analyzing Agents for Electronic Commerce," *Information Systems Management*, 16(1), Winter, 40–48.

Ward, E. 1999. "How to Build Community on your Site and Participate in Others," *Business Marketing*, June 1, 24.

Weber, T. 2000. "To Build Virtual Trust, Web Sites Develop 'Reputation Managers,'" *The Wall Street Journal*, July 17, B1.

Whiteman, L. 1999. "Novel Way to Market CDs: Auction them Over the Web," *American Banker*, 164(81), April 29, 7.

Wilder, C. 1998. "IT Joins the Bidding," *Information Week*, May 25, 94.

Wilder, C. and R. Thompson. 1998. "Saddle Up Them Browsers!" *Information Week*, September 7, 14.

THE ENVIRONMENT OF ELECTRONIC COMMERCE: INTERNATIONAL, LEGAL, ETHICS, AND TAX ISSUES

INTRODUCTION

In 1999, **Dell Computer** and **Micron Electronics**, two companies that sell personal computers through their Web sites, agreed to settle U.S. Federal Trade Commission (FTC) charges that they had disseminated misleading advertising to their customers and potential customers. The advertising in question was for computer leasing plans that both companies had offered on Web pages at their sites. The ads stated the price of the computer along with a monthly payment. Unfortunately for Dell and Micron, stating the monthly payment without disclosing full details of the lease plan is a violation of the Consumer Leasing Act of 1976. This law is implemented through a federal regulation that was written and that is updated periodically by the Federal Reserve Board. This regulation, called Regulation M, was designed to require banks and other lenders to fully disclose the terms of leases so that

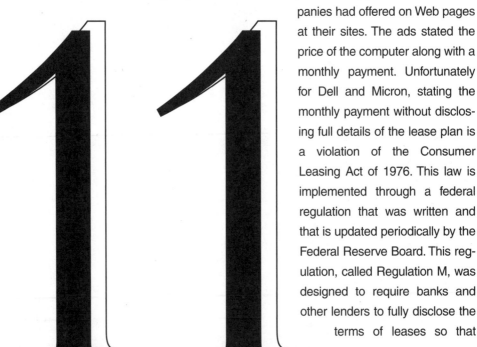

consumers have enough information to make informed financing choices when leasing cars, boats, furniture, and other goods.

Both Dell and Micron had included the required information on their Web pages, but FTC investigators noted that important details, such as the number of payments and fees due at the signing of the lease, were placed in a small typeface at the bottom of a long Web page. A consumer who wanted to determine the full cost of leasing a computer would need to scroll through a number of densely filled screens to obtain enough information to make the necessary calculations. In the settlement, both companies agreed to provide consumers with clear, readable, and understandable information in their lease advertising. The companies also agreed to record-keeping and federal monitoring activities designed to ensure their compliance with the terms of the settlement.

Dell and Micron are computer manufacturers. It apparently did not occur to them that they needed to become experts in Regulation M, generally considered to be a banking regulation. Companies that do business on the Web expose themselves, often unwittingly, to liabilities that arise from the environment of business today. That environment includes laws and ethical considerations that may be different from those with which the business is familiar. In the case of Dell and Micron, they were unfamiliar with the laws and ethics of the banking industry. The banking industry has a different culture than that of the computer industry—it is unlikely that a bank manager would have made such a mistake. As companies expand their markets on the Web, they encounter unfamiliar laws and different ethical frameworks, especially when they begin doing business in other countries.

LEARNING OBJECTIVES

In this chapter, you will learn about:

- International electronic commerce
- Laws that govern electronic commerce activities
- Ethics issues that arise for companies conducting electronic commerce
- Taxes that are levied on electronic commerce

INTERNATIONAL NATURE OF ELECTRONIC COMMERCE

Since the Internet connects computers all over the world, any business that engages in electronic commerce instantly becomes an international business. When companies use the Web to create a corporate image, build a brand, sell or purchase products or services, conduct auctions, or build a community, they are automatically operating in a global environment. A now-famous cartoon that appeared in *The New Yorker* magazine is shown in Figure 11-1. It illustrates a kind of anonymity that extends to all aspects of a Web presence.

"*On the Internet, nobody knows you're a dog.*"

Figure 11-1 *This cartoon from* The New Yorker *illustrates anonymity on the Web*

For example, a U.S. bank can establish a Web site that offers services throughout the world. No potential customer visiting the site will know by browsing through the site's pages just how large or well-established the bank is. Since Web site visitors will not become customers unless they trust the company behind the site, a plan for establishing credibility is essential. Sellers on the Web cannot assume that visitors will know that the site is operated by a trustworthy business.

Customers' lack of inherent trust in "strangers" on the Web is logical and to be expected; after all, people have been doing business with their neighbors—not strangers—for thousands of years. When businesses grew to become large corporations with multinational operations, their reputations grew commensurately. Before a company could do business in dozens of countries, it had to prove its trustworthiness by satisfying customers for many years as it grew. Businesses on the Web must find ways to overcome this well-founded tradition of distrusting strangers, because today a company can incorporate one day and, through the Web, be doing business the next day with people in almost every country in the world.

An important element of business trust is anticipating how the other party to a transaction will act in specific circumstances. For example, a potential buyer might like to know how the seller would react to a claim by the buyer that the seller has

388

misrepresented the quality of the goods sold. Part of this knowledge derives from the buyer and seller sharing a common language and common customs. Another part derives from having a common legal structure for resolving disputes. The combination of language and customs is often called **culture**. Most researchers agree that culture varies across national boundaries and, in many cases, varies across regions within nations. Businesses engaging in electronic commerce must be aware of the differences in language and customs that make up the culture of any region in which they intend to do business. The barriers to international electronic commerce include language, culture, and infrastructure issues.

Language Issues

Most companies have realized that the only way to do business effectively in other cultures is to adapt to those cultures. The phrase "think globally, act locally" is often used to describe this approach. The first step that a Web business usually takes to reach potential customers in other countries, and thus in other cultures, is to provide local language versions of its Web site. This may mean translating the Web site into another language or regional dialect. Researchers have found that customers are far more likely to buy products and services from Web sites in their own language, even if they can read English well. Only 370 million of the world's 6 billion people learned English as their native language.

Researchers estimate that about 80% of the content available on the Internet today is in English, but more than 40% of current Internet users do not read English. International Data Corporation predicts that by 2003, more than two-thirds of Internet users will be outside the United States and that 60% of Web use and 48% of electronic commerce sales will involve at least one party located outside the United States.

The most-used non-English languages for U.S. companies are Spanish, German, Japanese, French, and Chinese. Following closely behind is a second tier of languages that includes Italian, Korean, Portuguese, Russian, and Swedish. These language choices by U.S. businesses match closely the native languages of Internet users. As of September 2000, the languages most used on the Internet (other than English) were Chinese (8%), Japanese (7%), German (6%), Spanish (5%), Korean (4%), French (3%), Italian (3%), and Portuguese (3%). English is the native language of just under half of the people on the Internet. The Web site of **Global Reach**, a consulting firm that offers Web site globalization services, maintains current information about languages on the Web.

389

Some languages require multiple translations for separate dialects. For example, the Spanish spoken in Spain is different from that spoken in Mexico, which is different from that spoken elsewhere in Latin America. People in parts of Argentina and Uruguay use yet a fourth dialect of Spanish. Many of these dialect differences are spoken inflections, which are not important for Web site designers (unless, of course, their sites include audio or video elements); however, a significant number of differences occur in word meanings and spellings. You may be familiar with these types of differences, since they occur in the U.S. and British dialects of English. The U.S. spelling of *gray* becomes *grey* in Great Britain, and the meaning of *bonnet* changes from a type of hat in the United States to an automobile hood in Great Britain. Chinese has two main systems of writing: one used in mainland China, and another used in Hong Kong and Taiwan.

Most companies that translate their Web sites translate all of their pages. However, some sites have thousands of pages with much targeted content, and the cost of translating all pages might be prohibitive for some firms. The decision whether to translate a particular page should be made by the corporate department responsible for each page's content. The home page should have versions in all supported languages, as should all first-level links to the home page. Beyond that, pages that are devoted to marketing, product information, and establishing brand should be given a high translation priority. Some pages, especially those devoted to local interests, might be maintained only in the relevant language. For example, a weekly update on local news and employment opportunities at a company's plant in Frankfurt probably only needs to be maintained in German.

In Chapter 2, you learned how Web browsers and Web servers communicate with each other. Recall that the request message that a Web browser (client) sends to the Web server when it establishes a connection can include a request header. That request header contains information about the browser software, including the browser's default language setting. The Web server can detect the default language setting of the browser and automatically redirect the browser to the set of Web pages created in that language.

Another approach is to include links to multiple language versions on the Web site's home page. The Web site visitor must select one of the languages by clicking the appropriate link. Web sites that use this approach must make sure that the links are identifiable to visitors who read only those languages. Thus, the links should show the name of each language in that language. Many Web sites use country flags to indicate language. This practice can lead to errors and unintentional ill will. For example, a Bolivian visitor who must click the flag of Spain to navigate to the Spanish language pages might have difficulty identifying the Spanish flag and might resent that the Bolivian flag was not presented as a choice. The **Europages** home page, shown in Figure 11-2, includes identifiable links to pages in German, English, Spanish, French, Italian, and Dutch.

Firms that provide Web page translation services for companies include **Alis Technologies**, **Berlitz**, **LexFusion**, **Rubric, Ltd.**, **Transparent Language**, and **Worldpoint Interactive**. These firms will translate Web pages and maintain them for a fee that is usually between 25 and 50 cents per word. Languages that are complex or that are spoken by relatively few people are generally more expensive to translate than other languages.

Idiom Technologies sells software that automates the process of maintaining Web pages in multiple language versions. Idiom's WorldServer software tracks text that needs to be translated and inserts the translations into all sites that include that language. A human translator still must perform the translation. By automating the process, however, companies can reduce the cost of maintaining multiple-language Web sites. WorldServer places XML tags in the text of each translated Web site. These tags identify each text element with a corresponding text element at the company's main Web site. When the text in a page at the main Web site changes, the software sends out a notification of the change. The notification shows which pages need to be updated and tracks the exact location of the change in every page needing updating. When the human translation is complete, the software automatically inserts the translated text in the correct locations.

Lernout & Hauspie is another firm that offers a variety of useful translation and text-to-speech software products for Web site designers. The firm notes that different approaches can be appropriate for translating the different types of text that might appear on an electronic commerce site. For key marketing messages, the touch of a human translator is essential to capture subtle meanings. For more routine transaction processing functions, automated software translation may be an acceptable alternative. Software translation can reach speeds of 400,000 words per hour, so even if the translation is not perfect, businesses might find it preferable to a human who can translate between 400 and 600 words per hour.

name of the language shown in the language

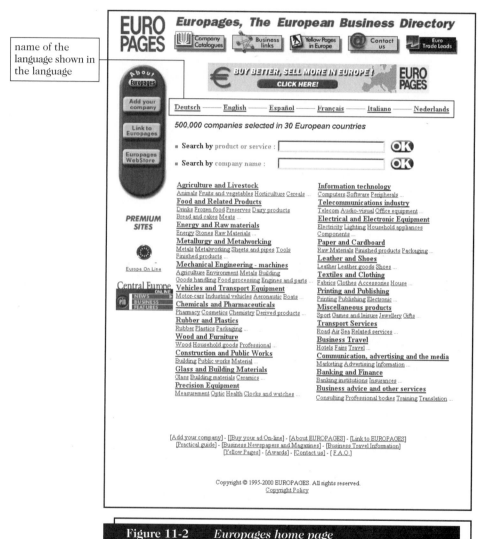

Figure 11-2 *Europages home page*

The translation services and software manufacturers that work with electronic commerce sites do not generally use the term "translation" to describe what they do. They prefer the term **localization**, which means a translation that considers multiple elements of the local environment, such as business and cultural practices, in addition to local dialect variations in the language. The cultural element is very important, since it can affect—and sometimes completely change—the user's interpretation of text.

The Environment of Electronic Commerce: International, Legal, Ethics, and Tax Issues

Culture Issues

Virtual Vineyards (now doing business as Wine.com), a company that sells wine and specialty food items on the Web, was perplexed. The company was getting an unusually high number of complaints from customers in Japan about short shipments. Virtual Vineyards sold most of its wine in case (12 bottles) or half-case quantities. Thus, to save on operating costs, it stocked shipping materials only in case, half-case, and two-bottle sizes. After investigation, the company determined that many of its Japanese customers ordered only one bottle of wine, which was shipped in a two-bottle container. To these Japanese customers, who consider packaging to be an important element of a high-quality product such as wine, it was inconceivable that anyone would ship one bottle of wine in a two-bottle container. They were e-mailing to ask where the other bottle was, notwithstanding the fact that they had ordered only one bottle.

Some errors stemming from subtle language and cultural standards have become classic examples that are regularly cited in international business training sessions. For example, General Motors could not understand why its Chevrolet Nova model was not selling well in Latin America until someone pointed out that *no va* means "it will not go" in Spanish. Pepsi's "Come Alive" advertising campaign fizzled in China because its message came across as "Pepsi brings your ancestors back from their graves." Another company sold baby food in jars adorned with the picture of a very cute baby. The jars sold well everywhere they had been introduced except in parts of Africa. The mystery was solved when the manufacturer learned that food containers in those parts of Africa always carry a picture of their contents.

On the Web, designers must be very careful when choosing icons that represent common actions. For example, in the United States, a shopping cart is a good symbol to use when building an electronic commerce site. However, Europeans use shopping *baskets* when they go to a store and may never have seen a shopping *cart*. In the United States, people often form a hand signal (the index finger touching the thumb to create a circle) that indicates "OK," or "everything is just fine." A Web designer might be tempted to use this hand signal as an icon to indicate that the transaction is completed or the credit card is approved, unaware that in countries such as Brazil, this hand signal is a very offensive gesture.

The cultural overtones of simple design decisions can be dramatic. In India, for example, it is inappropriate to use the image of a cow in a cartoon or other comical setting. Potential customers in Muslim countries can be offended by an image that shows human arms or legs uncovered. Even colors or Web page design elements can be troublesome. A Web page that is divided into four segments or that includes large white elements can be offensive to a Japanese visitor. Both the number four and the color white are symbols of death in that culture.

The design of a Web site built to attract customers from outside the United States should do more than avoid offending those visitors; it should also entice them. Web designers can do this by reflecting the visual preferences of the culture in which the site will operate. A site that strongly reflects a cultural design preference is the Swedish home page of **Bokus.com**, the largest online bookstore in Scandinavia. Figure 11-3 shows the Bokus.com. home page.

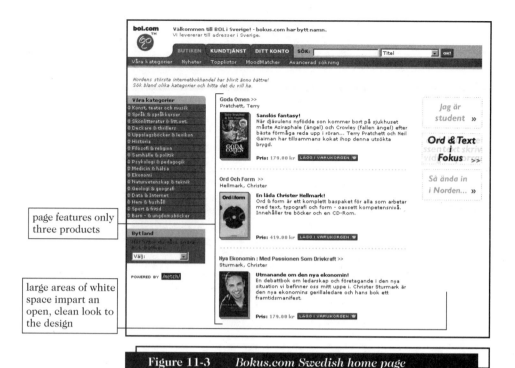

page features only three products

large areas of white space impart an open, clean look to the design

Figure 11-3 Bokus.com Swedish home page

The Bokus.com home page is quite different from that of most U.S. online bookstores. It is far less cluttered and has a crisp, clean design that is very typical of Scandinavian graphic arts, furniture design, and architecture. This page features three products; U.S. sites rarely feature fewer than five products on their home pages. The Bokus.com page also has many fewer hyperlinks and much more white space than the typical U.S. retail Web site. The Bokus.com design is not necessarily better, it is just different in ways that are likely to make Swedes feel more comfortable using the site.

Softbank, a major Japanese firm that invests in Internet companies, has devised a way to introduce electronic commerce to a reluctant Japanese population. Japanese shoppers have resisted the U.S. version of electronic commerce because they generally prefer to pay in cash or by cash transfer instead of by credit card, and they have a high level of apprehension about doing business online. In 1999, Softbank created a joint venture with **7-Eleven**, **Yahoo! Japan**, and Tohan (a major Japanese book distributor) to sell books and CDs on the Web. This venture, called eS-Books, allows customers to order items on the Internet, and then pick them up and pay for them in cash at the local 7-Eleven convenience store. By adding an intermediary—the exact opposite of the strategy used by U.S. firms—that satisfies the needs of the Japanese customer, Softbank has brought business-to-consumer electronic commerce to Japan.

Nike, a major U.S.-based maker of sports products, realized that it had to create special Web pages to attract the millions of its customers who live outside the United States. One such effort is the **Nike Football** site shown in Figure 11-4.

| Figure 11-4 | *Nike football English-language home page* |

This soccer imagery is not what most U.S visitors would expect to see when visiting a "football" site! Since Nike already had a site that was designed for its U.S. audience, it used the Nike Football site to appeal to European visitors by capitalizing on its involvement with the Euro 2000 soccer championships. Nike promoted its football pages separately in television and newspaper advertising campaigns related to the Euro 2000 games. The site was so successful that it was expanded to include pages devoted to Brazilian soccer. Nike plans to continue to develop pages directed at unique combinations of particular sports in particular countries in appropriate languages for those sport-country combinations.

Some parts of the world have cultural environments that are extremely inhospitable to electronic commerce initiatives. For example, a report issued in 1999 by **Human Rights Watch** stated that many countries in the Middle East and North Africa have been reluctant to allow their citizens free access to the Internet. The report notes that many governments in this part of the world regularly prevent free expression by their people and have taken specific steps to prevent the exchange of information outside of state controls. Saudi Arabia, Yemen, and the United Arab Emirates all use proxy servers to filter content. Jordan has imposed taxes that put the cost of Internet access beyond the means of most Jordanians. Jordan also has laws that prohibit publications in any media that conflict with the values of an Islamic nation. An organization devoted to the international promotion of democracy and civil liberties, **Freedom House**, offers a number of downloadable publications on its site that include in-depth reports on Internet censorship activities of governments throughout the world.

In contrast, Algeria, Morocco, and the Palestinian Authority have not limited online access or content. In most North African and Middle Eastern countries, officials have publicly denounced the Internet for carrying materials that are sexually explicit, anti-Islam, or that cast doubts on the traditional role of women in their societies. In many of these countries, Internet technology is so at odds with existing traditions, cultures, and laws that electronic commerce is unlikely to exist in these countries at any significant level in the near future.

Other countries, such as the People's Republic of China and Singapore, are wrestling with the issues presented by the growth of the Internet as a vehicle for doing business. These countries have a tradition of controlling their citizens' access to information from outside the country, but they want their economies to reap the benefits of electronic commerce. China has created a complex set of registration requirements and regulations that govern any business that engages in electronic commerce. These regulations are enforced by the state police, not by an administrative agency. For example, ISPs must register all of their customers with the police. Singapore has also adopted a number of restrictive rules and policies. Only time will tell whether these attempts will effectively control Internet activity yet allow business transactions to be conducted successfully.

Some countries, although they do not ban electronic commerce entirely, have strong cultural requirements that have found their way into the legal codes that govern business conduct. In France, an advertisement for a product or service must be in French. Thus, a business in the United States that advertises its products on the Web and that is willing to ship goods to France must provide a French version of its pages. Many U.S. electronic commerce sites include in their Web pages a list of the countries from which they will accept orders through their Web sites.

The official language of the Canadian province of Quebec is French. Quebec provincial law requires street signs, billboards, directories, and advertising created by Quebec businesses to be in French. In 1999, the government of Quebec fined photographer Michael Calomiris and ordered him either to remove his English-language Web site or to add a French translation of the pages to the site. Calomiris had been advertising his photographs for sale on his Quebec-based Web site and had targeted his ads to the U.S. market. He paid the fine and has appealed the government's decision. He is maintaining his Web site in English while his appeal is pending.

Infrastructure Issues

Businesses that successfully meet the challenges posed by language and culture issues still face the challenge posed by variations and inadequacies in the infrastructure that support the Internet throughout the world. Internet infrastructure includes the computers and software connected to the Internet and the communications networks over which the message packets travel. In many countries other than the United States, the telecommunications industry is either government-owned or heavily regulated by the government. In many cases, regulations in these countries have inhibited the development of the telecommunications infrastructure or limited the expansion of that infrastructure to a size that cannot reliably support Internet data packet traffic.

Local connection costs through the existing telephone networks in many countries are very high compared to U.S. costs for similar access. This can have a profound effect on the behavior of electronic commerce participants. For example, in Europe, where Internet connection costs can be quite high, few businesspeople would spend time surfing the Web to shop for a product. They will use a Web browser only to navigate to a specific site that they know will offer the product they want to buy. Thus, to be successful in selling to European businesses, a company must advertise its Web presence in traditional media instead of relying on high placement on search engine results pages.

The Organization for Economic Cooperation and Development's (OECD) Directorate of Science, Technology, and Industry has issued a number of **Statements on Information and Communications Policy** that deal with telecommunications infrastructure development issues throughout the world. These OECD statements have provided guidance for businesses and governments as they have begun building the technology capabilities that will support international electronic commerce in the future.

In 1998, business and government leaders in several European countries began pushing for flat-rate telephone line Internet access charges. Europeans pay for the time they are using the telephone line, even for local calls, despite recent moves to deregulate the telecommunications industries in those countries. In a **flat-rate access** system, the consumer or business pays one monthly fee for unlimited telephone line usage. Activists in Germany, Ireland, and Spain have all argued that flat-rate access has been the key to the success of electronic commerce in the United States.

The paperwork and often-convoluted processes that accompany international transactions are targets for technological solutions. Most firms that conduct business internationally rely on a complex array of freight forwarding companies, customs brokers, international freight carriers, and importers to navigate the maze of paperwork that must be completed at every step of the transaction to satisfy government and insurance requirements. The multiple flows of information and transfers of physical objects that occur in a typical international trade transaction are illustrated in Figure 11-5.

As you can see in Figure 11-5, the information flows can be very complex. Domestic transactions usually include only the seller, the buyer, their respective banks, and one freight carrier. International transactions almost always require physical handling of goods by several freight carriers, storage in a freight forwarder's facility before international shipment, and storage in a port or bonded warehouse facility in the destination country. This handling and storage requires monitoring by government customs offices in addition to the monitoring by seller and buyer that occurs in domestic transactions. International transactions usually require the coordinated efforts of customs brokers and freight forwarding agencies because the regulations and procedures governing international transactions are so complex.

In a report issued in late 2000, Forrester Research analyst Michael Putnam noted that 46% of the businesses he interviewed stated that they turned away all international orders because they did not have the processes in place to handle such orders. Some of the businesses in the study estimated they were losing more than $10 million in international business each year. This problem is shared by businesses around the globe. Not only are U.S. businesses having difficulty reaching their international markets, but businesses in other countries are having as much trouble reaching the U.S. market.

The United Nations estimated that the cost of handling paperwork for international transactions was $420 billion in 1998, or approximately 6% of the total $7 trillion spent in worldwide international trade that year. Companies such as **NextLinx**,

Syntra, and **Vastera** sell software designed to automate much of the international trade process. This is a difficult task because each country has its own paper-based forms and procedures with which international shippers must comply. Even countries that have automated some of the procedures often use computer systems that are incompatible with those of other countries. In 1999, a consortium of 120 banks and logistics firms founded **Bolero International**, an association dedicated to replacing the paper maze with a set of interoperable electronic commerce applications.

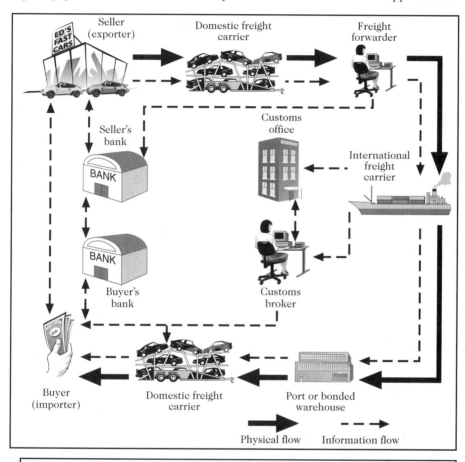

Figure 11-5 *An international trade transaction*

THE LEGAL ENVIRONMENT OF ELECTRONIC COMMERCE

Businesses that operate on the Web must comply with the same laws and regulations that govern the operations of all businesses. If they do not, they face the same set of penalties—including fines, reparation payments, court-imposed dissolution, and even jail time for officers and owners—that any business faces.

Businesses operating on the Web face two complicating factors as they try to follow the law. First, the Web extends a company's reach beyond traditional boundaries. As you learned in the previous section, a business that uses the Web immediately becomes an international business. Thus, a company can become subject to many more laws more quickly than a traditional brick-and-mortar business tied to one specific physical location. Second, the Web increases the speed and efficiency of business communications. As you learned in Chapter 8, customers often have much more interactive and complex relationships with the companies they buy from on the Web than they do with traditional merchants. Further, the Web creates a network of customers who often have significant levels of interaction with each other. Web businesses that violate the law or breach ethical standards can face rapid and intense reactions from many customers and other stakeholders who become aware of the businesses' activities.

Borders and Jurisdiction

Territorial borders in the physical world serve a useful purpose in traditional commerce: They mark the range of culture and reach of applicable laws very clearly. When people travel across international borders, they are made aware of the transition in many ways. For example, exiting one country and entering another requires a formal examination of documents, namely a person's passport. In addition, both language and the currency usually change upon entry into a new country. Each of these experiences, and countless others, are manifestations of the differences in legal rules and cultural customs in the two countries. In the physical world, geographic boundaries almost always coincide with legal and cultural boundaries. The limits of acceptable ethical behavior and the laws that are adopted in a geographic area are the result of the influences of the area's dominant culture. The relationships among a society's culture, laws, and ethical standards appear in Figure 11-6.

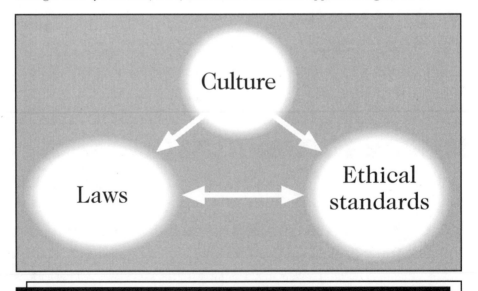

Figure 11-6   *Culture determines laws and ethical standards*

The geographic boundaries on culture are logical—for most of our history, we humans have been unable to travel great distances to learn about other cultures. Legal scholars define the relationship between geographic boundaries and legal boundaries in terms of four elements: power, effects, legitimacy, and notice.

Power

Power, in the form of control over physical space and the people and objects that reside in that space, is a defining characteristic of statehood. For laws to be effective, a government must be able to enforce them. Effective enforcement requires the power both to exercise physical control over residents, if necessary, and to impose sanctions on those who violate the law. The ability of a government to exert control over a person or corporation is called **jurisdiction**.

Laws in the physical world do not apply to people who are not located in or who do not own assets in the geographic area that created those particular laws. For example, the United States cannot enforce its copyright laws on a citizen of Japan who is doing business in Japan and who owns no assets in the United States. Any assertion of power by the United States over such a Japanese citizen would conflict with the Japanese government's recognized monopoly on using force with its citizens. Japanese citizens who bring goods into the United States to sell, however, would be subject to applicable U.S. copyright laws.

The level of power asserted by a government is limited to that which is accepted by the culture that exists within its geographic boundaries. Ideally, geographic boundaries, cultural groupings, and legal structures will all coincide. When they do not, internal strife and civil wars often erupt.

Effects

Laws in the physical world are grounded in the relationship between physical proximity and the effects of a person's behavior. Personal or corporate actions have stronger effects on people and things that are nearby than on those that are far away. Government-provided trademark protection is a good example of this. For instance, the Italian government can provide and enforce trademark protection for a business named Caruso's Ristorante located in Rome. The effects of another restaurant using the same name are strongest in geographic areas in Rome, close to Rome, and in other parts of Italy. If someone were to open a restaurant in Kansas City and call it Caruso's Ristorante, the restaurant in Rome would experience few, if any, negative effects from having this restaurant use its trademarked name. Thus, the effects of the trademark violation are controlled by the law in Italy because of the limited range within which such a violation has an effect.

The characteristics of laws are determined by the local culture's acceptance of or reluctance to various kinds of effects. For example, certain communities in the United States require that houses be built on lots that are at least five acres. Other communities prohibit outdoor advertising of various kinds. The local cultures in these communities make the effects of such restrictions acceptable.

Legitimacy

Most people agree that the legitimate right to create and enforce laws derives from the mandate of those who are subject to those laws. In 1970, the **United Nations** passed a resolution that affirmed this idea of governmental legitimacy. The resolution made clear that the people residing within a set of recognized geographic boundaries are the ultimate source of legitimate legal authority for people and actions within those boundaries. Thus, **legitimacy** is the idea that those subject to laws should have some role in formulating them.

Some cultures allow their governments to operate with a high degree of autonomy and unquestioned authority. China and Singapore are countries in which national culture permits the government to exert high levels of unchecked authority. Other cultures, such as those of the Scandinavian countries, place strict limits on governmental authority.

Notice

Physical boundaries are a convenient and effective way to announce the ending of one legal or cultural system and the beginning of another. The physical boundary, when crossed, provides notice that one set of rules has been replaced by a different set of rules. People can obey and perceive a law or cultural norm as fair only if they are notified of its existence. Borders provide this notice in the physical world. The legal systems of most countries include a concept called constructive notice. Persons receive **constructive notice** that they have become subject to new laws and cultural norms when they cross an international border, even if they are not specifically warned of the changed laws and norms by a sign or a border guard's statement. Thus, ignorance of the law is not a sustainable defense, even in a new and unfamiliar jurisdiction.

Jurisdiction on the Internet

Defining, establishing, and asserting jurisdiction are much more difficult on the Internet than they are in the physical world, mainly because traditional geographic boundaries do not exist. For example, a Swedish company that engages in electronic commerce may have a Web site that is entirely in English and a URL that ends in ".com," thus not indicating to customers that it is a Swedish firm. The server that hosts this company's Web page could be in Canada and the people who maintain the Web site might work from their homes in Australia.

If a Mexican citizen buys a product from the Swedish firm and is unhappy with the goods received, that person might want to file a lawsuit against the seller firm. However, the world's physical border–based systems of law and jurisdiction do not help this Mexican citizen determine where to file the lawsuit. The Internet does not provide anything like the obvious international boundary lines in the physical world. Thus, the four considerations that work so well in the physical world—power, effects, legitimacy, and notice—do not translate very well to the virtual world of electronic commerce.

Governments that want to enforce laws regarding business conduct on the Internet must establish jurisdiction over that conduct. A **contract** is a promise or set of promises between two or more legal entities—persons or corporations—that provides for an exchange of value (goods, services, or money) between or among them. A **tort** is an intentional or negligent action taken by a legal entity that causes harm to another legal entity. People or corporations that wish to enforce their rights based on either contract or tort law must file their claims in courts with jurisdiction to hear their case. A court has sufficient jurisdiction in a matter if it has both subject-matter jurisdiction and personal jurisdiction.

Subject-Matter Jurisdiction

Subject-matter jurisdiction is a court's authority to decide a particular type of dispute. For example, in the United States, federal courts have subject-matter jurisdiction over issues governed by federal law (for example, bankruptcy, copyright, patent, and federal tax matters), and state courts have subject-matter jurisdiction over issues governed by state laws (for example, professional licensing and state tax matters). If the parties to a contract are both located in the same state, a state court has subject-matter jurisdiction. The rules for determining whether a court has subject-matter jurisdiction are very clear and easy to apply. Very few disputes arise over subject-matter jurisdiction.

Personal Jurisdiction

Personal jurisdiction is, in general, determined by the residence of the parties. A court has personal jurisdiction over a case if the defendant is a resident of the state in which the court is located. In such cases, the determination of personal jurisdiction is straightforward. However, an out-of-state person or corporation can also voluntarily submit to the jurisdiction of a particular state court by agreeing to do so in writing or by taking certain actions in the state.

One of the most common ways that people voluntarily submit to a jurisdiction is by signing a contract that includes a statement, known as a **forum selection clause**, that the contract will be enforced according to the laws of a particular state. That state then has personal jurisdiction over the parties that signed the contract regarding any enforcement issue that arises from the terms of that contract.

In the United States, individual states have laws that can create personal jurisdiction for their courts. The details of these laws, called **long-arm statutes**, vary from state to state, but generally create personal jurisdiction over nonresidents who transact business or commit tortious acts in the state. For example, suppose that an Arizona resident drives recklessly while in California and, as a result, causes a collision with another vehicle that is driven by a California resident. Due to his tortious behavior in the state of California, the Arizona resident can expect to be called into a California court. In other words, California courts have personal jurisdiction over the matter.

Businesses should be aware of jurisdictional considerations when conducting electronic commerce over state and international lines. In most states, the extent to which these laws apply to companies doing business over the Internet is unclear. Since these procedural laws were written before electronic commerce existed, their application to Internet transactions continues to evolve as more and more disputes arise from online commercial transactions. The trend in this evolving law is that the more business activities a company conducts in a state, the more likely a court will assert personal jurisdiction over that company through the application of a long-arm statute.

One exception to that general rule, however, occurs in the case of tortious acts. A business can commit a tortious act by selling a product that causes harm to a buyer. The tortious act can be negligent, in which the seller unintentionally provides a harmful good, or it can be an intentional tort, in which the seller knowingly or recklessly causes injury to the buyer. The most common examples of business-related intentional torts are defamation, misrepresentation, fraud, and theft of trade secrets. Although case law is rapidly developing in this area also, courts tend to invoke their respective states' long-arm statutes much more readily in the case of tortious acts than in other business cases. If the matter involves an intentional tort or a criminal act, courts will more liberally assert jurisdiction.

Jurisdiction in International Commerce

Jurisdiction issues that arise in international business are even more complex than the rules governing personal jurisdiction across state lines within the United States. The exercise of jurisdiction across international borders is governed by treaties between the countries engaged in the dispute. In general, U.S. courts determine personal jurisdiction for foreign companies and persons in much the same way that those courts interpret the long-arm statutes in domestic matters. Non-U.S. corporations and individuals can be sued in U.S. courts if they conduct business or commit tortious acts in the United States. Similarly, foreign courts can enforce decisions against U.S. corporations or individuals through the U.S. court system if those courts can establish jurisdiction over the matter.

Jurisdictional issues are complex and change rapidly. Any business that intends to conduct electronic commerce should consult an attorney who is well-versed in these procedural issues. The **John Marshall Law School's Center for Information Technology and Privacy Law** Web page includes links to current cases, law review articles, and other updated resources related to electronic commerce legal issues. The Center provides a collection of materials related to cyberspace law, as shown in Figure 11-7.

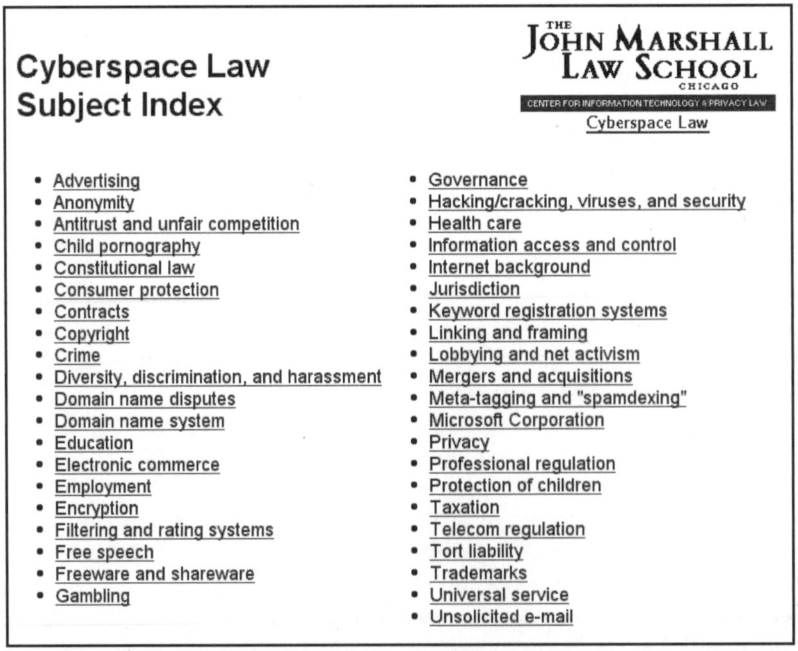

Figure 11-7   *John Marshall Law School Cyberspace Law site*

Contracting and Contract Enforcement in Electronic Commerce

Any contract includes three essential elements: an offer, an acceptance, and consideration. The contract is formed when one party accepts the offer of another party. An **offer** is a commitment with certain terms made to another party, such as a declaration of willingness to buy or sell a product or service. An offer can be revoked as long as no payment, delivery of service, or other consideration has been accepted. An **acceptance** is the expression of willingness to take an offer, including all of its stated terms. **Consideration** is the bargained-for exchange of something valuable, such as money, property, or future services. When a party accepts an offer based on the exchange of valuable goods or services, a contract has been created. An **implied contract** can also be formed by two or more parties that act as if a contract exists, even if no contract has been written and signed.

People enter into contracts on a daily, and often hourly, basis. Every kind of agreement or exchange between parties, no matter how simple, is a type of contract. For example, every time a consumer buys an item at the supermarket, the elements of a valid contract are met:

- The store offers an item at a stated price.
- The consumer accepts this offer by indicating a willingness to buy the product for the stated price.
- The store exchanges its product for another valuable item: the consumer's payment.

Contracts are a key element of traditional business practice and they are equally important on the Internet. Offers and acceptances can occur when parties exchange e-mail messages, engage in electronic data interchange (EDI), or fill out forms on Web pages. These Internet communications can be combined with traditional methods of forming contracts, including the exchange of paper documents, faxes, and verbal agreements made over the telephone or in person. An excellent resource for many of the laws concerning contracts, especially as they pertain to U.S. businesses, is the Cornell Law School Web site, which includes the full text of the **Uniform Commercial Code** (UCC).

When a seller advertises goods for sale on a Web site, that seller is not making an offer, but is inviting offers from potential buyers. If a Web ad was a legal offer to form a contract, the seller could easily become liable for delivery of more goods than it has available to ship. When a buyer submits an order, which is an offer, the seller can accept that offer and create a contract. If the seller does not have the ordered items in stock, the seller has the option of refusing the buyer's order outright or counteroffering with a decreased amount. The buyer then has the option to accept the seller's counteroffer.

Making a legal acceptance of an offer is quite easy to do in most cases. When enforcing contracts, courts tend to view offers and acceptances as actions that occur within a particular context. If the actions are reasonable under the circumstances, courts tend to interpret those actions as offers and acceptances. For example, courts have held that actions—including mailing a check, shipping goods, shaking hands, nodding one's head, taking an item off a shelf, or opening a wrapped package—are all, in some circumstances, legally binding acceptances of offers. Although the case law is limited regarding acceptances made over the Internet, it is reasonable to assume that courts would view clicking a button on a Web page, entering information in a Web form, or downloading a file to be legally binding acceptances.

Written Contracts on the Web

In general, contracts are valid even if they are not in writing or signed. However, certain categories of contracts are not enforceable unless the terms are put into writing and signed by both parties. In 1677, the British Parliament enacted a law that specified the types of contracts that had to be in writing and signed. Following this British precedent, every state in the United States today has a similar law, called a **Statute of Frauds**. Although these state laws vary slightly, each Statute of Frauds specifies that contracts for the sale of goods worth over $500 and contracts that require actions that cannot be completed within one year must be created by a signed writing. Fortunately for businesses and people who want to form contracts using electronic commerce, a writing does not require either pen or paper.

Most courts will hold that a **writing** exists when the terms of a contract have been reduced to some tangible form. An early court decision in the 1800s held that a telegraph transmission was a writing. Later courts have held that tape recordings of spoken words, computer files on disks, and faxes are writings. Thus, the parties to an electronic commerce contract should find it relatively easy to satisfy the writing requirement. Courts have been similarly generous in determining what constitutes a signature. A **signature** is any symbol executed or adopted for the purpose of authenticating a writing. Courts have held names on telegrams, telexes, faxes, and Western Union Mailgrams to be signatures. Even typed names or names printed as part of a letterhead have served as signatures. It is reasonable to assume that a symbol or code included in an electronic file would constitute a signature. As you learned in Chapter 6, the United States now has a law that explicitly makes digital signatures legally valid for contract purposes.

Firms conducting international electronic commerce do not need to worry about the signed writing requirement in most cases. The main treaty that governs international sales of goods, Article 11 of the **United Nations Convention on Contracts for the International Sale of Goods**, requires neither a writing nor a signature to create a legally binding acceptance.

Warranties

Most firms conducting electronic commerce have little trouble fulfilling the requirements needed to create enforceable, legally binding contracts on the Web. One area that deserves attention, however, is the issue of warranties. Any contract for the sale of goods includes implied warranties. A seller implicitly warrants that the goods that it sells are fit for the purposes for which they are normally used. If the seller knows specific information about the buyer's requirements, acceptance of an offer from that buyer may result in an additional implied warranty of fitness, which suggests that the goods are suitable for the specific uses of that buyer. Sellers can also create explicit warranties by providing a specific description of the additional warranty terms. It is also possible for a seller to create explicit warranties, often unintentionally, by making general statements in brochures or other advertising materials about product performance or suitability for particular tasks.

Sellers can avoid some implied warranty liability by making a warranty disclaimer. A **warranty disclaimer** is a statement that the seller will not honor some or all implied warranties. Any warranty disclaimer must be conspicuously made in writing, which means it must be easily noticed in the body of the written agreement. On a Web page, sellers can meet this requirement by putting the warranty disclaimer in larger type, a bold font, or a contrasting color. To be legally effective, the warranty disclaimer must be stated obviously and must be easy for a buyer to find on the Web site.

Authority to Form Contracts

As explained previously in this section, a contract is formed when an offer is accepted for consideration. Problems can arise when the acceptance is issued by an imposter or someone who does not have the authority to bind the company to a contract. In electronic commerce, the online nature of acceptances can make it relatively easy for identity forgers to pose as others. As you learned in Chapter 5, masquerading, which is the creation of false online identities and e-mail messages that appear to be from a different source than the actual origin of the message, is not too difficult to accomplish.

Fortunately, the Internet technology that makes forged identities so easy to create also provides the means to avoid being deceived by a forged identity. Digital signatures are an excellent way to establish identity in online transactions. If the contract is for any significant amount, the parties should require each other to use digital signatures to avoid identity problems. In general, courts will not hold a person or corporation whose identity has been forged to the terms of the contract; however, if negligence on the part of the person or corporation contributed to the forgery, a court may hold the negligent party to the terms of the contract. For example, if a company was careless about protecting passwords and allowed an imposter to enter the company's system and accept an offer, a court might hold that company responsible for fulfilling the terms of that contract.

Determining whether an individual has the authority to commit a company to an online contract is a greater problem than forged identities in electronic commerce. This issue, called **authority to bind**, can arise when an employee of a company accepts a contract and the company later asserts that the employee did not have such authority. For large transactions in the physical world, businesses check public information on file with the state of incorporation or ask for copies of corporate certificates or resolutions to establish the authority of persons to make contracts for their employers. These methods are available to parties engaged in online transactions; however, they can be time-consuming and awkward. Good electronic solutions are digital signatures and certificates from a certification authority (see Chapter 6). Some digital signatures and certificates can attest to the title and capacity of a person holding a particular public key, in addition to establishing that person's identity.

Web Site Content

In Chapter 5, you learned about copyright law and how it relates to materials that appear on Web pages. A number of other legal issues can arise regarding the Web page content of electronic commerce sites, including trademark infringement, deceptive trade practices, regulation of advertising claims, and defamation.

405

Trademark Infringement

The owners of registered trademarks have often invested a considerable amount of money in the development and promotion of their trademarks. Web site designers must be very careful not to use any trademarked name, logo, or other identifying mark without the express permission of the trademark owner. For example, a company Web site that included a photograph of its president who happened to be holding a can of Pepsi could violate Pepsi's trademark rights. Pepsi can argue that the appearance of its trademarked product on the Web site implies an endorsement of the president or the company by Pepsi.

Deceptive Trade Practices

Computer graphics, audio, and video technology allows Web site designers to do many creative and interesting things. Manipulations of existing pictures, sounds, and video clips can be very entertaining. If the objects being manipulated are trademarked, however, these manipulations can violate the trademark holder's rights. Fictional characters can be trademarked or otherwise protected. Many personal Web pages include unauthorized cartoon characters and scanned photographs of celebrities; often, these images are altered in some way. An electronic commerce Web site that uses an altered image of Mickey Mouse speaking in a modified voice is likely to hear from the Disney legal team.

Web sites that include links to other sites must be careful not to imply a relationship with the companies sponsoring the other sites unless such a relationship actually exists. For example, a Web design studio's Web page may include links to company Web sites that show good design principles. If those company Web sites were not created by the design studio, the studio must be very careful to state that fact. Otherwise, it would be easy for a visitor to assume that the linked sites were the work of the design studio.

In general, trademark protection prevents another firm from using the same or a similar name, logo, or other identifying characteristic in a way that would cause confusion in the minds of potential buyers of the trademark holder's products or services. For example, the trademarked name *Visa* is used by one company for its credit card services and another company for its type of synthetic fiber. This use is acceptable because the two products are very different. However, the use of very well-known trademarks can be protected for all products if there is a danger that the trademark might be diluted. Various state laws define **trademark dilution** as the reduction of the distinctive quality of a trademark by alternative uses. Trademarked names such as *Hyatt*, *Trivial Pursuit*, and *Tiffany*, and the shape of the Coca-Cola bottle have all been protected from dilution by court rulings. A Web site that sells gift-packaged seafood and claims to be the "Tiffany's of the Sea" risks a lawsuit from the famous jeweler claiming trademark dilution.

Advertising Regulation

In the United States, advertising is primarily regulated by the **Federal Trade Commission**. The FTC publishes regulations and investigates claims of false advertising. Its Web site, shown in Figure 11-8, includes helpful guidelines for businesses that want to comply with the law and avoid such claims.

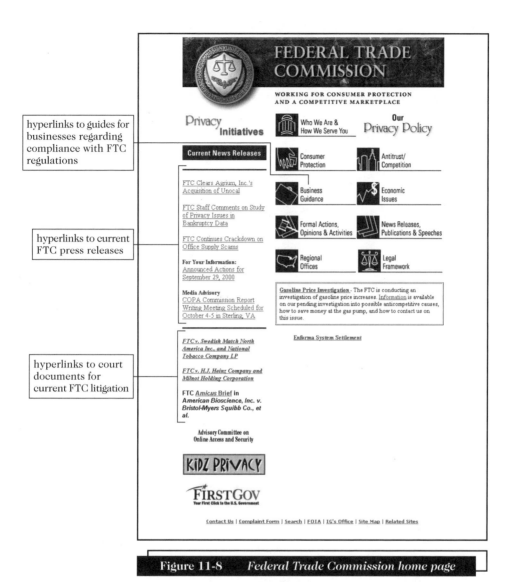

hyperlinks to guides for businesses regarding compliance with FTC regulations

hyperlinks to current FTC press releases

hyperlinks to court documents for current FTC litigation

Figure 11-8   *Federal Trade Commission home page*

Any advertising claim that can mislead a substantial number of consumers in a material way is illegal under U.S. law. In addition to conducting its own investigations, the FTC accepts referred investigations from organizations such as the Better Business Bureau. The FTC provides policy statements that can be helpful guides for designers creating electronic commerce Web sites. These policies include information on what is permitted in advertisements, and covers specific areas such as the following:

- Bait advertising
- Consumer lending and leasing
- Endorsements and testimonials
- Energy consumption statements for home appliances
- Guarantees and warranties
- Prices

Other federal agencies have the power to regulate online advertising in the United States. These agencies include the Food and Drug Administration (FDA), the Bureau of Alcohol, Tobacco, and Firearms (BATF), and the Department of Transportation (DOT). The FDA regulates information disclosures for food and drug products. In particular, any Web site that is planning to advertise pharmaceutical products will be subject to the FDA's drug labeling and advertising regulations. The BATF works with the FDA to monitor and enforce federal laws regarding advertising for alcoholic beverages and tobacco products. These laws require that every ad for such products includes statements that use very specific language. Many states also have laws that regulate advertising for alcoholic beverages and tobacco products. The state and federal laws governing advertising and the sale of firearms are even more restrictive. Any Web site that plans to deal in these products should consult with an attorney who is familiar with the relevant laws before posting any advertising for them online. The DOT works with the FTC to monitor the advertising of companies over which it has jurisdiction, such as bus lines, freight companies, and airlines.

ETHICS ISSUES

Companies using Web sites to conduct electronic commerce should adhere to the same ethical standards that other businesses follow. If they do not, they will suffer the same consequences that all companies suffer: the damaged reputation and long-term loss of trust that can result in loss of business. In general, advertising on the Web should include only true statements and should not omit any information that could mislead potential purchasers or wrongly influence their impressions of a product or service. Even true statements have been held to be misleading when the ad omits important related facts. Any comparisons to other products should be supported by verifiable information.

Ethical considerations are important in determining advertising policy on the Web. Recall from Chapter 8 that buyers on the Web often communicate with each other. Reports of an ethical lapse that is rapidly passed among customers can seriously affect a company's reputation. In 1999, *The New York Times* ran a story that disclosed Amazon.com's arrangements with publishers for book promotions. Amazon.com was accepting payments of up to $10,000 from publishers to give their books editorial reviews and placement on lists of recommended books as part of a cooperative advertising program. When this news broke, Amazon.com issued a statement that it had done nothing wrong and that such advertising programs were a standard part of publisher-bookstore relationships. The outcry on the Internet in newsgroups and mailing lists was overwhelming. Two days later—before most mass media outlets had even reported the story—Amazon.com announced that it would end the practice and offer unconditional refunds to any customers who had purchased a promoted book. Amazon.com had done nothing illegal, but the practice appeared to be unethical to many of its customers and potential customers.

In early 1999, eBay faced a similar ethical dilemma. Several newspapers had begun running stories about sales of illegal items, such as assault weapons and drugs, on the eBay auction site. At this point in time, eBay was listing about 250,000 items each day. Although eBay would investigate claims that illegal items were up for auction on its site, eBay did not actively screen or filter listings before the auctions were placed on the site.

Even though eBay was not legally obligated to screen the items auctioned, and even though the screening would be fairly expensive to do, the executive team decided that screening for illegal and copyright-infringing items would be in the best long-run interest of eBay. The team decided that such a decision would send a signal about the character of the company to its customers and the public in general. The eBay executive team also decided to remove an entire category, firearms, from the site. Not all of eBay's users were happy about this decision—the sale of firearms, when done properly, was legal on the site. However, the eBay executive team again decided that presenting an overall image of an open and honest marketplace was so important to the future success of eBay that it chose to ban all firearm sales.

Defamation

A **defamatory** statement is a statement that is false and that injures the reputation of another person or company. If the statement injures the reputation of a product or service instead of a person, it is called **product disparagement**. In some countries, even a true and honest comparison of products may give rise to product disparagement. Since the difference between justifiable criticism and defamation can be hard to determine, commercial Web sites should avoid making negative evaluative statements about other persons or products.

Web site designers should be especially careful to avoid potential defamation liability by altering a photo or image of a person in a way that depicts the person unfavorably. In most cases, a person must establish that the defamatory statement caused injury. However, most states recognize a legal cause of action, called per se **defamation**, in which a court deems some types of statements to be so negative that injury is assumed. For example, the court will hold inaccurate statements alleging conduct potentially injurious to a person's business, trade, profession, or office as defamatory per se—the complaining party need not prove injury to recover damages. Thus, online statements about competitors should always be carefully reviewed before posting to determine whether they contain any elements of defamation.

Privacy Rights and Obligations

The issue of online privacy, which you learned about in Chapter 5, is continuing to evolve as the Internet and the Web grow in importance as tools of communication and commerce. Many legal and privacy issues remain unsettled and are hotly debated in various forums. The **Electronic Communications Privacy Act of 1986** is the main law governing privacy on the Internet today. Of course, this law was enacted before the general public began its wide use of the Internet. A more recent law, the **Children's Online Privacy Protection Act** of 1998, provides restrictions on data collection that must be followed by electronic commerce sites aimed at children.

In recent years, a number of legislative proposals have been advanced that specifically address online privacy issues in general, but thus far none has withstood constitutional challenges. In July 1999, the FTC issued a report that examined how well Web sites were respecting visitor's privacy rights. Although it found a significant number of sites without posted privacy policies, the report concluded that companies operating Web sites were developing privacy practices with sufficient speed and that no federal laws regarding privacy were required at that time. Privacy advocacy groups responded to the FTC report with outrage and calls for legislation. Thus, the near-term future of privacy regulation in the United States is unclear. The Direct Marketing Association (DMA), a trade association of businesses that advertise their products and services directly to consumers by using mail, telephone, Internet, and mass media outlets, has established a set of privacy standards for its members. However, critics note that past efforts by the DMA to regulate its members' activities have been less than successful.

Ethics issues are significant in the area of online privacy because laws have not kept pace with the growth of the Internet and the Web. The nature and degree of personal information that Web sites can record when collecting information about visitors' page-viewing habits, product selections, and demographic information can threaten the privacy rights of those visitors. Differences in cultures throughout the world have resulted in different expectations about privacy in electronic commerce. In Europe, for example, most people expect that information they provide to a commercial Web site will be used only for the purpose for which it was collected. Many European countries have laws that prohibit companies from exchanging consumer data without the express consent of the consumer. In 1998, the European Union adopted a **Directive on the Protection of Personal Data**. This directive codifies the constitutional rights to privacy that exist in most European countries and applies them to all Internet activities. In addition, the directive prevents businesses from exporting personal data outside the European Union unless the data will continue to be protected in accordance with provisions of the directive.

Until the legal environment of privacy regulation becomes more clear, electronic commerce sites should be conservative in their collection and use of customer data. Mark Van Name and Bill Catchings, writing in *PC Week* in 1998, outlined four principles for handling customer data that provide a good outline for Web site administrators. These principles include the following:

- Use the data collected to provide improved customer service.
- Do not share customer data with others outside your company without the customer's permission.
- Tell customers what data you are collecting and what you are doing with it.
- Give customers the right to have you delete any of the data you have collected about them.

TAXATION AND ELECTRONIC COMMERCE

Companies that do business on the Web are subject to the same taxes as any other company. However, even the smallest Web businesses can become subject to taxes in many states and countries instantly because of the Internet's worldwide scope. Traditional businesses may operate in one location and be subject to only one set of tax laws for years. By the time those businesses are operating in multiple states or countries, they have developed the internal staff and record-keeping infrastructure needed to comply with multiple tax laws. Firms that engage in electronic commerce must comply with these multiple tax laws from their first day of existence.

A government acquires the power to tax a business when that business establishes a connection with the area controlled by the government. For example, a business that is located in Kansas has a connection with the state of Kansas and is subject to Kansas taxes. If that company opens a branch office in Arizona, it forms a connection with Arizona and becomes subject to Arizona taxes on the portion of its business that occurs in Arizona. This connection between a taxpaying entity and a government is called **nexus**. The concept of nexus is similar in many ways to the concept of personal jurisdiction discussed earlier in this chapter. The activities that create nexus vary from state to state. Nexus issues have been frequently litigated and the case law is fairly complex. Determining nexus can be difficult when a company conducts only a few activities in or has minimal contact with the state. In such cases, it is advisable for the company to obtain the services of a professional tax advisor.

An online business is potentially subject to several types of taxes, including income taxes, transaction taxes, and property taxes. Income taxes are levied by national, state, and local governments on the net income generated by business activities. Transaction taxes, which include sales taxes, use taxes, and customs duties, are levied on the products or services that the company sells or uses. Customs duties are taxes levied by the United States on certain commodities when they are imported into the country. Property taxes are levied by states and local governments on the personal property and real estate used in the business. In general, the taxes that cause the greatest concern for Web businesses are income taxes and sales taxes.

Income Taxes

The **Internal Revenue Service** (IRS) is the U.S. government agency charged with administering the country's tax laws. A basic principle of the U.S. tax system is that any increase in a company's wealth is subject to federal taxation. Thus, any company whose U.S.-based Web site generates income is subject to U.S. federal income tax. Further, a Web site maintained by a company in the United States must pay federal income tax on income generated outside of the United States. To reduce the incidence of double-taxation of foreign earnings, U.S. tax law provides a credit for taxes paid to foreign countries. The IRS Web site appears in Figure 11-9.

The IRS has been subject to criticism in recent years for its heavy-handed tactics and intimidating auditors. The agency is making attempts to improve its operations and its image with taxpayers. Its Web site is a good example of this effort. The site uses a friendly newspaper motif to help calm visitors who might be desperate to find a tax form or ruling as they prepare their tax returns at the last minute. The site includes links to downloadable tax forms, copies of tax regulations and IRS publications, and the Taxpayer Advocate Service.

Most states levy an income tax on business earnings. If a company conducts activities in several states, it must file tax returns in all of those states and apportion its earnings in accordance with each state's tax laws. In some states, the individual cities, counties, and other political subdivisions within the state also have the power to levy income taxes on business earnings. Companies that do business in multiple local jurisdictions must apportion their income and file tax returns in each locality that levies an income tax. The number of taxing authorities in the United States exceeds 30,000.

hyperlink to Taxpayer
Advocate Service

hyperlink to
downloadable tax
forms and IRS
publications

Figure 11-9 *U.S. Internal Revenue Service home page*

Companies that sell through their Web sites do not, in general, establish nexus everywhere their goods are delivered to customers. Usually, a company can accept orders and ship from one state to many other states and avoid nexus by using a contract carrier such as FedEx or United Parcel Service to deliver goods to customers.

Sales Taxes

Most states levy a sales tax on goods sold to consumers. Businesses that establish nexus with a state must file sales tax returns and remit the sales tax they collect from their customers. If a business ships goods to customers in other states, it is not required to collect sales tax from those customers unless the business has established nexus with the customer's state. However, the customer in this situation is required to file a use tax return and pay the amount that the business would have collected as sales tax if it had been a local business. Few consumers file use tax returns and few states enforce their use tax laws with regularity.

Larger businesses use complex software to manage their sales tax obligations. Not only are the sales tax rates different in the 7500 U.S. sales tax jurisdictions (which include states, counties, cities, and other sales tax authorities), but the rules about which items are taxable differ. For example, New York's sales tax law provides that large marshmallows are taxable (because they are "snacks") but small marshmallows are not taxable (because they are "food").

Some purchasers are exempt from sales tax, such as certain charitable organizations and businesses buying items for resale. Thus, to determine whether a particular item is subject to sales tax, a seller must know where the customer is located, what the laws of that jurisdiction say about taxability and tax rate, and the taxable status of the customer.

Summary

Businesses face many challenges posed by differences in language, culture, and infrastructure when conducting electronic commerce across international borders. Due to the speed and vast availability of online information, companies must quickly establish credibility with online customers in order to succeed in new and different markets.

Translation of Web pages by companies familiar with the culture of the target country helps to avoid some of the problems that can arise from doing business across international borders. Some countries require that companies doing business within their borders make their Web pages available in the local language.

Ethics issues can arise even when no laws have been broken. Since laws and ethics standards derive from local cultures, and local cultures vary significantly around the world, businesses must work hard to become aware of those cultural, ethical, and legal differences—especially those that exist in their target markets. Strategies and systems that would fail in the United States may be prerequisites for electronic commerce success in other countries.

Variations and inadequacies of the infrastructure that supports the Internet worldwide can make it challenging to conduct electronic commerce in certain countries. Where adequate infrastructure exists, ownership of the networks can impact visitors' ability to access the Internet.

The relationship between geography and culture is historically and legally intertwined. For most of our history, people have not been able to travel

413

great distances to learn about other cultures. The relationship between geographic boundaries and legal boundaries is based on four elements: power, effects, legitimacy, and notice.

As in traditional commerce, contracts are a part of doing business on the Web and are established through various types of offers and acceptances. Any contract for the electronic sale of goods or services includes implied warranties. Contracts can be invalidated when one of the parties to the transaction is an imposter; however, forged identities are becoming easier to detect through electronic security tools such as digital signatures.

Seemingly innocent inclusion of photographs, whether manipulated or not, and other elements of a Web page can lead to infringement of trademarks. Electronic commerce sites must be careful not to imply relationships that do not actually exist. Negative evaluative statements about entities, even when true, are best avoided given the subjective nature of defamation and product disparagement.

Collecting information and tracking consumer habits raises questions of ethics regarding online privacy. Some countries are far more restrictive than others in terms of what type of information collection is acceptable and legal.

Companies that conduct electronic commerce are subject to the same laws and taxes as other companies, but the nature of doing business on the Web can expose companies to a large number of laws and taxes sooner than traditional companies usually face them. Although some legal issues are straightforward, others are difficult to interpret and follow because of the newness of electronic commerce and the unsettled nature of applicable law. The large number of government agencies that have jurisdiction and the power to tax makes it essential that companies doing business on the Web understand the potential liabilities of doing business with customers in those jurisdictions.

Key Terms

Acceptance	Long-Arm Statute
Authority to Bind	Nexus
Consideration	Offer
Constructive Notice	Per Se Defamation
Contract	Personal Jurisdiction
Culture	Product Disparagement
Defamatory	Signature
Flat-Rate Access	Statute of Frauds
Forum Selection Clause	Subject-Matter Jurisdiction
Implied Contract	Tort
Jurisdiction	Trademark Dilution
Legitimacy	Warranty Disclaimer
Localization	Writing

Review Questions

1. Explain the difference between language translation and language localization in fewer than 200 words.
2. In a paragraph, describe the advantages of a flat-rate telecommunications access system for countries that want to encourage electronic commerce.
3. What is the difference between subject-matter jurisdiction and personal jurisdiction? Keep your explanation under 300 words.

4. Define product disparagement. Describe a situation that would be an example of product disparagement. Limit your answer to two paragraphs.

5. In 300 words or fewer, explain nexus. Why is it an important concept in state taxation?

Exercises

1. Use the **AltaVista Translation** Web site to translate the following business messages from English to one of the foreign languages available on that site. Translate each message back into English. Write a short memo that summarizes the problems you think an electronic commerce Web site that uses translation software might experience. Translate the following messages:

   - The flight has been delayed for several hours and your shipment of components will not arrive as scheduled.
   - We would be happy to bid on your proposal; however, we will need the drawings of subassembly #24 and the supervising mechanical engineer's quality control report by next Thursday.
   - Our company offers the latest and greatest hot deals on wheels. We would love to send you a brochure that explains why our brakes, wheels, and suspension components will do the job for you effectively and economically.

2. Use **Northern Light** or your favorite Web search engine to obtain a list of Web pages that include the word "warranty." Visit the Web pages on the search results list until you find a page that includes a warranty or guarantee statement for products that the site is offering for sale. Print the page and turn it in with your answers to the following questions:

   - Is the warranty statement on a separate page?
   - Could you read the entire warranty statement without scrolling your browser window?
   - Does the statement include a warranty disclaimer? If so, is the disclaimer conspicuous?
   - Does the statement make any explicit claims about the suitability of the product for specific purposes?
   - Does the statement deny warranty coverage if the product is used in specific ways?

3. Visit the Better Business Bureau's **BBBOnLine** Web site and the **TRUSTe** Web site. Examine each site to determine what types of privacy policies a member company must have to qualify for these two programs. Evaluate the two programs from the standpoint of a consumer who is interested in privacy protection—that is, determine which privacy program you would prefer to see in a Web site with which you are doing business. Summarize your findings in a memo of about 200 words.

For Further Study and Research

Alsop, S. 1999. "Copyright Protection is for Dinosaurs," *Fortune*, 139(8), April 26, 399–400.

Angwin, J. 2000. "Credit Card Scams Bedevil E-Stores," *The Wall Street Journal*, September 19, B1.

Ardito, S., P. Eiblum, and R. Daulong. 1999. "Realistic Approaches to Enigmatic Copyright Issues," *Online*, 23(3), May–June, 91–95.

Balkin, R. 1999. "AltaVista's Automatic Translation Program," *Database*, 22(2), April–May, 56–57.

Beams, C. 1999. "The Copyright Dilemma Involving Online Service Providers: Problem Solved...for Now," *Federal Communications Law Journal*, 51(3), June, 823–847.

Beckman, D. and D. Hirsch. 2000. "Web Worries of Dot-Com Lawyers," *ABA Journal*, June, 82.

Betts, M., C. Sliwa, and J. DiSabatino. 2000. "Global Web Sites Prove Challenging," *Computerworld*, 34(34), August 21, 17.

Bingi, P., A. Mir, and J. Khamalah. 2000. "The Challenges Facing Global E-Commerce," *Information Systems Management*, 17(4), Fall, 26–34.

Bond, R. and C. Whiteley. 1998. "Untangling the Web: A Review of Certain Secure E-Commerce Legal Issues," *International Review of Law, Computers & Technology*, 12(2), July, 349–370.

Boyle, M., J. Peterson, W. Sample, T. Schottenstein, and G. Sprague. 1999. "The Emerging International Tax Environment for Electronic Commerce," *Tax Management International Journal*, 28(6), June 11, 357–382.

Brandweek. 2000. "Trends in Trademarks 2000," 41(24), June 12, 68–70.

Brilmayer, L. 1989. "Consent, Contract, and Territory," *Minnesota Law Review*, 74(1), 11–12.

Brown, W. 2000. "Modern Technology Speaks with a Global Tongue; and It Certainly Isn't French," *The Daily Telegraph*, February 2, 10.

Carlo-Casellas, J. 1999. "Translating Policies into Spanish Requires Care," *National Underwriter*, 103(1), January 4, 21, 26.

Carney, D. 2000. "E-Exchanges May Keep Trustbusters Busy," *Business Week*, May 1, 52.

Chissick, M. and A. Kelman. 1998. *E-Commerce*. London: Sweet & Maxwell.

Clark, P. 2000. "E-Hub Will Court Power of Attorneys," *B to B*, 85(13), August 28, 3.

Clausing, J. 1999. "Study Says Most Children's Web Sites Are Lax on Privacy," *The New York Times*, July 20. Available online at: (http://www.nytimes.com/library/tech/99/07/cyber/articles/20privacy-day.html).

Corgel, J. 2000. "International Business: E-Business in Europe," *Vital Speeches of the Day*, 66(20), August 1, 637–640.

Davenport, T. 2000. "E-Commerce Goes Global," *CIO*, 13(20), August 1, 52–54.

Dempsey, G. and R. Sussman. 1999. "A Hands-On Guide for Multilingual Web Sites," *Communication World*, 16(6), June–July, 45–47.

DePalma, A. 2000. "Getting There Is Challenge for Latin America E-Tailing," *The New York Times*, August 17, 4.

Digital Millennium Copyright Act. 1998. Public Law No. 105–304, 112 Statutes 2860.

DiSabatino, J. 2000. "Globalization," *Computerworld*, 34(28), July 10, 46.

Doernberg, R. and L. Hinnekens. 1998. *Electronic Commerce and International Taxation*. Cambridge, MA: Kluwer Law International.

Echikson, W., C. Matlack, and D. Vannier. 2000. "American E-Tailers Take Europe by Storm," *Business Week*, August 7, 54–56.

The Economist. 2000. "Business Ethics: Doing Well by Doing Good," 355(8167), April 22, 65–67.

Einhorn, B., A. Webb, and P. Engardio. 2000. "China's Tangled Web: Will Beijing Ruin the Net by Trying to Control It?" *Business Week*, July 17, 28–30.

Emond, M. 2000. "Preparing for the E-Planet," *E-com*, 2(4), Available online at: (http://www.e-commag.com/printresources/v2n4/v2n4037.htm).

Federal Trade Commission (FTC). 1999. *Self-Regulation and Privacy Online: A Report to Congress*. Washington: FTC.

Fitzloff, E. and E. Schwartz. 1999. "Internet Sites to Leap Across Linguistic Gap," *InfoWorld*, 21(6), February 8, 1, 28.

Flynn, L. 2000. "Whose Name Is It Anyway? Arbitration Panels Favoring Trademark Holders in Disputes Over Web Names," *The New York Times*, September 4, C3.

Friedman, M. 1999. "Photographer Fights Quebec Language Law," *Computing Canada*, 25(24), June 18 1, 4.

Gantz, J. 1999. "E-Commerce: Here's What You Need to Know," *Computerworld*, 33(17), April 26, 34.

Gleckman, H. 2000. "The Tempest Over Taxes: The 'Too Complex' Excuse Won't Last for Long," *Business Week*, February 7, EB32–EB33.

416

Gleckman, H. and D. Carney. 2000. "Watching Over the World Wide Web: The Internet and the Rise of Globalization Are Creating New Pressure to Develop a Commercial Code that's Recognized from Kuala Lumpur to Kansas City," *Business Week*, August 28, 195–196.

Goldstein, E. 1999. *The Internet in the Mideast and North Africa: Free Expression and Censorship*. Washington: Human Rights Watch.

Gurley, W. 2000. "Like It or Not, Every Startup Is Now Global," *Fortune*, June 26, 324.

Guttman, R. 2000. "Erkki Liikanen: European Commissioner for Enterprise and the Information Society," *Europe*, May, 11–13.

Haddock, F. 2000. "European E-volution," *Global Finance*, 14(4), April, 39–40.

Hance, O. 1997. *Business and Law on the Internet*. Translated from French by S. Balz. New York: McGraw-Hill.

Hardesty, D. 1999. *Electronic Commerce Taxation and Planning*. Boston: Warren, Gorham & Lamont.

Harrington, L. 2000. "Point-and-Click to Anyplace on Earth," *Transportation & Distribution*, 41(8), August, 93–96.

Harvard Law Review. 1999. "The Criminalization of Copyright Infringement in the Digital Era," 112(7), May, 1705–1722.

Heckman, J. 2000. "Trademarks Protected Through New Cyber Act," *Marketing News*, 34(1), January 3, 6–7.

Heilemann, J. 2000. "David Boies: The Wired Interview," *Wired*, October. Available online at: (http://www.wired.com/wired/archive/8.10/boies.html).

Hong, V. 2000. "'Brussels 1' Angers EC Businesses," *The Industry Standard*, December 1. Available online at: (http://www.thestandard.com/article/display/0,1151,20531,00.html).

Hurt, E. 2000. "FTC Wins Internet's Respect," *Business 2.0*, October 13. Available online at: (http://www.business2.com/content/channels/technology/2000/10/13/21123).

Janal, D. 1999. "Thirty Essential Steps to Take Right Now to Prevent Online Crime," *Communication World*, 16(4), March, 34–36.

Kahin, B. and C. Nesson (eds.). 1997. *Borders in Cyberspace*. Cambridge, MA: MIT Press.

Keeler, D. 2000. "Taxation Slips Through the Net," *Global Finance*, 14(6), June, 60–61.

King, J. 1999. "Idiom App Speaks Your Language," *Computerworld*, 33(22), May 31, 66.

Lapres, D. 2000. "Legal Dos and Don'ts of Web Use in China," *China Business Review*, 27(2), March–April, 26–28.

Lawson, J. 1999. *The Complete Internet Handbook for Lawyers*. Chicago: American Bar Association.

Le Seac'h, M. and A. Klotz. 1999. "Corporate Translating: Handle with Care," *Business and Economic Review*, 45(2), January–March, 12–14.

Lessig, L. 2000. *Code and Other Laws of Cyberspace*. New York: Basic Books.

Loro, L. 1999. "Marketers Look to Stave Off Net Taxes," *Advertising Age's Business Marketing*, 84(4), April, 1, 31.

Los Angeles Times. 1999. "Hey, Web, Wake Up," May 14, B6.

McCarthy, B. 2000. "All E-Business Is Global," *Informationweek*, June 5, 204.

McClintock, M., N. Maguire, J. Kilby, and D. Barlow. 2000. "Electronic Commerce," *International Tax Review*, July–August, 9–13.

McCune, J. 1999. "English Written Here," *Management Review*, 88(2), February, 12.

Meller, P. 2000. "Europe Passes Stiff E-Commerce Law," *The Industry Standard*, December 1. Available online at: (http://www.thestandard.com/article/display/0,1151,20526,00.html).

Messmer, E. 1999. "Teaching the Web to Speak to Everyone," *Network World*, 16(2), May 24, 29–30.

Morrow, J. 2000. "Study: Fraud No Threat to E-Commerce," *E-Commerce Times*, September 20. Available online at: (http://www.ecommercetimes.com/news/articles2000/000920-1.shtml).

Moschella, D. 1999. "Consumers Being Forgotten in the Copyright Debate," *Computerworld*, 33(25), June 21, 35.

Murray, J. 2000. "E-Contracts Present Courts with Special Legal Challenges," *Purchasing*, 129(3), August 24, 119–120.

Oliva, R. and S. Prabakar. 1999. "Copyright Perils Can Lurk on the Business Web," *Marketing Management*, 8(1), Spring, 54–57.

417

Pantazis, A. 1999. "Zeran v. America Online, Inc.: Insulating Internet Service Providers from Defamation Liability," *Wake Forest Law Review*, 34(2), Summer, 531–555.

Porter, K. and S. Bradley. 1999. *eBay, Inc.* Case #9-700-007. Cambridge, MA: Harvard Business School.

Posch, R. 1999. "What Is Fair Use?" *Direct Marketing*, 62(1), May, 26–28.

Reagle, J. 1999. "The Platform for Privacy Preferences," *Communications of the ACM*, 42(2), February, 48–51.

Retsky, M. 1999. "Protect Your Property Intellectually," *Marketing News*, 33(13), June 21, 10–11.

Rich, J. 2000. "Latin America Is a Difficult—But Attractive—Region for E-Marketplaces that Are Trying to Enlist Small Businesses," *The New York Times*, September 25, C4.

Roberts, B. 2000. "Ready, Fire, Aim," *Electronic Business*, 26(7), July, 80–88.

Samborn, H. 1999. "Small World, Big Questions," *ABA Journal*, 85(2), February, 78.

Samuelson, P. 1999. "Good News and Bad News on the Intellectual Property Front," *Communications of the ACM*, 42(3), March, 19–24.

Schwartau, W. 2000. "Safe Passage," *Network World*, 17(9), February 28, 95–97.

Sclafane, S. 1998. "Web Sites Create World Wide Exposures," *National Underwriter*, 102(6), February 9, 14, 31.

Segal, J. 2000. "Cybertraps for HR," *HRMagazine*, 45(6), June, 217–231.

Shaller, D. 2000. "E-mail, the Internet, and Other Legal and Ethical Nightmares," *Strategic Finance*, August, 82(2), 48–52.

Shannon, P. 2000. "Including Language in your Global Strategy for B2B E-commerce," *World Trade*, 13(9), September, 66–68.

Shari, M. 2000. "Cutting Red Tape in Singapore," *Business Week*, September 18, 92.

Skipton, C. 1999. "Think Globally, Act Locally," *New Media*, 9(6), June, 58.

Sliwa, C. 2000. "Boo.com Makeover Draws Skeptical Reactions at Unveiling," *Computerworld*, 34(32), August 7, 6.

Smedinghoff, T. (ed.). 1996. *Online Law: The SPA's Legal Guide to Doing Business on the Internet*. Reading, MA: Addison-Wesley Developers Press.

Steinberg, J. 2000. "An Unusual Commute Leads to a Different Kind of Partnership," *The New York Times*, September 20, H4.

Swire, P. and R. Litan. 1998. *None of your Business: World Data Flows, Electronic Commerce, and the European Privacy Directive*. Washington: Brookings Institution Press.

Thornton, J. 2000. "Should Drugs Ads Be Legal?" *Marketing*, May 4, 28–29.

Towle, H. 2000. "No Guiding Light," *CIO*, 13(21), August 15, 72–74.

United Nations. 1970. "Declaration on Principles of International Law Concerning Friendly Relations and Cooperation Among States in Accordance with the Charter of the United Nations," *General Assembly Resolution*, #2625, 35th Session.

Van Name, M. and B. Catchings. 1998. "Practical Advice about Privacy and Customer Data," *PC Week*, 15(27), July 6, 38.

Vijayan, J. and K. Ohlson. 2000. "Standards Issue Mars E-Signatures," *Computerworld*, 34(28), July 10, 1, 16.

Walker, P. 2000. "Watch Out for the Web," *Credit Management*, March, 24–25.

Wallraff, B. 2000. "What Global Language?" *The Atlantic Monthly*, 286(5), 52–66.

Whitaker, B. 2000. "The Web Makes Going Global Easy, Until You Try to Do It," *The New York Times*, H20.

Wiley, L. 1999. "Proposed Revisions to European Copyright Laws Cause a Stir," *E Media Professional*, 12(4), April, 16–17.

Williams, J., J. Clark, C. Clark, and J. Noe. 1999. "What a Tangled Web: The Legal and Public Relations Dangers of Operating a Web Site," *Information Strategy*, 15(3), Spring, 6–12.

Winston, J. "Copyright on the Internet," *Target Marketing*, 22(6), June, 40.

Wright, B. 1995. *The Law of Electronic Commerce: EDI, E-Mail, and Internet: Technology, Proof, and Liability*. Boston: Little, Brown.

418

PLANNING FOR ELECTRONIC BUSINESS: RESOURCE AND IMPLEMENTATION ISSUES

INTRODUCTION

AlliedSignal was a diversified manufacturing and technology business selling products in the aerospace, automotive, chemicals, fibers, and plastics industries. The company had more than 70,000 employees and annual sales exceeding $15 billion. Although some of AlliedSignal's products used new technologies or helped other firms create new technologies, many of them were commodity items that were manufactured and sold just as they had been for decades. In 1999, AlliedSignal's CEO, Larry Bossidy, called together the heads of the company's business units for a one day conference. He invited Michael Dell, chairman and CEO of **Dell Computers**, and John Chambers, CEO of **Cisco Systems**, to speak about their companies' electronic commerce implementation successes.

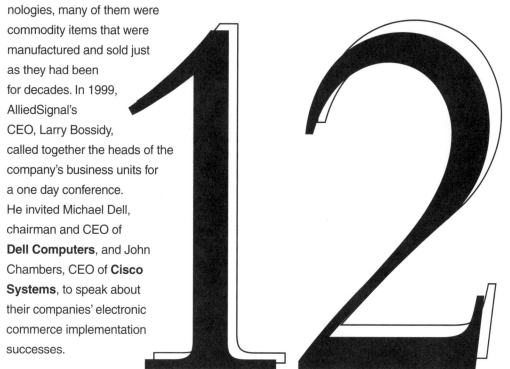

At the end of the day, Bossidy gave the business unit heads their marching orders. They were to take what they had learned and create a strategy for implementing electronic commerce in their business units—in two months. Bossidy told the room full of rather stunned managers that, although most of their business units were at or near the top of their industries, the Internet would change everything. He believed that the kinds of electronic commerce strategies that had worked so well for Dell and Cisco in the computer industry would also work in AlliedSignal's businesses. He wanted to make sure that AlliedSignal was the first to exploit those strategies and any other strategies that the business managers could devise. In two months, each manager reported back with a strategy that included multiple electronic commerce projects, such as Web sites for selling products, providing customer service, improving corporate infrastructure, managing supply chains, coordinating logistics, holding auctions, and creating virtual communities. These plans were evaluated in the company's annual strategic planning process, and the best ones were chosen for funding and immediate implementation. In a matter of months, one of the largest industrial enterprises in the world had drastically altered its course, setting sail for the uncharted waters of electronic commerce.

Late in 1999, AlliedSignal changed its name to Honeywell, after merging with that company. Further changes were in store for the business; it was purchased by **General Electric** in late 2000. General Electric was one of the first large manufacturing companies to embrace electronic commerce. It is likely that the moves toward electronic commerce initiated by AlliedSignal made that company more attractive to General Electric.

The ability of companies to plan, design, and implement cohesive electronic commerce strategies will make the difference between success and failure for the majority of them. The tremendous leverage that firms can gain on the Web by being the first to enter a market or do business a new way has caught the attention of top executives in many industries. The keys to successful implementation of any information technology project are planning and execution. This chapter will provide some useful guidelines for those who will manage the planning, implementation, and continuing operations of electronic commerce initiatives.

In this chapter, you will learn about:

- Identifying the value of electronic commerce initiatives
- Aligning implementation plans with strategies
- Deciding which electronic commerce project elements to outsource
- Selecting Web hosting services
- Using incubators and fast venturing techniques to launch Internet business initiatives
- Using formal project management techniques to plan and control electronic commerce activities
- Staffing electronic commerce activities

PLANNING THE ELECTRONIC COMMERCE PROJECT

A successful business plan for an electronic commerce initiative should include activities that will:

- Identify the initiative's specific objectives
- Link those objectives to business strategies (identified in Chapters 8, 9, and 10)
- Manage the implementation of those business strategies
- Oversee the continuing operations of the initiative once it is launched

In setting the objectives for an electronic commerce initiative, managers should consider the strategic role of the project, its intended scope, and the resources available for executing it.

Identifying Objectives

Businesses undertake electronic commerce initiatives for a wide variety of reasons. Common objectives that a business might hope to accomplish through electronic commerce could include increasing sales in existing markets, opening new markets, serving existing customers better, identifying new vendors, coordinating more efficiently with existing vendors, or recruiting employees more effectively.

Resource decisions for electronic commerce initiatives should consider the expected benefits and expected costs of meeting the objectives. These decisions should also consider the risks inherent in the electronic commerce initiative and compare them to the risks of inaction—a failure to act could concede a strategic advantage to competitors.

Linking Objectives to Business Strategies

Businesses can use **downstream strategies**, which are tactics that improve the value that the business provides to its customers. Alternatively, businesses can pursue

421

upstream strategies that focus on reducing costs or generating value by working with suppliers or inbound logistics.

You have already learned about many of the things that companies are doing on the Web. Although the Web is a tremendously attractive sales channel for many firms, companies can use electronic commerce in a variety of ways to do much more than selling: They can use the Web to complement their business strategies and improve their competitive positions. As described in earlier chapters of this book, electronic commerce opportunities can inspire businesses to undertake activities such as:

- Building brands
- Enhancing existing marketing programs
- Selling products and services
- Selling advertising
- Improving after-sale service and support
- Purchasing products and services
- Managing supply chains
- Operating auctions
- Creating virtual communities and Web portals

Although the success of each of these activities is measurable to some degree, many companies have undertaken these activities on the Web without setting specific, measurable goals. In the mid-1990s—the early days of electronic commerce—businesses that had good ideas could start a business activity on the Web and not face competition. Successes and failures were measured in broad strokes. A company would either become the Amazon.com or the eBay of its industry or it would disappear—either slipping into bankruptcy or being acquired by another company.

As electronic commerce is now beginning to mature, more companies are taking a closer look at the benefits and costs of their electronic commerce projects. Measuring both benefits and costs is becoming more important. A good business plan will set specific objectives for benefits to be achieved and costs to be incurred. In many cases, a company will create a pilot Web site to test an electronic commerce idea, and then release a production version of the site when it works well. These companies must specify clear goals for the pilot test so that they know when the site is ready to scale up.

Measuring Benefit Objectives

Many companies create Web sites to build their brands or enhance existing marketing programs. These companies can set goals in terms of increased brand awareness, as measured by market research surveys and opinion polls. Companies that sell goods or services on their sites can measure sales volume in units or dollars. A complication that occurs in measuring either brand awareness or sales is that the increases can be caused by other things that the company is doing at the same time or by a general improvement in the economy. A good marketing staff or outside consulting firm can help a company sort out the specific causes and effects of marketing and sales programs. Firms may need these groups to help set and evaluate these kinds of goals for electronic commerce initiatives.

Companies that want to use their Web sites to improve customer service or after-sale support might set goals of increased customer satisfaction or reduced costs of providing customer service or support. For example, **Philips Lighting** wanted to use the Web to provide an ordering system for its smaller customers that did not use EDI. The primary goal for this initiative was to reduce the cost of processing smaller orders. Philips had identified that over half the cost of processing smaller orders was handling inventory availability and order status requests. Customers who placed small orders often called or sent faxes asking for this information. In 1999, Philips built a pilot Web site and invited a number of its smaller customers to try it. The company found that customer service phone calls from the test group of customers dropped by 80%. Based on that measurable increase in efficiency, Philips decided to invest in additional hardware and personnel to staff a version of the Web site that could handle virtually all of its smaller customers. The reduction in the cost of handling small orders justified the additional investment.

Companies can use a variety of similar measurements to assess the benefits of other electronic commerce initiatives. Supply chain managers can measure supply cost reductions, quality improvements, or faster deliveries of ordered goods. Auction sites can set goals for the number of auctions, the number of bidders and sellers, the dollar volume of items sold, the number of items sold, or the number of registered participants. The ability to track such numbers is usually built into auction site software. Virtual communities and Web portals measure the number of visitors and try to measure the quality of their visitors' experiences. Some sites use online surveys to gather these data; however, most settle for approximations provided by measuring the length of time that each visitor remains on the site and how often visitors return. A summary of benefits and measurements that companies can make to assess the value of those benefits appears in Figure 12-1.

Electronic commerce initiatives	Common measurements of benefits provided
Build brands	Surveys or opinion polls that measure brand awareness
Enhance existing marketing programs	Change in per-unit sales volume
Improve customer service	Customer satisfaction surveys, the number of customer complaints
Reduce cost of after-sale support	Quantity and type (telephone, fax, e-mail) of support activities
Improve supply chain operation	Cost, quality, and on-time delivery of materials or services purchased
Hold auctions	Quantity of auctions, bidders, sellers, items sold, registered participants; dollar volume of items sold
Provide portals and virtual communities	Number of visitors, number of return visits per visitor, and duration of average visit

Figure 12-1 *Measuring the benefits of electronic commerce initiatives*

Planning for Electronic Business: Resource and Implementation Issues

No matter how a company measures the benefits provided by its Web site, it usually tries to convert the raw activity measurements to dollars. Having the benefits measured in dollars lets the company compare benefits to costs and compare the net benefit (benefits minus costs) of a particular initiative to the net benefits provided by other projects. Although each activity provides some value to the company, it is often difficult to measure that value in dollars. Usually, even the best attempts to convert benefits to dollars yield only rough approximations.

Measuring Cost Objectives

At first glance, the task of identifying and estimating costs may seem much easier than the task of setting benefits objectives. However, many managers have found that information technology project costs can be as difficult to estimate and control as the benefits of those projects. Since Web development uses relatively new hardware and software technologies, managers have little experience on which they can draw to make estimates. Most changes in the cost of hardware are downward, but the increasing sophistication of software provides an ever-increasing demand for more of the newer, cheaper hardware. This often yields a net increase in overall hardware costs. Even though electronic commerce initiatives tend to be completed within a shorter time frame than many other information technology projects, the rapid changes in Web technology can destroy a manager's best-laid plans very quickly.

In addition to hardware and software costs, the project budget must include the costs of hiring, training, and paying the personnel who will design the Web site, write or customize the software, create the content, and operate and maintain the site. As more companies build electronic commerce sites, people who have the skills necessary to do the work are commanding increasingly higher compensation.

Based on data collected in separate recent surveys, International Data Corporation and the GartnerGroup both estimated that the cost for a large company to build and implement an adequate entry-level electronic commerce site was about $1 million. About 79% of this cost was labor-related; 10% was the cost of software and 11% was the cost of hardware. GartnerGroup added that it would take between $2 million and $5 million to build a site that would compare favorably to leading sites. International Data Corporation noted that 10 of the top 100 electronic commerce sites had spent over $10 million for development and implementation.

A 1999 *Advertising Age* survey of smaller companies showed that they were spending an average of $78,000 to build new electronic commerce sites, an increase of 75% over the previous year's average. Experts expect the cost of entry-level Web sites to continue to increase as more companies establish themselves on the Web and as expensive features such as shopping carts and search engines become standard on even the most basic sites. The GartnerGroup estimates that establishing a basic electronic commerce operation on the Web today will cost a company between $300,000 and $1 million. Figure 12-2 summarizes the estimated cost of creating a Web business at three different levels: a basic entry level, a level comparable to most existing Web competitors, and a level that makes the Web site stand out as truly different from competitors' sites.

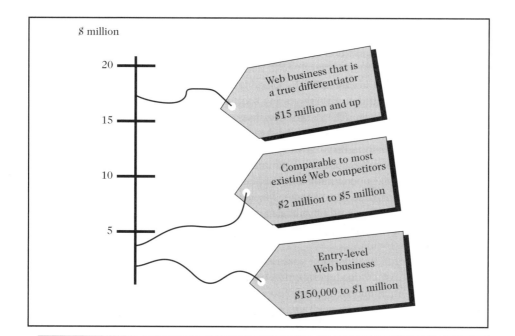

Figure 12-2 *Starting a Web business: three price tags*

The initial cost of building an electronic commerce site is not the whole story, unfortunately. Since Web technology continues to evolve at a rapid pace, most businesses will want to take advantage of what that technology offers to remain competitive. Most experts agree that the annual cost to maintain and improve a site once it is up and running—whether it is a small site or a large site—will be between 50% and 100% of its initial cost.

As an increasing number of traditional businesses create Web versions of their physical stores, the cost to build an online business that is a true differentiator—a site that stands out and offers something new to customers—will continue to increase. Much of the cost in such a Web site is for elements that make a major difference in how well the site works, but are not readily apparent to a site visitor. For example, Kmart's Web business site, **BlueLight.com**, cost more than $140 million to create. The site's home page, shown in Figure 12-3, is certainly well-designed and highly functional, but the typical visitor would never guess how much this site cost to build. Much of the site's cost is hidden—it was incurred to build connections to Kmart's vast inventory and logistics databases.

425

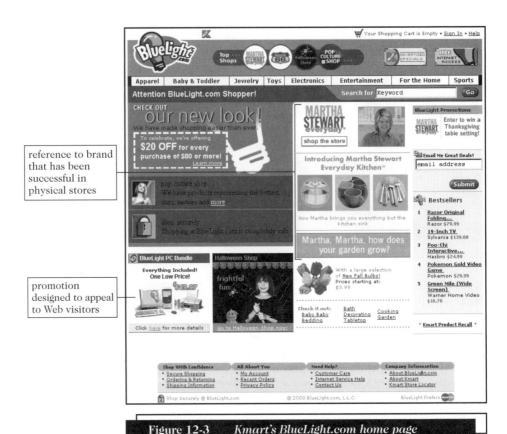

reference to brand that has been successful in physical stores

promotion designed to appeal to Web visitors

Figure 12-3   Kmart's BlueLight.com home page

Comparing Benefits to Costs

Most companies have procedures that call for an evaluation of any major expenditure of funds. These major investments in equipment, personnel, and other assets are called **capital projects** or **capital investments**. The techniques that companies use to evaluate proposed capital projects range from very simple calculations to complex computer simulation models. However, no matter how complex the technique, it always reduces to a comparison of benefits and costs. If the benefits exceed the cost of a project by a comfortable margin, the company invests in the project.

A key part of creating a business plan for electronic commerce initiatives is the process of identifying potential benefits (including intangibles such as employee satisfaction and company reputation), identifying the costs required to generate those benefits, and evaluating whether the benefits exceed the costs. Companies should evaluate each element of their electronic commerce strategies using this cost/benefit approach. A simplified representation of the cost/benefit approach appears in Figure 12-4.

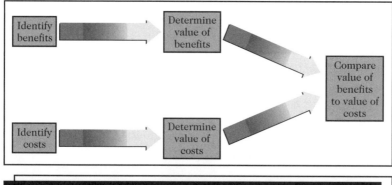

Figure 12-4 *Cost/benefit evaluation of electronic commerce strategy elements*

You may have learned techniques for capital project evaluation, such as the payback method or the net present value method, in your accounting or finance courses. These evaluation approaches provide a quantitative expression of a comfortable benefit-to-cost margin for a specific company. They can also mathematically adjust for the reduced value of benefits that the investment will return in future years (benefits received in future years are worth less than those received in the current year). Managers often use the term **return on investment** (ROI) to describe any capital investment evaluation technique, even though ROI is the name of only one of these techniques.

Although most companies evaluate the anticipated value of electronic commerce initiatives in some way before approving them, many companies see these projects as absolutely necessary investments. Thus, they might not subject them to the same close examination and rigid requirements as they do other capital projects. These companies fear being left behind as competitors stake their claims in the online marketspace. The value of early positioning in a new market is so great that many companies are willing to invest very large amounts of money with no near-term prospects of profit.

Newspaper Web sites are a very good example of this desire to establish a foothold in the online marketspace. Profitable electronic commerce initiatives in the newspaper business, such as Gannet's **USA Today** and Dow Jones' **Wall Street Journal Interactive Edition** sites, are few. *Editor & Publisher* magazine estimated that newspaper Web sites lost a total of $80 million in 1998 alone. Despite the losses, most newspaper companies believe that they cannot afford to ignore the long-term potential of the Web and feel compelled to make whatever investment is required to move into the online world.

Strategies for Web Site Development

When companies began establishing their presences on the Web, the typical Web site was a static brochure that was not updated frequently with new information and seldom had any capabilities for helping the company's customers or vendors transact business. As Web sites have become the home not only of transaction processing but also of automated business processes of all kinds, these Web sites have become important parts of companies' information systems infrastructures. The evolution of Web site functions—from the static brochures of the early days of electronic commerce, to transaction processing tools, to today's automated homes for business processes of all kinds—appears in Figure 12-5.

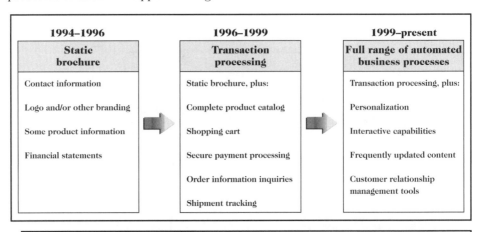

1994–1996	1996–1999	1999–present
Static brochure	**Transaction processing**	**Full range of automated business processes**
Contact information	Static brochure, plus:	Transaction processing, plus:
Logo and/or other branding	Complete product catalog	Personalization
Some product information	Shopping cart	Interactive capabilities
Financial statements	Secure payment processing	Frequently updated content
	Order information inquiries	Customer relationship management tools
	Shipment tracking	

Figure 12-5   *Increasing complexity of Web site functions*

This transformation occurred rapidly—taking only a year or two in most companies—and very few businesses have caught up with the changes in terms of how they develop and manage Web sites. Thus, the purposes and scope of Web sites have increased greatly, but few businesses today manage them as the dynamic business applications they have become. The tools that companies have developed over the years to manage software development projects are designed to help those companies meet the needs of their current customers and operate more effectively within existing value chains.

Many large and medium-sized companies have found it extremely difficult to develop new information systems and Web sites that work with such systems to create new markets or reconfigure their supply chains. In the past, companies that have had success in exploring new ways of working with their customers and suppliers by reconfiguring supply chains have had the luxury of time—in many cases, years—to complete those reconfigurations. However, the speed at which the Internet has changed markets and marketing channels throughout entire industry value chains precludes lengthy reconfigurations. Now companies that want to successfully adapt to the changed business environment of the information age must explore alternatives to traditional systems development methods.

Internal Development vs. Outsourcing

Although many companies would like to think that they can avoid electronic commerce site development problems by outsourcing the entire project, savvy leaders realize that they cannot. No matter what kind of electronic commerce initiative a company is contemplating, the initiative's success depends on how well it is integrated into and supports the activities in which the business is already engaged. However, few companies are large enough or have sufficient in-house expertise to launch an electronic commerce project without some external help. Even Wal-Mart, with annual sales of more than $150 billion, did not undertake its 2000 Web site relaunch alone. The key to success is finding the right balance between outside and inside support for the project. Hiring another company to provide the outside support for the project is called **outsourcing**.

The Internal Team

The first step in determining which parts of an electronic commerce project to outsource is to create an internal team that is responsible for the project. This team should include people with enough knowledge about the Internet and its technologies to know what kinds of things are possible. Team members should be creative thinkers who are interested in taking the company beyond its current boundaries, and they should be people who have distinguished themselves in some way by doing something very well for the company. If they are not already recognized by their peers as successful individuals, the project may suffer from lack of credibility.

Some companies make the mistake of appointing as electronic commerce project leader a technical wizard who does not know much about the business and is not well-known throughout the company. Such a choice can greatly increase the likelihood of failure. Business knowledge, creativity, and the respect of the firm's line managers are all much more important than technical expertise in establishing successful electronic commerce.

Measuring the achievements of this internal team is very important. The measurements do not have to be monetary. Achievement can be expressed in whatever terms are appropriate to the objectives of the initiative. Customer satisfaction, number of sales leads generated, and reductions in order-processing time are examples of metrics that can provide a sense of the team's level of accomplishment. The measurements should show how the project is affecting the company's ability to provide value to the consumer. John Stoiber of the consulting firm **Value Technology** advises that companies set aside between 5% and 10% of a project's budget for quantifying the project's value and measuring the achievement of that value.

Increasingly, companies are recognizing the value of the intellectual capital they have built up in the form of employees' knowledge about the business and its

429

processes. In the past, many companies ignored the value of their human assets because they do not appear in the accounting records or financial statements. Leif Edvinsson has pioneered the use of human capital measures at Skandia Group, a large financial services company in Sweden. In addition to acknowledging employees' competencies, Edvinsson's measures include the value of customer loyalty and business partnerships as part of a company's intellectual capital. This networking approach to evaluating intellectual capital shows promise as a tool for assessing and tracking the value of internal teams and their connections to external consultants. Although these measurements are just now being adapted for use in measuring systems development efforts, we have included references to books by Edvinsson and Max Boisot, another proponent of human capital measurement, in the "For Further Study and Research" section at the end of this chapter.

The internal team should hold ultimate and complete responsibility for the electronic commerce initiative, from the setting of objectives to the final implementation and operation of the site. The internal team will decide which parts of the project to outsource (and to whom those parts will be outsourced) and what consultants or partners the company will need to hire for the project. Consultants, outsourcing providers, and partners can be very important early in the project because they often develop skills and expertise in new technologies before most information systems professionals do.

Early Outsourcing

In many electronic commerce projects, the company outsources the initial site design and development to launch the project quickly. The outsourcing team then trains the company's information systems professionals in the new technology before handing the operation of the site over to them. This approach is called **early outsourcing**. Since operating an electronic commerce site can rapidly become a source of competitive advantage for a company, it is best to have the company's own information systems people work closely with the outsourcing team and develop ideas for improvements as early as possible in the life of the project.

Late Outsourcing

In the more traditional approach to information systems outsourcing, the company's information systems professionals do the initial design and development work, implement the system, and operate the system until it becomes a stable part of the business operation. Once the company has gained all the competitive advantage provided by the system, the maintenance of the electronic commerce system can be outsourced so that the company's information systems professionals can turn their attention and talents to developing new technologies that will provide further competitive advantage. This approach is called **late outsourcing**. Although for years late outsourcing has been the standard for allocating scarce information systems talent to projects, electronic commerce initiatives lend themselves more to the early outsourcing approach.

Partial Outsourcing

In both the early outsourcing and late outsourcing approaches, a single group is responsible for the entire design, development, and operation of a project group—either inside or outside the company. This typical outsourcing pattern works well for many information systems projects. However, electronic commerce initiatives can benefit from a partial outsourcing approach, too. In **partial outsourcing**, which is also called **component outsourcing**, the company identifies specific portions of the project that can be completely designed, developed, implemented, and operated by another firm that specializes in a particular function.

Many smaller Web sites outsource their e-mail handling and response function. Customers expect rapid and accurate responses to any e-mail inquiry they make of a Web site with which they are doing business. Many companies like to send an automatic order confirmation via e-mail as soon as the order or credit card payment is accepted. A number of companies provide e-mail auto-response functions on an outsourcing basis.

Another common example of partial outsourcing is an electronic payment system. As you learned in Chapter 7, many vendors are willing to provide complete customer payment processing. These vendors provide a site that takes over when customers are ready to pay and returns the customers to the original site after processing the payment transaction.

One of the most common elements of electronic commerce initiatives that companies outsource using this approach is the Web hosting activity that you learned about in Chapter 4. Internet service providers (ISPs) offer Web hosting services to companies that want to operate electronic commerce sites but that do not want to invest in the hardware and staff needed to create their own Web servers. ISPs are usually willing to accommodate requests for a variety of service levels. Small businesses can rent space on an existing server at the ISP's location. Larger companies can purchase the server hardware and have the ISP install and maintain it at the ISP's location. The ISP provides the continuous staffing and expertise needed to keep an electronic commerce site up and running 24 hours a day, seven days a week (this kind of service is often called **24/7 operation**). Most ISPs offer a wide range of services, including personal Web access for individuals. Some ISPs specialize in services to business. These larger ISPs cater to companies that want to operate electronic commerce sites. They usually offer wider bandwidth connections to the Internet than smaller ISPs and offer more reliable continuous service.

A number of ISPs and other firms offer services beyond basic Internet connectivity to companies that want to do business on the Web. Many of these services were described earlier as candidates for partial outsourcing strategies and include automated e-mail response, transaction processing, payment processing, security, customer service and support, order fulfillment, and product distribution. Recall from Chapter 6 that a company that offers these services can be called a commerce service provider (CSP) or, if the service uses specific application software (such as an automated e-mail response service), an application service provider (ASP).

431

Selecting a Hosting Service

The internal team should be responsible for selecting the ISP that will provide the site's hosting service. For smaller electronic commerce projects, teams can consult an ISP directory such as **The List**, which appears in Figure 12-6. The List site includes a good search engine that helps visitors choose an ISP, Web hosting service, or ASP that meets their needs from the thousands of listings on the site.

Figure 12-6 *The List ISP, Web hosting service, and ASP directory*

For larger Web site implementations, the team will want to obtain the advice of consultants or other firms that rate ISPs and CSPs, such as **Keynote Systems** and the *Directory of Internet Service Providers* published by ***Boardwatch Magazine***. The most important factors to evaluate when selecting a hosting service include:

- Functionality
- Reliability
- Bandwidth and server scalability
- Security
- Backup and disaster recovery
- Cost

Companies that sell hosting services provide different features and different levels of service. The functionality offered by a service provider can include credit card processing and the ability to link to existing databases that store customer and product information. Almost all hosting services offer site visitor tracking, but as you learned in Chapter 4, the capabilities of different tracking software packages are not all equal. Some tracking software provides much more detailed information and easier-to-use report generators than other tracking software. You should determine the functionality offered by a hosting service and carefully evaluate whether that functionality will be sufficient to meet the needs of your Web site.

The service should offer a guarantee that limits possible downtime. Electronic commerce buyers expect hosting services to be up and running 24 hours a day, every day. Of course, no hosting service can promise never to fail, but it can provide staffing and backup hardware that minimizes reliability problems. Coordination of this function with the service provider can be very important. Usually, a business must have some around-the-clock staff available or on-call to work with the service provider when an interruption occurs.

The bandwidth of the service's connection to the Internet must be sufficient to handle the peak transaction loads that its customers require. Sometimes a service provider will sign up new accounts faster than it can expand the bandwidth of its connections, resulting in access bottlenecks. A guarantee that specifies bandwidth availability or server response times is worth negotiating into a service provider contract. If you expect your site's traffic to increase rapidly, it is important that your service provider can increase rapidly the server capacity and the bandwidth provided. In general, larger hosting services can scale up more easily than smaller hosting services. Again, it is worth negotiating some scalability into the service provider contract in such situations.

433

Since the company's information on customers, products, pricing, and other data will be placed in the hands of the service provider, the vendor's security policies and practices are very important. You learned about electronic commerce security issues and practices in Chapters 5 and 6. The service provider should specify the types of security it provides and how it implements security. No matter what security guarantees the service provider offers, the company should monitor the security of the electronic commerce operation through its own personnel or by hiring a security consulting firm. Security consultants can periodically test the system and can launch attacks on the security features used by the service provider to determine whether they are easily breached.

The hosting service should be able to guarantee close to 100% reliability by having a workable disaster recovery plan in place. In addition to having off-site data backup or mirroring, the hosting service should have a way to restore your site very quickly in the case of a natural disaster.

Service providers offer many different pricing plans for different levels of service. Knowing what types of server hardware and software your site will require and having a good estimate of the range of transaction loads the site is likely to generate can help in negotiating a price for the hosting service.

New Methods for Implementing Partial Outsourcing

In the past five years, new ways of implementing the partial outsourcing strategy have evolved specifically for Web businesses. The next two sections describe two of the more popular of these methods.

Incubators

An **incubator** is a company that offers start-up companies a physical location with offices, accounting and legal assistance, computers, and Internet connections at a very low monthly cost. Sometimes, the incubator will offer seed money, management advice, and marketing assistance. In exchange, the incubator receives an ownership interest in the company, typically between 10% and 50%.

When the company grows to the point that it can obtain venture capital financing or launch a public offering of its stock, the incubator sells all or part of its interest and reinvests the money in a new incubator candidate. One of the first Internet incubators was **idealab!**, which helped companies such as **GoTo.com**, **etoys.com**, and **CarsDirect.com** get their start.

Some companies have created internal incubators. A number of companies used internal incubators in the past to develop technologies that the companies planned to use in their main business operations. Most of these programs, such as the Kodak internal venturing program of the 1980s, were unsuccessful and ultimately were shut down. Companies such as Kodak found it difficult to maintain an entrepreneurial spirit when everyone knew that the technology being developed would ultimately be taken away and controlled by the parent company. More recently, companies such as Matsushita Electric's U.S. Panasonic division have started internal incubators to help launch new companies that will grow to become important strategic partners. The companies launched in the incubator will retain their individual management teams and the assets they develop. The prospects for these strategic partner incubators appear to be much brighter than those of the old-style technology development incubators.

Fast Venturing

Often, large companies struggle to emulate the entrepreneurial spirit of smaller companies as they launch their Internet business initiatives. Many of these companies are trying to expand the internal incubator model and create an effective support system for new business and technology ideas, such as electronic commerce initiatives. One proposed approach that is becoming popular is called fast venturing.

434

In **fast venturing**, an existing company that wants to launch an electronic commerce initiative joins external equity partners and operational partners that can offer the experience and skills needed to develop and scale up the project very rapidly. Equity partners are usually banks or venture capitalists that sometimes offer money, but are more likely to offer experience gained from guiding other start-ups that they have funded. Operational partners are firms, such as systems integrators, consultants, and Web portals, that have experience in moving projects along and scaling up prototypes. The roles of each participant in fast venturing are described in Figure 12-7.

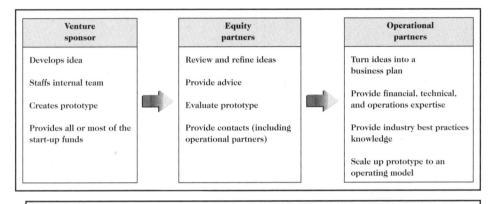

Figure 12-7   Fast venturing

The venture sponsor is the existing company that wants to launch the electronic commerce initiative. The equity partners are entities that have provided start-up money to new ventures in the past and have developed knowledge about operating new ventures. The equity partners provide advice based on this knowledge to the venture sponsor, which typically has little experience in developing new ventures. The operational partners are people and companies that have built Web business sites and provide expertise in the technologies and business practices needed to create a successful operating electronic commerce site.

MANAGING ELECTRONIC COMMERCE IMPLEMENTATIONS

The best way to manage any complex business software implementation is to use formal project management techniques. Project management was developed by the U.S. military and the defense contractors that worked with the military in the 1950s and the 1960s to develop weapons and other large systems. Not only was defense spending increasing in those years, but individual projects were becoming so large that it became impossible for managers to maintain control without some kind of assistance.

Project Management

Project management is a collection of formal techniques for planning and controlling the activities undertaken to achieve a specific goal. The project plan includes criteria for cost, schedule, and performance: it helps project managers make intelligent trade-off decisions regarding these three criteria. For example, if it becomes necessary for a project to be completed early, the project manager can compress the schedule by either increasing the project's cost or decreasing its performance.

Today, project managers use specific application software called **project management software** to help them manage projects. Project management software products, such as **Microsoft Project** and **Primavera Project Planner**, give managers an array of built-in tools for managing resources and schedules. The software can generate charts and tables that show, for example, which parts of the project are critical to its timely completion, which parts can be rescheduled or delayed without changing the project completion date, and where additional resources might be most effective in speeding up the project.

In addition to managing the people and tasks of the internal team, project management software can help the team manage the tasks assigned to consultants, technology partners, and outsourced service providers. By examining the costs and completion times of tasks as they are completed, project managers can learn how the project is progressing and continually revise the estimated costs and completion times of future tasks.

Information systems development projects have a well-deserved reputation for running out of control and ultimately failing. They are much more likely to fail than other types of projects, such as building construction projects. The main causes for information systems project failures are rapidly changing technologies, long development times, and changing customer expectations. Because of this vulnerability, many teams rely on project management software to help them achieve project goals.

Although electronic commerce certainly uses rapidly changing technologies, the development times for most electronic commerce projects are relatively short—often they are accomplished in under six months. This gives both the technologies and the expectations of users less time to change. Thus, electronic commerce initiatives are, in general, more successful than other types of information systems implementations.

You can learn more about project management by reading the references listed in the "For Further Study and Research" section at the end of this chapter or by clicking the Online Companion link for the **Project Management Institute**, a not-for-profit organization devoted to the promotion of professional project management practices.

Staffing the Operation

Regardless of whether the internal team decides to outsource parts of the design and implementation activity, it must determine the staffing needs of the electronic commerce initiative. The general areas of staffing that are most important to the success of an electronic commerce initiative include:

- Business management
- Application specialists
- Customer service staff
- Systems administration
- Network operations staff
- Database administration

The business management function should include internal staff. The business manager should be a member of the internal team that sets the objectives for the project. The **business manager** is responsible for implementing the elements of the business plan and reaching the objectives set by the internal team. If revisions to the plan are necessary as the project proceeds, the business manager develops specific proposals for plan modifications and additional funding and presents them to the internal team and top management for approval.

The business manager should have experience and knowledge related to the business activity that is being implemented in the electronic commerce site. For example, if business managers are assigned to a retail consumer site, they should have experience managing a retail sales operation. In the future, a company might want to hire an experienced electronic commerce business manager from another company; however, electronic commerce is too new for many managers to have obtained experience yet. Thus, most companies try to find internal candidates for this position.

In addition to including the business manager, the business management function in large electronic commerce initiatives may include other individuals who carry out specialized functions, such as project management or account management, that the business manager does not have time to handle personally. The account manager keeps track of the various Web sites in use by the project.

Most larger projects will have a test version, a demonstration version, and a production version of the Web site located on different servers. The test version is the "under construction" version of a Web site. Since most sites are frequently updated with new features and content, the test version gives the firm a place to make sure that each new feature works before exposing it to customers. The demonstration version has features that have passed testing and must be demonstrated to an internal audience (for example, the marketing department) for approval. The production version is the full operating version of the site that is available to customers and other visitors. The account manager supervises the location of specific Web pages and related software installations as they are moved from test to demonstration to production.

As more vendors provide packaged software solutions for electronic commerce, such as those described in Chapter 4, companies will need information systems staff that can install and maintain the software. Most large businesses have **applications specialists** who maintain accounting, human resources, and logistics software. Similarly, electronic commerce sites that buy software to handle catalogs, payment processing, and other features will need applications specialists to maintain the software. Although the installation of these software packages can be outsourced, most companies will want to train their own staff to serve in this function when the site becomes operational.

The Web offers businesses a unique opportunity to reach out to their customers. Thus, business-to-consumer and business-to-business sites that want to capitalize on that opportunity must include a customer relationship management function. **Customer service** personnel help design and implement customer relationship management in the electronic commerce operation. They can, for example, issue and administer passwords, design customer interface features, handle customer e-mail and telephone requests for service or follow-up action, and conduct telemarketing for the site. Some companies outsource parts of their customer relationship management operation to independent call centers. A **call center** is a company that handles customer telephone calls and e-mails for other companies. Using a call center often makes sense for smaller companies that do not have the volume of customer inquiries to justify creating their own internal call center operation. Some call centers work with a variety of businesses; others focus on one specialty area. For example, a specialized call center might contract with software manufacturers to provide installation help for their software products. Call center employees who are skilled in helping customers install one software package are often able to learn how to support other software packages very quickly.

A systems administrator who understands the server hardware and operating system is an essential part of a successful electronic commerce implementation. The **systems administrator** is responsible for the system's reliable and secure operation. If the site operation is outsourced to an ISP or CSP, the vendor will provide this function. If the site is hosted by the company, it will need to devote at least one person to this job. In addition, the internal system administrator needs sufficient staff to maintain full 24/7 operation and site security. These **network operations** staff functions include load estimation and load monitoring, resolving network problems as they arise, designing and implementing fault-resistance technologies, and managing any network operations that are outsourced to ISPs, CSPs, or telephone companies.

Every electronic commerce site will require some kind of **database administration** function to support activities such as transaction processing, order entry, inquiry management, or shipment logistics. These activities require either an existing database into which the site is being integrated or a separate database established for the electronic commerce initiative. It is important to have a database administrator who can effectively manage the design and implementation of this function.

Post-Implementation Audits

After an electronic commerce site is successfully launched, most of the project's resources are devoted to maintaining and improving the site's operations. However, an increasing number of businesses are realizing the value of a post-implementation audit. A **post-implementation audit** is a formal review of a project after it is up and running.

The post-implementation audit gives managers a chance to examine the objectives, performance specifications, cost estimates, and scheduled delivery dates that were established for the project in its planning stage and compare them to what actually happened. In the past, most project reviews focused on identifying individuals to blame for cost overruns or missed delivery dates. Since many external forces in technology projects can overwhelm the best efforts of managers, this blame identification approach was generally unproductive as well as uncomfortable for the managers on the project.

A post-implementation audit allows the internal team, the business manager, and the project manager to raise questions about the project's objectives and provide their "in-the-trenches" feedback on strategies that were set in the project's initial design. By agreeing beforehand not to lay blame, the company obtains valuable information that it can use in planning future projects and gives the participants a meaningful learning experience.

Summary

This chapter provided an overview of key elements that are typically included in business plans for electronic commerce implementations. The first step is setting objectives. Specific objectives derive from the initiative's overall goals and include planned benefits and planned costs. The benefit and cost objectives should be stated in measurable terms, such as dollars or quantities. Before undertaking an electronic commerce project, most companies will evaluate its estimated costs and benefits.

Businesses use a number of evaluation techniques; however, most businesses calculate projects' return on investment to gauge their value. Many companies have undertaken electronic commerce projects without evaluating their costs and benefits in detail because they fear being left out of the Internet's marketspace.

Companies must decide how much, if any, of an electronic commerce project they will outsource. The first step in determining an outsourcing strategy is to form an internal team that includes knowledgeable individuals from within the company. The internal team develops the specific project objectives and is responsible for meeting those objectives. The internal team designs an outsourcing strategy, selects a hosting service (or decides to have the company host its own Web server), and supervises the staffing of the project.

Project management is a formal way to plan and control specific tasks and resources used in a project. It provides project managers with a tool they can use to make intelligent trade-offs among the project schedule, cost, and performance. Electronic commerce initiatives are usually completed within a short time frame and thus are less likely to run out of control than other information systems development projects.

The company must staff the electronic commerce initiative regardless of whether portions of the project are outsourced. Critical staffing areas include business management, application specialists, customer service staff, systems administration, network operations staff, and database administration. A good way for all participants to learn from project experiences is to conduct a post-implementation audit that compares project objectives to the actual results.

Key Terms

24/7 Operation
Applications Specialist
Business Manager
Call Center
Capital Investment
Capital Project
Component Outsourcing
Customer Service
Database Administration
Downstream Strategies
Early Outsourcing
Fast Venturing

Incubator
Late Outsourcing
Network Operations
Outsourcing
Partial Outsourcing
Post-Implementation Audit
Project Management
Project Management Software
Return on Investment
Systems Administrator
Upstream Strategies

Review Questions

1. Name three benefit objectives that a business might decide to measure in an electronic commerce business plan.

2. Why have some firms approved electronic commerce projects without taking a close look at their return on investment numbers?

3. Why is late outsourcing seldom used in electronic commerce projects?

4. In fewer than 200 words, name and briefly describe four factors that a company should evaluate when selecting an ISP or CSP to provide Web hosting services.

5. Why should the head of the business management function of an electronic commerce initiative be an employee of the company implementing the project? Limit your answer to 250 words.

Exercises

1. The Grover Cams Company manufactures cams and other components for diesel engines. As Webmaster for Grover, you have created an attractive Web site that includes information about the company's history, its financial statements, and digitized depictions of the company's main products. You have been talking with your manager, Chief Information Officer Tom Buckles, for several months about adding electronic commerce features to the Web site that will allow your smaller customers to order directly from Grover instead of through their local distributors. Tom finally created a capital budget proposal for the Web site expansion and submitted it to Grover's board of directors. The board always calculates and evaluates a capital project's return on investment before approving it. Tom came back from the board meeting looking unhappy. The board told him that the project did not provide a high enough financial return to approve it. However, the board realized that electronic commerce initiatives could be important to Grover's future strategic position in the business; thus it would be willing to consider nonmonetary factors as a basis for approving the project. Tom would like to take the project back to the board next month, but he does not have a good sense of what nonmonetary factors might persuade the board to approve the project. He wants you to write a memo that outlines some of those factors and explains why they are important to Grover's future strategic position. In addition to considering the discussion in this chapter, you may want to use the Online Companion and draw on resources at *Business Week's ebiz*, *CIO's* **Electronic Commerce Research Center, E-Commerce at** *The Industry Standard,* **Electronic Commerce Guide**, or **ZDNet's E-Business** as you prepare your memo.

2. You are working for International Delicacies, which has become successful selling unusual food and other gift items through its mail-order catalog. Your manager, Jagdish Singh, is interested in exploring electronic commerce options for the company. He wants you to be a member of the internal team for the project and has asked you to identify some potential ISPs or CSPs to which the company could outsource the Web hosting portion of the project. Since the project is at a very early stage, the team has not specified the exact nature of the Web hosting function; however, the team would like to get a rough idea of what the setup costs and monthly charges might be. Use **The List**, which is internet.com's online directory of ISPs and CSPs, to find three Web hosting service providers in your area. Follow the directory's links to your chosen service providers to learn more about them. Write a memo to Jagdish that summarizes the services offered by each Web hosting service provider and

that compares the costs charged by each for those services. Based on your research, evaluate each service provider and recommend whether the team should further consider each service provider as a potential outsourcing partner by the team.

3. As manager of networks and computing operations for Fashion Land, a retailer of women's clothing and accessories, you have seen the business grow from seven stores in Kansas City to over 100 stores located throughout the Midwest. Fashion Land's marketing research team has found that many members of its target customer group—females between the ages of 15 and 35—are becoming regular users of the Web. The researchers have found that these customers would not want to buy major clothing items on the Web; however, they would like to buy accessories.

Alone, or in a team assigned by your instructor, do the following:

- Design a business plan for Fashion Land's electronic commerce initiative. The plan should include a list of specific objectives and the costs and benefits of accomplishing each objective. The plan should also include recommendations regarding what to outsource, what Web hosting services will be needed, and what staff should be hired.

- Prepare a memo that outlines the major hardware, software, security, payment processing, advertising, international, legal, and ethics issues that might arise in the development of this electronic commerce site.

For Further Study and Research

Abdel-Hamid, T. and S. Madnick. 1991. *Software Project Dynamics: An Integrated Approach*. Englewood Cliffs, NJ: Prentice Hall.

Abdel-Hamid, T., K. Sengopta, and C. Sweet. 1999. "The Impact of Goals on Software Project Management: An Experimental Investigation," *MIS Quarterly*, 23(4), December, 531–555.

Alter, A. 2000. "IT Leaders Require Strategy, Flexibility," *Computerworld*, 34(19), May 8, 36.

Andrews, W. 1999. "In Commerce, It's Build vs. Buy," *Internet World*, 5(26), August 1, 42.

Barker, R. 2000. "When to Outsource," *Business 2.0*, March 1. Available online at: (http://www.business2.com/content/magazine/indepth/2000/03/01/11125).

Berlind, D. 1999. "The Harris Interactive Poll: An E-Commerce To-Do List," *ZDNet Enterprise*, June 23. Available online at: (http://www.zdnet.com/enterprise/e-business/stories/0,5918,2281595,00.html).

Boisot, M. 1999. *Knowledge Assets: Securing Competitive Advantage in the Information Economy*. New York: Oxford University Press.

Booker, E. 1999. "Bigger Not Always Better—Despite Reorgs, Large SIs Still Seen as Lacking E-Biz Acumen," *Internetweek*, September 20, 1.

Brache, A. and J. Webb. 2000. "The Eight Deadly Assumptions of E-Business," *Journal of Business Strategy*, 21(3), May–June, 13–17.

Brooks, F. 1995. *The Mythical Man-Month: Essays on Software Engineering, Anniversary Edition*. Reading, MA: Addison-Wesley.

Carmichael, M. 1999. "E-Commerce Sites Return Investment," *Advertising Age's Business Marketing*, 84(4), April, 38.

Conlon, G. and J. Winchester. 1999. "Formula for Success," *Sales & Marketing Management*, 151(2), February, 76.

Cope, J. 2000. "Revving Up E-Commerce," *Computerworld*, 34(19), May 8, 26–27.

Dugan, S. 2000. "Where Will E-Business Take You?" *InfoWorld*, 22(14), April, 49–52.

Edvinsson, L. and M. Malone. 1997. *Intellectual Capital: Realising your Company's True Value by Finding its Hidden Brainpower*. New York: HarperCollins.

Elango, B. 2000. "Do You Have an Internet Strategy?" *Information Strategy*, 17(1), Fall, 32–38.

Eng, S. 2000. "Hatching Schemes," *The Industry Standard*, November 27, 174–175.

Fogarty, K. 2000. "Muddling Through to a Disaster," *Computerworld*, 34(16), April 17, 42.

Fryer, B. 2000. "Consulting's Next Big Thing," *Computerworld*, 34(1), January 3, 94–95.

Gantz, J. 1999. "E-Commerce: Here's What You Need to Know," *Computerworld*, 33(17), April 26, 34.

GartnerGroup. 1999. *Survey Results: The Real Cost of E-Commerce Sites*. Stamford, CT: GartnerGroup.

Gido, J. and J. Clements. 1999. *Successful Project Management*. Cincinnati, OH: South-Western College Publishing.

Girishankar, S. 1999. "Sizzling Start-Ups," *Internetweek*, February 8, 39–45.

Glass, R. 1997. *Software Runaways: Lessons Learned from Massive Software Project Failures*. Upper Saddle River, NJ: PTR Prentice Hall.

Goldratt, E. 1997. *Critical Chain*. Great Barrington, MA: North River Press.

Grackin, A. 2000. "Internet-Driven Business Models," *Manufacturing Systems*, 18(6), June, 56.

Greene, A. 2000. "The Missing Element of an E-Strategy," *Manufacturing Systems*, 18(1), January, 44.

Griffin, J. 2000. "Rethinking Internet Valuation," *Business 2.0*, June 13. Available online at: (http://www.business2.com/content/insights/opinion/2000/06/01/10989).

Hellweg, E. and S. Donahue. 2000. "The Smart Way to Start an Internet Company," *Business 2.0*, March 1. Available online at: (http://www.business2.com/content/magazine/indepth/2000/03/01/13754).

Heun, C. 2000. "No Web Bargains for Kmart," *InformationWeek*, August 21, 18.

Horowitz, A. 1999. "Emerging Enterprises Set Their Own IT Agendas," *Information Week*, June 14, 52.

Hudgins-Bonafield, C. 1998. "The Electronic Crane: E-Commerce Infrastructure Builds Upward," *Network Computing*, 9(23), December 15, 44–48.

Kalakota, R. and M. Robinson. 1999. *E-Business: Roadmap for Success*. Reading, MA: Addison-Wesley.

Kambil, A., E. Eselius, and K. Monteiro. 2000. "Fast Venturing: The Quick Way to Start a Web Business," *Sloan Management Review*, 41(4), Summer, 55–67.

Kara, D. 1999. "Sourcing Solutions for Wired World Emerging," *Software Magazine*, 19(1), June, 60–71.

Karpinski, R. 1999. "Cost Conscious Configurator," *Internetweek*, April 12, 9.

Keen, P. 2000. "Back to Process," *Computerworld*, 34(19), May 8, 50.

Keen, P. 2000. "Six Months—or Else," *Computerworld*, 34(15), April 10, 48.

Keil, M., P. Cule, K. Lyytinen, and R. Schmidt. 1998. "A Framework for Identifying Software Project Risks," *Communications of the ACM*, 41(11), November, 76–83.

Keil, M. and D. Robey, 1999. "Turning Around Troubled Software Projects: An Exploratory Study of the De-Escalation of Commitment to Failing Courses of Action," *Journal of Management Information Systems*, 15(4), 63–87.

Kerzner, H. 2000. *Advanced Project Management: Best Practices*. New York: John Wiley & Sons.

Kling, A. 2000. "Penny Wise, Project Foolish," *PlanetIT*, August 18. Available online at: (http://www.PlanetIT.com/docs/PIT20000818S0013).

Lewis, D. 2000. "Some Retailers De-emphasize Web Payback," *Internetweek*, October 19. Available online at (http://www.internetweek.com/lead/lead101900.htm).

Maitra, A. 1999. *Project Management for Internet Businesses*. New York: John Wiley & Sons.

McConnell, S. 1996. *Rapid Development: Taming Wild Software Schedules*. Redmond, WA: Microsoft Press.

McCune, J. 1998. "Making Web Sites Pay," *Management Review*, 87(6), June, 36–38.

Melymuka, K. 2000. "Born to Lead Projects," *Computerworld*, 34(13), March 27, 62–63.

Neuwirth, R. 1998. "Race into Cyberspace Gushes $80M Red Ink," *Editor & Publisher*, 131(51), December 19, 12–13.

O'Kelly, C. 2000. "Planning and Implementation of Internet Based Technologies," *Accountancy Ireland*, 32(1), February, 24–25.

Radcliff, D. 2000. "The Web's Master Builders," *Computerworld*, 34(11), March 13, 81.

Ramsey, C. 2000. "Managing Web Sites as Dynamic Business Applications," *Intranet Design Magazine*, June. Available online at: (http://idm.internet.com/articles/200006/wm_index.html).

Randall, L. 1999. "Average E-Commerce Web Site Costs US$1 Million," *Computing Canada*, 25(24), June 18, 11.

Rogers, A. 1999. "Up-Front Web Costs Are Half the Story," *Computer Reseller News*, June 7, 3.

Ruud, M. and J. Deutz. 1999. "Moving your Company Online," *Management Accounting*, 80(8), February, 28–32.

Schultz, K. 1998. "Winners Craft Creative Solutions for Business," *InfoWorld*, 20(39), September 28, 73–74.

Schwalbe, K. 2000. *Information Technology Project Management*. Cambridge, MA: Course Technology.

Siebel, T. and P. House. 1999. *Cyber Rules: Strategies for Excelling at E-Business*. New York: Currency-Doubleday.

Slayton, J. 2000. "Recruiter Beware," *The Industry Standard*, September 11. Available online at: (http://www.thestandard.com/article/display/0,1151,18304,00.html).

Spieler, G. 2000. "Ordering Out: Outsourcing E-Commerce Services," *GartnerGroup*, January 25. Available online at: (http://www.gartner.com/public/static/hotc/00075938.html).

Stavropoulos, A. and S. Jurvetson. 2000. "Does Your Idea Make Sense? Seven Questions to Ask Yourself Before Chasing Your Dot-Com Dreams," *Business 2.0*, March 1. Available online at: (http://www.business2.com/content/magazine/indepth/2000/03/01/20720).

Stewart, M. 1998. "My Big Bet on the Net," *Newsweek*, 132(23), December 7, 53.

Stewart, T. 1997. *Intellectual Capital: The New Wealth of Nations*. New York: Doubleday Dell.

Stewart, T. 1999. "Larry Bossidy's New Role Model: Michael Dell," *Fortune*, 139(7), April 12, 166–167.

Stoiber, J. 1999. "Maximizing IT Investments," *CIO Enterprise Magazine*, July 15. Available online at (http://www.cio.com/archive/enterprise/071599_checks.html).

Tattum, L. 2000. "Chemical Industry E-Strategies and Implementations," *Chemical Week*, July 26, S11–S15.

Tebbe, M. 1999. "EBay's Outages Highlight Problems with I-Commerce," *InfoWorld*, 21(25), June 21, 34.

Treese, G. and L. Stewart. 1998. *Designing Systems for Internet Commerce*. Reading, MA: Addison-Wesley.

Vijayan, J. 1999. "Underlying Tech Key to E-Success," *Computerworld*, 33(25), June 21, 1, 113.

Violino, B. 2000. "Payback Time for E-Business— Net Projects No Longer Too 'Strategic' for ROI," *Internetweek*, May 1, 1.

The Wall Street Journal. 1999. "Spending Campaign Is Set for Newspaper's Web Site," June 22, B16.

Walsh, B. 1998. "Building a Business Plan for an E-Commerce Project," *Network Computing*, 9(17), September 15, 69–73.

Wexler, J. 2000. "Lands of Opportunity," *Computerworld*, 34(26), June 26, 72–73.

Wilder, C. 1999. "Customized Web Sites," *InformationWeek*, February 2, 124.

Wilder, C. 1999. "ROI: E-Business Strategic Investment," *InformationWeek*, May 24, 48–56.

Willard, C. 2000. "Taking Techies to their Limits," *Computerworld*, 34(26), June 26, 74–75.

Wysocki, B. 2000. "U.S. Incubators Help Japan Hatch Ideas," *The Wall Street Journal*, June 12, A1.

Yourdon, E. 2000. "Success in E-Projects," *Computerworld*, 34(34), August 21, 36.

Yourdon, E. and P. Becker. 1997. *Death March: The Complete Software Developer's Guide to Surviving "Mission Impossible" Projects*. Upper Saddle River, NJ: Prentice Hall.

443

Glossary

24/7 Operation The operation of a site or service 24 hours a day, seven days a week.

A

Acceptance An expression of willingness to take an offer, including all of its stated terms.

Access Control List (ACL) A list of resources and the usernames of people who are permitted access to those resources within a computer system.

Accredited Standards Committee X12 (ASC X12) A committee that develops and maintains uniform EDI standards in the United States.

Acquiring Bank Synonymous with merchant bank, which is a bank that does business with merchants who want to accept credit cards.

Active Content Programs that are embedded transparently in Web pages that cause action to occur.

Active Server Pages (ASP) Applications that generate dynamic content within Web pages using either Jscript code or Visual Basic.

Active Wiretapping An integrity threat that exists when an unauthorized party can alter a message stream of information.

ActiveX An object, or control, that contains programs and properties that are put in Web pages to perform particular tasks.

Ad View A Web site visitor page request that contains an advertisement.

Addressable Media Advertising efforts sent to a known addressee; these include direct mail, telephone calls, and e-mail.

Affiliate Marketing An advertising technique in which one Web site (called an "affiliate") includes descriptions, reviews, ratings, or other information about products that are sold on another Web site. The affiliate site includes links to the selling site, which pays the affiliate site a commission on sales made to visitors that arrived from a link on the affiliate site.

Agent A program that performs information gathering, information filtering, and/or mediation on behalf of a person or entity.

American National Standards Institute (ANSI) The coordinating body for electrical, mechanical, and other technical standards in the United States.

Anchor Tag The HTML tag used to specify hyperlinks.

Anonymous Electronic Cash Electronic cash that cannot be traced back to the person who spent it.

Anonymous FTP A protocol that allows users to access limited parts of a remote computer using FTP without having an account on the remote computer.

Applet A program that executes within another program; it cannot execute directly on a computer.

Application Construction The use of Web editors and extensions to produce Web pages.

Application Program Interface (API) A set of routines, protocols, and tools for building software applications.

Application Server A middle-tier software and hardware combination that lies between the Internet and a corporate backend server.

Application Service Provider A Web-based site that provides management of applications such as spread sheets, human resources management, or e-mail to companies for a fee.

446

Applications Specialist
The member of an electronic commerce team who is responsible for maintenance of software that performs a specific function, such as catalog, payment processing, accounting, human resources, and logistics software.

Archiving Saving a log on a storage device.

Ascending-Price Auction A type of auction in which bidders publicly announce their successively higher bids until no higher bid is forthcoming, also called an English auction.

Asymmetric Digital Subscriber Line (ADSL) Internet connections using the DSL protocol with bandwidths from 16 to 640 Kbps upstream and 1.5 to 9 Mbps downstream.

Asymmetric Encryption Synonymous with public-key encryption, which is the encoding of messages using two mathematically related but distinct numeric keys.

Asynchronous Transfer Mode (ATM) Internet connections with bandwidths of up to 622 Gbps.

Attachment A data file (document, spreadsheet, or other) that is appended to an e-mail message.

Auctioneer The person who manages an auction.

Authority to Bind The ability or authority of an individual to commit his or her company to a contract.

Automated Clearing House (ACH) One of several systems set up by banks or government agencies, such as the U.S. Federal Reserve Board, that process high volumes of low dollar amount electronic fund transfers.

B

Backbone The main network of connections that carry most of the traffic on the Internet.

Backdoor An electronic hole in electronic commerce software left open by accident or intentionally.

Bandwidth The amount of data that can be transmitted in a fixed amount of time. Also, the number of simultaneous site visitors that a Web site can accommodate without degrading service.

Banner Exchange Site (BES) Web sites that help electronic merchants promote their stores with advertising on other members' Web sites.

Benchmarking Testing that compares hardware and software performances.

Bid An offer of a certain price made on an item that is up for auction.

Bidder A potential buyer at an auction; one who places bids.

Bot Synonymous with spider, which is the first part of a search engine. It automatically and frequently searches the Web to find pages and updates its database of information about old Web sites.

Browser Synonymous with Web browser, which is software that lets users read HTML documents and move from one HTML document to another using hyperlinks.

Buffer An area of a computer's memory that is set aside to hold data read from a file or database.

Bulk Mail Synonymous with spam, which is electronic junk mail.

Business Manager The member of an electronic commerce team who is responsible for implementing the elements of the business plan and reaching the objectives set by the internal team. The business manager should have experience in and knowledge of the business activity being implemented in the site.

Business Processes The activities in which businesses engage as they conduct commerce.

Business-to-Business Transactions conducted between businesses on the Web.

Business-to-Consumer Transactions conducted between shoppers and businesses on the Web.

C

Cache A high-speed memory area set aside to store Web pages.

Call Center A company that handles customer telephone calls and e-mails for other companies.

Cannibalization The loss of traditional sales of a product to its electronic counterpart.

Capital Investment A major outlay of funds made by a company to purchase fixed assets such as property, a factory, or equipment.

Capital Project Synonymous with capital investment.

Card Not Present The condition that occurs when the merchant's location and the purchaser's location are different.

Cascading Style Sheets Utilities that allow designers to apply many predefined page display styles to Web pages.

Catalog On electronic commerce sites, a listing of goods or services that may include photographs and descriptions, often stored in a database.

Cause Marketing An affiliate marketing program that benefits a charitable organization.

Certification Authority (CA) An agency that issues digital certificates to organizations or individuals.

Channel A Web page or category of information in a particular area of interest that is automatically delivered to a user's computer.

Channel Conflict Sales activities on a company's Web site that interfere with its existing sales outlets.

Charge Card A card with no preset spending limit. The entire amount charged to the card

must be paid in full each month.

Cipher Text Text that comprises a seemingly random assemblage of bits. Cipher text is what messages become after they are encrypted.

Circuit A specific route between source and destination along which data travels.

Circuit Switching A way of connecting computers or other devices that uses a centrally controlled single connection. In this method, which is used by telephone companies to provide voice telephone service, the connection is made, data are transferred, and the connection is terminated.

Clear Text Text that has not been encrypted or encoded in any way.

Click Synonymous with click-through.

Click-Through The loading of an advertiser's Web page that results from a visitor clicking on a banner advertisement on another Web page.

Click-Through Count The number of visitors that click on a Web advertisement link and go to the advertiser's Web site.

Client A networked PC or workstation on which users run applications.

Client/Server Model A network in which each computer on the network is either a client or a server.

Client-Side Electronic Wallet An electronic wallet that stores a consumer's information on the consumer's own computer.

Closed Loop System A payment card arrangement involving a consumer, a merchant, and a payment card company (such as American Express or Discover) that processes transactions between the consumer and merchant without involving banks.

Closing Tag The second half of a two-sided HTML tag; it is identified by a slash (/) that precedes the tag's name.

Clustering The tendency of processes in a given business to include activities that are similar to the activities in which other businesses engage.

Collision The occurrence of two messages resulting in the same hash value; the probability of this happening is extremely small.

Colocated Hosting Self-hosting wherein the server is owned by the online store but is located at the Web host's site, and the Web host provides maintenance based on the level of service the online business requires.

Commerce A negotiated exchange of valuable objects or services between two or more parties. Commerce includes all activities that each party undertakes to complete the transaction.

Commerce Service Provider (CSP) A Web host service that also provides commerce hosting services on their computer.

Commerce Suite A family of electronic commerce software consisting of all the software needed to establish an electronic commerce Web site.

Commodity A product or service that has become so standardized and well-known that buyers cannot detect a difference in the offerings of various sellers, and they decide to buy based on price.

Common Gateway Interface (CGI) A protocol that allows Web servers to interact dynamically with other software packages to create custom Web pages.

Component Outsourcing Synonymous with partial outsourcing.

Computer Forensics The field responsible for the collection, preservation, and analysis of computer-related evidence.

Computer Forensics Expert An individual hired to access client computers to locate information that can be used in legal proceedings.

Computer Security The protection of computer resources from various types of threats.

Contract An agreement between two or more legal entities that provides for an exchange of value between or among them.

Conversion Rate Used in advertising to calculate the percentage of recipients that respond to an ad or promotion.

Consideration The bargained-for exchange of something valuable, such as money, property, or future services.

Constructive Notice The idea that citizens should know that when they leave one area and enter another, they become subject to the laws of the new area.

Cookie Bits of information about Web site visitors created by Web sites and stored on client computers.

Cookie Blocker A third-party program that prevents cookie storage selectively.

Copy Control An electronic mechanism for providing a fixed upper limit to the number of copies that one can make of a digital work.

Copyright A legal protection of intellectual property.

Cost Per Thousand (CPM) An advertising pricing metric that equals the dollar amount paid to reach 1000 people in an estimated audience.

Countermeasure A physical or logical procedure that recognizes, reduces, or eliminates a threat.

Crawler Synonymous with spider, which is the first part of a search engine.

Credit Card A card with a preset spending limit based on the card holder's credit limit. A minimum monthly payment must be made against the balance on the card, and interest is charged on the unpaid balance.

Cryptography The science that studies encryption, which is the hiding of messages so that only the sender and receiver can read them.

Culture The combination of language and customs that are unique to a particular population.

Customer Relationship Management Synonymous with technology-enabled relationship management, which is the obtaining and use of detailed customer information.

Customer Service The persons within an electronic commerce team that are responsible for managing customer relationships in the electronic commerce operation.

Cyber vandalism The electronic defacing of an existing Web site page.

Cybersquatting The practice of registering a domain name that is the trademark of another person or company with the hope that the trademark owner will pay huge amounts of money for the domain rights.

Cycling Replacing the oldest log with the newest log.

D

Data Encryption Standard (DES) An encryption standard adopted by the U.S. government for encrypting sensitive information.

Data Mining Looking for hidden patterns in data.

Database Administration The function within an electronic commerce team that is responsible for defining the data elements in the database design and the operation of the database management software.

Dead Link A Web link that when clicked displays an error message instead of a Web page.

Decrypted Information that has been decoded. The opposite of encrypted.

Decryption Program A procedure to reverse the encryption process, resulting in the decoding of an encrypted message.

Dedicated Hosting Web hosting wherein a firm has an individually dedicated server that is owned and administered by a service provider who dictates terms of usage.

Defamatory Statement A statement that is false and injures the reputation of a person or company.

Delay Threat The disruption of normal computer processing.

Denial of Service Threat Synonymous with necessity threat, which is the disruption of normal computer processing.

Denial Threat The prohibition of normal computer processing.

Descending-Price Auction Synonymous with Dutch auction, which is an open auction in which bidding starts at a high price and drops until a bidder accepts the price.

Digital Certificate An attachment to an e-mail message or data embedded in a Web page that verifies the identity of a sender or Web site.

Digital ID Synonymous with digital certificate.

Digital Signature An encryption message digest.

Digital Subscriber High-grade telephone service (superior to POTS) offered by some telephone companies.

Digital Subscriber Line Synonymous with digital subscriber.

Digital Subscriber Loop (DSL) Synonymous with digital subscriber.

Direct Connection EDI The form of EDI in which EDI translator computers at each company are linked directly to each other through modems and dial-up telephone lines or leased lines.

Discretionary Access Control Control mechanisms that specify which users have access to which computer files and resources.

Disintermediation The removal of an intermediary from a value chain.

Doing Business Synonymous with commerce, which is a negotiated exchange of valuable objects or services between two or more parties.

Domain Name The address of a Web page, it can contain two or more word groups separated by periods. Components of domain names become more general from right to left.

Domain Name Ownership Change The changing of owner information maintained by a public domain registrar in the registrar's database to reflect the new owner's name and business address. This usually only happens when safeguards are not in place.

Dotted Quad The representation of an IP address; it appears as up to four separate numbers delineated by periods (for example, 126.204.89.56).

Double Auction A type of auction in which buyers and sellers each submit combined price-quantity bids to an auctioneer. The auctioneer matches the sellers' offers (starting with the lowest price, then going up) to the buyers' offers (starting with the highest price, then going down) until all of the quantities are sold.

Double Spending The spending of the same unit of electronic cash twice by submitting the same electronic currency to two different vendors.

Download To receive a file from another computer.

Downstream The connection that occurs when information is sent to a user's computer from an ISP.

Downstream Strategies Tactics that improve the value that a business provides to its customers.

Dutch Auction A type of open auction in which bidding starts at a high price and drops until a bidder accepts the price.

Dynamic Content Nonstatic information constructed in response to a Web client's request.

Dynamic Page A Web page whose content is shaped by a program in response to a user request.

E

E-Business (Electronic Business) Software Commerce software that provides tools for both business-to-consumer and business-to-business commerce, it is often designed to interface with existing back office systems.

E-Card An electronic greeting card.

E-Zine An electronic magazine.

Early Outsourcing The hiring of an external company to do initial electronic commerce site design and development. The external team then trains the original company's information systems professionals in the new technology, eventually handing over complete responsibility of the site to the internal team.

Eavesdropper A person or device who is able to listen in on and copy Internet transmissions.

EBCDIC Extended Binary Coded Decimal Interchange Code, an eight-bit code system developed by IBM that represents characters as numbers.

EDI for Administration, Commerce, and Transport (EDIFACT) The 1987 publication that summarizes the United Nations' standard transaction sets for international EDI.

EDI-Capable Banks Banks that are able to exchange payment and remittance through value-added networks.

EDI-Compatible Firms that are able to exchange data in specific standard electronic formats with other firms.

Electronic Business (E-Business) Another term for electronic commerce; sometimes used to mean business-to-business electronic commerce.

Electronic Cash A form of electronic payment that is anonymous and can be spent only once.

Electronic Commerce (E-Commerce) Business activities conducted using electronic data transmission via the Internet and the World Wide Web.

Electronic Customer Relationship Management (eCRM) Synonymous with technology-enabled relationship management.

Electronic Data Interchange (EDI) Exchange between businesses of computer-readable data in a standard format.

Electronic Funds Transfer (EFT) Electronic transfer of account exchange information over secure private communications networks.

Electronic Mail (E-mail) Messages that are sent from one user to another (or multiple recipients) using particular mail programs and protocols.

Electronic Wallet A software utility that holds electronic cash, credit card information, owner identification and address information, and provides these data automatically at electronic commerce sites.

Encryption The coding of information using a mathematical-based program and secret key, it makes a message illegible to casual observers or those without the decoding key.

Encryption Algorithm The logic that implements an encryption program.

Encryption Program A program that transforms clear text into cipher text.

English Auction A type of auction in which bidders publicly announce their successively higher bids until no higher bid is forthcoming.

Enterprise Resource Planning (ERP) Business software that integrates all facets of a business, including planning, manufacturing, sales, and marketing.

Entity Body An optional, though almost always present, part of a message from a client that contains the HTML page requested by the client and passes bulk information to the server.

Escrow Service An independent third party who holds an auction buyer's payment until the buyer receives the purchased item and is satisfied that it is what the seller represented it to be.

Extensible Markup Language (XML) A language that describes the semantics of a page's contents and defines data records on a page.

Extensible System Any system that can be easily enhanced without voiding earlier work done on the system.

Extranet A network system that extends a company's intranet and allows it to connect with the networks of business partners or other designated associates.

F

Fast Venturing The joining of an existing company that wants to launch an electronic commerce initiative with external equity partners and operational partners who provide the experience and skills needed to develop and scale up the project very rapidly.

Fair Use The approved limited use of copyright material when certain conditions are met.

File Transfer Protocol (FTP) A protocol that enables users to transfer files via the Internet.

Financial VANS (FVANS) Value-added networks that are not banks but can translate financial transaction sets into ACH formats and transmit them to banks that are not EDI capable.

Firewall A computer that provides a defense between one network (inside the firewall) and another network (outside the firewall, such as the Internet) that could pose a threat to the inside network. All traffic to and from the network must pass through the firewall. Only authorized traffic, as defined by the local security policy, is allowed to pass through the firewall, which itself is immune to penetration.

Firm A business engaged in commerce.

First-Price Sealed-Bid Auction A type of auction in which bidders submit their bids independently and privately, with the highest bidder winning the auction.

Flat File A single file containing all the data on a particular subject.

Flat-Rate Access The monthly fee paid by a consumer or business for unlimited telephone line usage.

Float Money deposited in a customer's account that earns interest for the merchant.

Forum Selection Clause A statement within a contract that dictates that the contract will be enforced according to the laws of a particular state; signing a contract with a forum selection clause constitutes voluntary submission to the jurisdiction named in the forum selection clause.

Full-Privilege FTP A protocol that allows users to upload files to and download files from a remote computer using FTP.

Gateway Server A firewall that filters traffic based on applications requested by clients on the trusted network.

Gopher A system predating the World Wide Web that displays a hierarchically structured list of files on both Web and Gopher (remote) servers. Gopher was developed at the University of Minnesota and named after its mascot.

Graphical User Interface (GUI) Computer program control functions that are displayed using pictures, icons, and other easy-to-use graphical elements.

Group Purchasing Site A type of auction Web site that negotiates with a seller to obtain lower prices on an item as individual buyers enter bids on that item.

H

Hash Algorithm A security utility that mathematically combines every character in a message to create a fixed-length number (usually 128 bits in length) that is a condensation, or fingerprint, of the original message.

Hash Coding The process used to calculate a number from a message.

Hash Value The number that results when a message is hash coded.

Hierarchical Business Organization Firms that include a number of levels with cumulative responsibility. These organizations are typically headed by a top-level president or officer. A number of vice presidents report to the president. A larger number of middle managers report to the vice presidents.

Hierarchical Hyperlink Structure A hyperlink structure in which the user starts from a home page and follows links to other pages in whatever order they wish.

Hit A Web page that is stored (indexed) in a search engine's database and contains text that matches a search expression.

Hops The number of hosts between two Internet-connected computers.

Hypertext A system of navigating between HTML pages using links.

Hypertext Link (Hyperlink) A pointer in an HTML document to another location within the same document or to a different HTML document.

Hypertext Markup Language (HTML) The language of the Internet; it contains codes attached to text that describe text elements and their relation to one another.

Hypertext Server Synonymous with Web server, which is a computer that is connected to the Internet and that stores files written in HTML that are publicly available through an Internet connection.

Hypertext Transfer Protocol (HTTP) The Internet protocol responsible for transferring and displaying Web pages.

I

Implied Contract The agreement between two parties stating that a contract exists, even if no contract has been written and signed.

Impression The loading of a banner ad on a Web page.

Incubator A company that offers start-up businesses a physical location with offices, accounting and legal assistance, computers, and Internet connections at a very low monthly cost.

Index A list containing every Web page found by a spider, crawler, or bot.

Indirect Connection EDI The form of EDI in which each company transmits and receives EDI messages through a value-added network.

Industry Value Chain The larger stream of activities in which a particular business unit's value chain is embedded.

Integrated Services Digital Network (ISDN) High-grade telephone service that uses the DSL protocol and offers bandwidths of up to 128 Kbps.

Integrity The category of computer security that addresses the validity of data; confirmation that data have not been modified.

Integrity Violation A security violation that occurs whenever a message is altered while in transit between sender and receiver.

Intellectual Property The ownership of ideas and control over the tangible or virtual representation of those ideas.

Intelligent Agent Synonymous with agent.

Interactive Message Access Protocol (IMAP) A newer e-mail protocol with improvements over POP.

Internet2 A newly developed successor to the Internet with bandwidths in excess of 1 Gbps.

Internet A global system of interconnected computer networks.

Internet Access Provider (IAP) Synonymous with Internet Service Provider.

Internet Commerce Another term for electronic commerce; sometimes used to refer to electronic commerce conducted on the Internet or World Wide Web instead of via private networks.

Internet Host A computer that is directly connected to the Internet.

Internet Protocol See TCP/IP.

Internet Service Provider (ISP) A company that sells Internet access rights directly to Internet users.

InterNIC One of the primary official World Wide Web domain name registration services.

Interoperable Software Software that will run transparently on a variety of hardware and software configurations.

Intranet An interconnected network of computers operated within a single company or organization.

IP Address The 32-bit number that represents the address of a particular location (computer) on the Internet.

J

Java Applet A Java application that runs on a browser.

Java Sandbox A security model that confines Java applet actions to a security model-defined set of rules.

Java Servlet An application that runs on a Web server and generates dynamic content.

Jurisdiction A government's ability to exert control over a person or corporation.

K

Key A number used to encode or decode messages.

Knowledge Management The intentional collection, classification, and dissemination of information about a company, its products, and its processes.

L

Late Outsourcing The hiring of an external company to maintain an electronic commerce site that has been designed and developed by an internal information systems team.

Leased Line A permanent telephone connection between two points; it is always active.

Legitimacy The idea that those subject to laws should have some role in formulating them.

Limited Edition A trial version of a software program; it can be used for free for a limited time period or for a certain number of uses.

Limited Use Synonymous with limited edition.

Linear Hyperlink Structure A hyperlink structure that resembles conventional paper documents whereby the user reads pages in serial order.

Liquidation Broker An agent that finds buyers for unusable and excess inventory.

Local Area Network (LAN) A network that connects workstations and PCs within a single physical location.

Localization A type of language translation that considers multiple elements of the local environment, such as business and cultural practices, in addition to local dialect variations in the language.

Log File A collection of data that shows information about Web site visitors' access habits.

Logical Security The protection of assets using nonphysical means.

Long-Arm Statute A state law that creates personal jurisdiction for courts.

Macro Virus A virus that is transmitted or contained inside a downloaded file attachment; it can cause damage to a computer and reveal otherwise confidential information.

Mail Bomb A security attack wherein many people (hundreds or thousands) each send a message to a particular address, exceeding the recipient's allowable mail limit and causing mail systems to malfunction.

Mail Server Programs and hardware used to manage and store e-mail over the Internet.

Mailing List An e-mail address that forwards messages to certain users who are subscribers.

Maintenance, Repair, and Operating (MRO) Commodity supplies, including general industrial merchandise and standard machine tools that are used in a variety of industries.

Managed Hosting A Web hosting service in which the service provider manages the operation and oversight of all servers and assigns a dedicated service manager.

Managed Server Hosting Synonymous with managed hosting.

Many-to-Many Communications A model of communications in which a number of entities communicate with a number of other entities.

Many-to-One Communications A model of communications in which a number of entities communicate with a single other entity.

Market A real or virtual space in which potential buyers and sellers come into contact with each other and agree on a medium of exchange (such as currency or barter).

Market Research Activities undertaken by firms—such as conducting surveys, engaging in conversations with customers, and holding focus groups—to help identify customer needs.

Market Segmentation The identification by advertisers of specific subsets of their markets that have common characteristics.

Marketspace A market that occurs in the virtual world instead of in the physical world.

Masquerading Pretending to be someone you are not (for example, by sending an e-mail that shows someone else as the sender) or representing a Web site as an original when it is an imposter.

Mass Media The method of contacting potential customers through the distribution of broadcast, printed, billboard, or mailed advertising materials.

Merchandising The combination of store design, layout, and product display intended to create an environment that encourages customers to buy.

Merchant Bank A bank that does business with merchants who want to accept credit cards.

Message Digest Synonymous with hash value, which is the number that results when a message is hash coded.

Meta Language A language that comprises a set of language elements and can be used to define other languages

META Tag A special HTML tag that contains keywords that represent Web page content; these are used by search engines to build indexes.

Micromarketing The practice of targeting very small and well-defined market segments.

Micropayments Internet payments for items costing very little—usually $1 or less.

Minimum Bid In an English auction, the price for an item at which the auctioning begins.

Minimum Bid Increment The amount by which one bid must exceed the previous bid.

Multipurpose Internet Mail Extension (MIME) An e-mail protocol that allows users to attach binary files to e-mail messages.

N

Name Changing A problem that occurs when someone registers purposely misspelled variations of well-known domain names. These variants sometimes lure consumers who make typographical errors in entering a URL.

Name Stealing Theft of a Web site's name that occurs when someone, posing as a site's administrator, changes the ownership of the domain name assigned to the site to another site and owner.

National Center for Supercomputing Applications (NCSA) Housed at the University of Illinois, Urbana-Champaign, the NCSA is one of the five original centers in the National Science Foundation's Supercomputer Centers Program. Mosaic, the first Internet browser program and predecessor to the Netscape browser, was invented at NCSA.

Necessity The category of computer security that addresses data delay or data denial threats.

Necessity Threat The disruption of normal computer processing or denial of processing.

Nesting Synonymous with clustering.

Net Bandwidth The actual speed information travels, taking into account traffic on the communication channel at any given time.

Network Access Points (NAPs) The four primary connection points for access to the Internet backbone in the United States.

Network Access Providers
The four companies
(Pacific Bell, Sprint,
Ameritech, MFS
Corporation) that are
primary providers of
Internet access rights;
they sell these rights to
smaller Internet service
providers.

**Network Control Protocol
(NCP)** Used by
ARPANET in the early
1970s to route messages
in its experimental wide
area network.

**Network Economic
Structure** A business
structure wherein firms
coordinate their strate-
gies, resources, and skill
sets by forming a long-
term, stable relationship
based on a shared
purpose.

Network Operations Web
site staff whose responsibilities include load estimation and monitoring,
resolving network prob-
lems as they arise,
designing and imple-
menting fault-resistance
technologies, and man-
aging any network opera-
tions that are outsourced
to ISPs, CSPs, or tele-
phone companies.

Nexus The association
between a taxpaying
entity and a governmen-
tal taxing authority.

Nonrepudiation
Verification that a particular transaction actually
occurred; this prevents
parties from denying a
transaction's validity or
its existence.

O

Offer A declaration of willingness to buy or sell a
product or service; it
includes sufficient
details to be firm, pre-
cise, and unambiguous.

One-Sided Tag HTML tags
that require only an
opening tag.

**One-to-Many
Communication Model**
A model of communica-
tions in which one entity
communicates with a
number of other entities.

**One-to-One
Communication Model**
A model of communica-
tions in which one entity
communicates with one
other entity.

One-Way Function An
algorithm that cannot be
converted back to its
original value.

Online Community
Synonymous with virtual
community, which is an
electronic gathering
place for people with
common interests.

Open Architecture The
philosophy behind the
Internet that dictates
that independent net-
works should not require
any internal changes to
be connected to the net-
work, packets that do
not arrive at their destinations must be retrans-
mitted from their source
network, routers do not
retain information about
the packets they handle,
and no global control
exists over the network.

Open Auction An auction
in which bids are pub-
licly announced (such as
an English auction).

**Open DataBase
Connectivity (ODBC)**
A database protocol that
makes it possible for a
program to access data
from an application,
regardless of which data-
base management
system is dispensing
the data.

Open EDI EDI conducted
on the Internet instead of
over private leased lines.

Open Loop System A payment card arrangement
involving a consumer
and his or her bank, a
merchant and its bank,
and a third party (such
as Visa or MasterCard)
that processes transac-
tions between the con-
sumer and merchant.

Opening Tag An HTML tag
that precedes the text
that a tag affects.

Open-Outcry Auction
Synonymous with open
auction.

**Operating Resource
Management System
(ORMS)** Software that
automates routine pur-
chasing decisions.

Operating Resources
Support components of
business that include
information technology,
telecommunications
equipment, professional
services, MRO supplies,
travel and entertainment
expenses, and office
equipment.

Operating System
Software (such as
Windows or UNIX) that
performs basic resource
allocation and dealloca-
tion tasks for a
computer.

Opt-In E-Mail The practice
of sending e-mail mes-
sages to people who have
requested information on
a particular topic or
about a specific product.

Outsourcing The hiring of another company to perform design, implementation, or operational tasks for an information systems project.

P

Packet Filter Firewall A firewall that examines all data flowing back and forth between a trusted network and the Internet.

Packet Switching A network system of data transmission in which files and messages are broken down into small units called "packets." These packets are labeled electronically with codes indicating origin and destination. Packets travel from computer to computer along the network and upon reaching their destination are reassembled.

Page View A page request made by a Web site visitor.

Partial Outsourcing The outsourcing of the design, development, implementation or operation of specific portions of an electronic commerce system.

Patch A small piece of code designed to correct a software bug.

Per Se Defamation A legal cause of action in which a court deems some types of statements to be so negative that injury is assumed.

Permission Marketing A marketing strategy that only sends specific information to persons who have indicated an interest in receiving information about the product or service being promoted.

Persistent Cookie A cookie that exists indefinitely.

Personal Contact A method of identifying and reaching customers that involves searching for, qualifying, and contacting potential customers.

Personal Identification Number A random assemblage of digits, chosen by the customer, that serves as a password for monetary transactions.

Personal Jurisdiction A court's authority to hear a case based on the residency of the defendant; a court has personal jurisdiction over a case if the defendant is a resident of the state in which the court is located.

Physical Security Tangible protection devices such as alarms, guards, fireproof doors, fences, and vaults.

Plain Old Telephone Service (POTS) The network connecting telephones; it provides a reliable data transmission bandwidth of about 56 Kbps.

Plug-In An application that helps a browser to display information (such as video or animation) but that is not part of the browser.

Port The endpoint to a TCP/IP connection devoted to a particular type of Internet traffic.

Portal A Web site that serves as a customizable home base from which users do their searching, navigating, and other Web-based activity.

Post Office Protocol (POP) The protocol responsible for retrieving e-mail from a mail server.

Post-Implementation Audit A formal review of a project after it is up and running.

Presence The public image conveyed by an organization to its stakeholders.

Primary Activities Activities that are required to do business: design, production, promotion, marketing, delivery, and support of products or services.

Privacy The protection of individual rights to nondisclosure of information.

Private Key A single key that is used to encrypt and decrypt messages. Synonymous with symmetric key.

Private-Key Encryption The encoding of a message using a single numeric key to encode and decode data, it requires both the sender and receiver of the message to know the key, which must be guarded from public disclosure.

Private Network A private, leased-line connection between two companies that physically links their individual computers or intranets.

Private Valuation The amount a bidder is willing to pay for an item that is up for auction.

Procurement The business activity that includes all purchasing activities plus the monitoring of all elements of purchase transactions.

Product Disparagement A statement that is false and injures the reputation of a product or service.

Project Management Formal techniques for planning and controlling activities undertaken to achieve a specific goal.

Project Management Software Application software that provides built-in tools for managing people, resources, and schedules.

Prospecting The part of personal contact selling in which the salesperson identifies potential customers.

Protocol A collection of rules for formatting, ordering, and error-checking data sent across a network.

Proxy Bid In an electronic auction, a predetermined maximum bid submitted by a bidder.

Proxy Server A firewall that communicates with the Internet on behalf of the trusted network.

Public Key One of a pair of mathematically related numeric keys, it is used to encrypt messages and is freely distributed to the public.

Public Network An extranet that allows the public to access its intranet or when two or more companies link their intranets.

Public-Key Encryption The encoding of messages using two mathematically related but distinct numeric keys.

Q

Query Synonymous with search expression, which is the key word on which search engines perform searches.

R

Rational Branding An advertising strategy that substitutes an offer to help Web users in some way in exchange for their viewing an ad.

Registrar An official Web domain name registration service.

Repeat Visits Subsequent visits a Web site visitor makes to a particular page.

Request Header The part of a message from a client to a server that contains additional information about the client and more information about the request.

Request Line The part of a message from a client to a server that contains a command, the name of the target resource (without the protocol or domain name), and the protocol name and version.

Reserve Price The minimum price a seller will accept for an item sold at auction.

Reserve Synonymous with reserve price.

Response Header Field In a client/server transmission, it follows the response header line and returns information describing the server's attributes.

Response Header Line The part of a message from a server to a client that indicates the HTTP version used by the server, status of the response, and an explanation of the status information.

Response Time The amount of time a server requires to process one request.

Return on Investment A method for evaluating the potential costs and benefits of a proposed capital investment.

Reverse Auction A type of auction in which sellers bid prices for which they are willing to sell items or services.

Ripper Software that stores music in digital format on a computer.

Ripping The act of extracting a track from a music CD and storing it in digital format on a computer.

Router A computer that determines the best way for data packets to move forward to their destination.

Routing Algorithm The program used by a router to determine the best path for data packets to travel.

S

Save Area The location of a computer where programs store critical information before control of that information is passed to another program.

Scalable A system's ability to be adapted to meet changing requirements.

Scrip Digital cash minted by a small number of third-party organizations.

Scripting Language Code Code that allows a downloaded Web page to execute programs on a user's computer.

Sealed-Bid Auction An auction in which bidders submit their bids independently and are usually prohibited from sharing information with each other.

Search Engine Web software that find other pages based on key word matching.

Search Expression The key word on which search engines perform searches, which can include instructions telling the search engine how to perform its search.

Second-Price Sealed-Bid Auction A type of auction in which bidders submit their bids independently and privately; the highest bidder wins the auction but pays only the amount bid by the second-highest bidder.

Secrecy The category of computer security that addresses the protection of data from unauthorized disclosure and confirmation of data source authenticity.

Secure Electronic Transaction (SET) A secure protocol that provides security for card payments as they traverse the Internet between merchant sites and processing banks.

Secure Envelope A security utility that encapsulates a message and provides secrecy, integrity, and client/server authentication.

Secure Sockets Layer (SSL) A protocol for transmitting private information securely over the Internet.

Security Policy A written statement describing assets to be protected, the reasons for protecting the assets, the parties responsible for protection, and acceptable and unacceptable behaviors.

Segment Also called a market segment; a subset of a company's potential customer pool that has common demographic characteristics.

Self Hosting A system of Web hosting in which the online business owns and maintains the server and all its software.

Server A powerful computer dedicated to managing disk drives, printers, or network traffic.

Server Console A console that is in the same room as the server and is directly attached to it.

Server-Side Electronic Wallet An electronic wallet that stores a customer's information on a remote server that belongs to a particular merchant or to the wallet's publisher.

Server Side Include (SSI) A type of HTML comment that directs the Web server to dynamically generate data for a Web page when it is requested.

Session Cookie A cookie that exists only until you shut down your browser.

Session Key A key used by an encryption algorithm to create cipher text from plain text during a single secure session.

Shared Hosting A Web hosting arrangement in which a corporate Web site is on a server that hosts other Web sites simultaneously and is controlled by a third-party service provider.

Shill Bidder An individual employed by a seller or auctioneer who makes bids on behalf of the seller, sometimes artificially inflating an item's price. Shill bidders may be prohibited by the rules of a particular auction.

Shopping Cart An electronic commerce utility that keeps track of selected items for purchase and automates the purchasing process.

Signature Any symbol executed or adopted for the purpose of authenticating a writing.

Signed Java Applet A Java applet that contains an embedded digital signature from a trusted third party; it is proof of the identity of the applet's source.

Signed Message or Code The status of a message or Web page when it contains an attached digital certificate.

Simple Mail Transfer Protocol (SMTP) A standardized protocol used by a mail server to format and administer e-mail.

Smart Card A plastic card with an embedded microchip that contains information about the card owner.

Sniffer Program A program that taps into the Internet and records information that passes through a router from the data's source to its destination.

Software Agent Synonymous with agent.

Spam Electronic junk mail.

Specialist The auctioneer in a sealed-bid double auction (such as the New York Stock Exchange). The specialist firm is often responsible for the matching of buy and sell orders and must use its own funds when necessary to maintain an orderly market.

Spider The first part of a search engine, it automatically and frequently searches the Web to find pages and updates its database of information about old Web sites.

Spoofing Synonymous with masquerading.

Stakeholders The various entities involved in a business; these include customers, suppliers, employees, stockholders, neighbors, and the general public.

Standard Generalized Markup Language (SGML) A computer language used to mark up documents independent of any software application, it contains an international standard that defines methods for representing electronic documents. SGML is the basis for HTML.

Static Page A Web page that displays unchanging information retrieved from disk.

Statute of Frauds State laws that specify that contracts for the sale of goods worth over $500 and contracts that require actions that cannot be completed within one year must be created by a signed writing.

Steganography The hiding of information (such as commands) within another piece of information.

Stickiness The ability of a Web site to keep visitors at its site and to attract repeat visitors.

Sticky The condition of having stickiness.

Stored Value Card Either an elaborate smart card or a simple plastic card with a magnetic strip that records the currency balance, such as a prepaid phone, copy, subway, or bus card.

Strategic Business Unit (Business Unit) A unit within a company that is organized around a specific combination of product, distribution channel, and customer type.

Subject Matter Jurisdiction A court's authority to decide a dispute between entities based on the issue of dispute.

Subscription The delivery of specific information to a user's computer. Users provide information about what information to deliver, amount of information, and schedule for updates. More specific than a channel.

Supply Alliances Long-term relationships among participants in the supply chain.

Supply Chain The part of an industry value chain that precedes a particular strategic business unit. It includes the network of suppliers, transportation firms, and brokers that combine to provide a material or service to the strategic business unit.

Supply Chain Management The process of taking an active role in working with suppliers and other participants in the supply chain to improve products and processes.

Supply Management Synonymous with procurement, which is the business activity that includes all purchasing activities plus the monitoring of all elements of purchase transactions.

Supporting Activities Secondary activities that back up primary business activities. These include human resource management, purchasing, and technology development.

Symmetric Encryption Synonymous with private-key encryption, which is the encoding of a message using a single numeric key to encode and decode data.

Systems Administrator A member of an electronic commerce team who understands the server hardware and software and is responsible for the system's reliable and secure operation.

T

T1 High-bandwidth telephone company connections that operate at 1.544 Mbps.

T3 High-bandwidth telephone company connections that operate at 44.736 Mbps.

Tags HTML codes inserted into documents that specify formatting and arrangement of page elements.

TCP/IP The set of protocols that provide the basis for the operation of the Internet. The TCP protocol includes rules that computers on a network use to establish and break connections. The IP protocol determines routing of data packets.

Technology-Enabled Relationship Management The business practice of obtaining detailed information about a customer's behavior, preferences, needs, and buying patterns and using that information to set prices, negotiate terms, tailor promotions, add product features, and provide other customized interactions.

Telnet A protocol that allows users to log on to a computer and access its contents from a remote location.

Terminal Emulation A program that allows access to files on a remote computer.

Thin Client The relatively low workload of a Web client, compared with that of a server.

Threat An act or object that poses a danger to assets.

Three-Tier Architecture A client/server architecture that builds on the two-tier architecture by adding applications and their associated databases that supply non-HTML information to the Web server on request.

Throughput The number of HTTP requests that a particular hardware and software combination can process in a unit of time.

Tier-One Suppliers The capable suppliers that work directly with and have long-term relationships with businesses.

Tier-Three Suppliers Suppliers that provide components and raw materials to tier-two suppliers.

Tier-Two Suppliers Suppliers that provide components and raw materials to tier-one suppliers.

Tort An action taken by a legal entity that causes harm to another legal entity.

Total Cost of Ownership (TCO) Includes all the costs associated with design and maintenance of a computer network system, hardware and software costs, technical support costs, and training costs.

Trademark Dilution The reduction of the distinctive quality of a trademark by alternative uses.

Trading Partners Businesses that engage in EDI with one another.

Transaction Costs The total of all costs incurred by a buyer and seller as they gather information and negotiate a transaction.

Transaction Processing Processes that occur as part of completing a sale; these include calculation of any discounts, taxes, or shipping costs and transmission of payment data (such as a credit card number).

Transaction Sets Formats for specific business data interchanges using EDI.

Transmission Control Protocol See TCP/IP.

Trial Visit The first visit a Web site visitor makes to a particular page.

Triple Data Encryption Standard (3DES) A robust version of the Data Encryption Standard used by the U.S. government that cannot be cracked even with today's supercomputers.

Trojan Horse A program hidden inside another program or Web page that masks its true purpose (usually destructive).

Trusted (Network) A network that is within a firewall.

Trusted Applet A Java applet that has full access to system resources on a client computer.

Two-Sided Tags HTML tags that require both an opening and a closing tag.

Two-Tier Architecture A client/server architecture in which only a client and server are involved in the requests and responses that flow between them over the Internet.

Uniform Resource Locator (URL) Names and abbreviations representing the IP address of a particular Web page. Contains the protocol used to access the page and the page's location. Used in place of dotted quad notations.

Unmanaged Hosting A system of server hosting in which the customer is responsible for maintaining and staffing all servers.

Unmanaged Server Hosting Synonymous with unmanaged hosting.

Untrusted (Network) A network that is outside a firewall.

Untrusted Applet A Java applet that is not known to be secure.

Upload To send a file to another computer.

Upstream The transfer of information from a client computer to another computer.

Upstream Strategies Tactics that focus on reducing costs or generating value by working with suppliers or inbound logistics.

URL Broker A business that sells or auctions domain names that it believes others will find valuable.

Usability Testing The testing and evaluation of a company's Web site for ease of use by visitors.

Usenet (User's News Network) One of the first mailing lists; it allows subscribers to read and post articles within topic areas.

Value-Added Bank A bank that offers value-added network services for nonfinancial transactions.

Value-Added Network (VAN) An independent company that provides connection and EDI transaction forwarding services to businesses engaged in EDI.

Value Chain A way of organizing the activities that each strategic business unit undertakes to design, produce, promote, market, deliver, and support the products or services it sells.

Value System Synonymous with industry value chain.

Vanilla Wafer A cookie created by a browser that contains little or no personal information.

VBScript A programming language that can create dynamic pages within HTML documents.

Vendor-Specific Scrip Scrip that only a particular merchant will accept.

Vertical Integration The practice of an existing firm replacing one of its suppliers with its own strategic business unit that creates the supplied product.

Vickrey Auction Synonymous with second-price sealed-bid auction. Named for William Vickrey, who won the 1996 Nobel Prize in Economics for his studies of the properties of this auction type.

Viral Marketing Tactics that rely on existing customers to tell other persons—the company's prospective customers—about the products or services they have enjoyed using.

Virtual Community An electronic gathering place for people with common interests.

Virtual Host Multiple servers that exist on a single computer.

Virtual Private Network (VPN) A network that uses public networks and their protocols to transmit sensitive data using a system called "tunneling" or "encapsulation."

Virtual Server Synonymous with virtual host.

Virus Software that attaches itself to another program and can cause damage when the host program is activated.

Visit The request of a Web site visitor for a page from a Web site.

Vortal Portals for vertical industries that are part of a value chain beginning with raw materials and ending with finished products.

W

Warranty Disclaimer A statement indicating that the seller will not honor some or all implied warranties.

Web Browser Software that lets users read HTML documents and move from one HTML document to another using hyperlinks.

Web Bug A tiny, invisible Web page graphic that provides a way for a Web site to place cookies.

Web Catalog Model A business model of selling goods and services on the Web wherein the seller establishes a brand image that conveys quality and uses the strength of that image to sell through catalogs mailed to prospective buyers. Buyers place orders by mail or by calling the seller's toll-free telephone number.

Web Client A computer that is connected to the Internet and used to download Web pages.

Web Community Synonymous with virtual community.

Web Directory A listing of hyperlinks to Web pages that is organized into hierarchical categories.

Web Hosting Provides the Internet access services of an ISP along with electronic commerce software, store space, and commerce expertise.

Web Mall A type of Web hosting that groups commerce sites in a portal-style directory.

Web Portal (Portal) A location on the Web that acts as a launching point for searching, navigating, and other Web-based activities.

Web Server A computer that is connected to the Internet and that stores files written in HTML that are publicly available through an Internet connection.

Winner's Curse A psychological phenomenon that causes bidders to become caught up in the excitement of competitive bidding and bid more than their private valuation.

Wire Transfer Synonymous with electronic funds transfer, which is the electronic transfer of account exchange information over secure private communications networks.

World Wide Web (Web) The subset of Internet computers that connects computers and their contents in a specific way, and that allows for easy sharing of data using a standard interface.

Worm A virus that replicates itself on other machines.

Writing A tangible representation of the terms of a contract.

Y

Yankee Auction A type of English auction that offers multiple units of an item for sale and that allows bidders to specify the quantity of items they want to buy.

Z

Zombie A program that secretly takes over another computer for the purpose of launching attacks on other computers. Zombie attacks cannot be traced to their creators.

Index

A

About.com, 104
access control list (ACL), 229
access controls, 90, 227–29
Accredited Standards Committee
    X12 (ASC X12), 334–35, 343
ACH. *See* Automated Clearing
    House (ACH)
ACL. *See* access control list (ACL)
ACLU, 286–87
ACM Digital Library, 313
acquiring banks, 273
active content
    digital certificates and, 202–6
    Internet Explorer and, 206–8
    launching, 170
    monitoring, 202–10
    Netscape Navigator and,
        209–10
    process, 169–70
    security threats from,
        168–70, 202
    types of, 168
Active Server Pages (ASP),
    94, 99, 100
active wiretapping, 180–81
ActiveX controls, 168, 174
addressable media, 295
Adleman, Leonard, 214
administration, 28
ADSL. *See* Asymmetric Digital
    Subscriber Line (ADSL)
AdSubtract, 201
Advanced Encryption Standard
    (AES), 215–16
Advanced Rotocraft
    Technology, 309
advertising, 7
    costs of, 298–99
    effectiveness of, 298–99
    employment, 316
    ethical issues, 408
    legal issues, 403
    newspaper publishers, 315
    regulation of, 406–8
    Web, 96
advertising-subscription mixed
    models, 317–18

advertising-supported business
    models, 314–16
ad views, 298
affiliate marketing, 306–7
after-sale service and support, 27
agents, intelligent, 108–11
agricultural commodities, 24
Airborne, 122
Alertbox Web site, 292
algorithms, 214, 217–19
Alis Technologies, 390
AllAdvantage.com, 306
Allaire Corporation, 92
Allegiance Telecom, 330
AlliedSignal, 419–20
Amazon.com, 1–2, 104, 123,
    295, 306, 307, 408
    auction site, 366
    1-Click, 261
American Civil Liberties Union
    (ACLU), 286–87
American Express Blue smart
    card, 268
American National Standards
    Institute (ANSI), 334
American Red Cross, 287
American Society of Mechanical
    Engineers, 310
America Online (AOL), 46
Ameritrade, 281–82
Amnesty International, 287
Analog Web server log file
    analyzer, 91
anchor tags, 55–56
Andreessen, Marc, 20
Anonymizer, 179–80
anonymous browser service,
    179–80
anonymous electronic cash,
    247–48
anonymous File Transfer
    Protocol (FTP), 90
anonymous FTP, 41
ANSI. *See* American National
    Standards Institute (ANSI)
antivirus software, 211
Apache HTTP Server, 96, 98–99,
    111, 118

Apartments.com, 364
API. *See* Application Program
    Interface (API)
applets, 168
application construction, 93
Application Program Interface
    (API), 85, 98, 99
application servers, 84–85
application service providers
    (ASPs), 211, 431
applications specialists, 437
archiving logs, 98
Ariba, 78–79, 350
ARIS MusiCode, 198
ARIS Technology, 198
ARPANET, 36, 37, 38, 45
Art.com, 309–10
Artuframe, 309
ascending-price auctions, 359
ASCII text, 41–42
ASC X12. *See* Accredited
    Standards Committee X12
    (ASC X12)
ASP. *See* Active Server
    Pages (ASP)
ASPs. *See* application service
    providers (ASPs)
Association for Computer
    Machinery (ACM), 313
Asymmetric Digital Subscriber
    Line (ADSL), 71, 72
asymmetric encryption, 214
Asynchronous Transfer Mode
    (ATM), 72
At Home Corporation, 307
ATM. *See* Asynchronous Transfer
    Mode (ATM)
Atrieva, 331
attachments, to e-mail, 45
    computer viruses and, 176
    security and, 176–77
    viruses in, 193–94, 211
AT&T IP Services, 342
Auction Block, 369–70
AuctionBot, 110
Auction Guide, 372
AuctionInsider, 372
Auction IT, 370–71